CONTENTS

PREFACE

We have intended to present an introduction to the physical interpretation of phenomena which govern hydrological events related to soil or the upper most mantle of the earth's crust. The text is based upon our teaching and research experience. The book can serve either as the first reading for future specialists in soil physics or soil hydrology. Or, it can be a source of basic information on soil hydrology for specialists in other branches, e.g. in agronomy, ecology, environmental protection, forestry, geomorphology, hydrologic science, meteorology and water management. We assume that those specialists do not intend to conduct research in soil hydrology but they may wish to successfully use its tools in their own academic domains.

The first requirement of such a book is its simplicity without neglecting all of the complexity of soils and their porous systems as they react with the atmosphere, biosphere and hydrosphere. With mathematical derivations in the majority of cases being reduced to the level of calculus, we were obliged in some instances to deviate from the usually rigorous derivations. However, if the reader has an aversion to mathematics, basic information on any particular problem can be gained even without going into the details of its mathematical development.

According to our experience, the level of information presented in this book is adequate for a qualified use of programs and models in soil hydrology. If the reader does not intend to passively accept ready-made software, and has intentions of being more academically creative, we recommend more detailed study of information readily available in related journals and other publications.

Physical interpretation and mathematical formulation of such complex events as rain infiltration or evapotranspiration cannot be done without a certain degree of simplification. We emphasize here that this simplification leads only to an approximation to reality. Similar simplifying approaches can be applied in neighboring branches of investigation. By removing some simplifying assumptions, a more exact but usually a more complicated but still approximate solution of the soil hydrological problem is formulated.

Inasmuch as the text of our book is purposely not a monograph, the references cited are our subjective selections from the voluminous literature. Indeed, many titles to which we did not refer may be of the same importance as our selected references for a given problem . The same consideration applies to the theories presented in the text – not all are explained nor quoted in our book. Even if we had tried to subordinate our subjective choice to criteria based upon further development and broad applicability, we are aware of the fact that we would have neglected some which will eventually be extended to fundamental contributions in the future. Those contributions are the "surprises" that make research and scientific investigation so attractive to many creative brains.

For further, more detailed reading, the literature can be found in our references.

We acknowledge the assistance of our colleagues in reading, completing and critiquing parts of the manuscript. We thank namely to Professor Němeček from the Agricultural University in Prague, Dr. V. Novák from the Institute of Hydrology in Bratislava and Dr. M. Parlange from the University of California, Davis. Last but not least, we thank Mrs. Margot Rohdenburg, the publisher, for her patience in dealing with our slow writing and the completion of the book.

Miroslav Kutílek
Prague, Czech Republic

Donald R. Nielsen
Davis, California, U.S.A.

1 SOILS IN HYDROLOGY

Soil is a very thin layer of the earth when its thickness is compared to the dimensions of the atmosphere or geosphere. Even the average depth of water in the oceans is orders of magnitude greater than that of soil. In spite of its slim dimension, soil is indispensable for life on continents. Without soil, our planet would not be green and all life would be restricted to the oceans.

Hydrologically, soil is an important reservoir of fresh water. Owing to the action of this reservoir, non continuous rainfall or snow are transformed into a continuous flow of water to the roots of plants. Together with ground waters, soil transforms discontinuous precipitation into continuous discharges recognized as streams and rivers. Without soil, the face of our planet would have continents wrinkled by dry wadis conducting water only during and immediately after each rainfall event. The retention capacity of soil able to sequester rain water is approximately the same as the capacity of all lakes. Moreover, the amount of water in soil equals one-third of all fresh water in lakes (including artificial reservoirs) and is one order of magnitude larger than that in riverbeds. On the other hand, the volume of ground waters is substantially greater than that in soil. Soil water together with ground water exceeds by more than two orders of magnitude all surface fresh water (Dyck and Peschke, 1983).

Transport of water soluble materials occurring naturally or anthropogenically is linked to hydrologic processes. For example, without water as a transporting agent, anthropogenically derived chemical pollutants would remain at the location of their deposition provided they were not carried into the atmosphere by diffusion or the wind. The distance and rate of their motion on and below the soil surface depends upon hydrologic events. Because soil is a chemically and biologically active medium, pollutants are partly retained, released and transformed with the boundary of the pollution transposed in both time and space. Similarly, naturally occurring constituents within soil are mobilized and transported as a result of precipitation and irrigation. The quality of continental water resources in space and time is therefore greatly influenced by soil hydrology.

1.1 SOILS

With the qualities and properties of soil not being uniform across continents, we note that soils are classified according to their morphological features, genesis and soil forming processes. Because properties of soils vary across the landscape, a relationship observed at one location may indeed not be applicable for other locations. Hence, measured data and deduced relationships are location specific and should be interpreted with information regarding the particular soil type or pedotop according to an appropriate soil taxonomy.

1.1.1 Soil Genesis

When fluid magma from the depth of the earth reaches the earth's surface, it solidifies under new conditions of pressure and temperature. After this relatively rapid process, the freshly formed rock, not completely in equilibrium with its new environment, progresses slowly toward a new equilibrium but never actually reaching it. Primary minerals weather and transform in the presence of microbial organisms to slowly form substances required for soil genesis. Similarly, whenever subsurface rocks from the depth of the earth or unconsolidated materials such as sand or loess are exposed to the atmosphere, the process of soil genesis is initiated.

The weathering of rocks is the disturbance of the crystal lattices of minerals constituting the rocks. The susceptibility of minerals to weathering depends upon the characteristics of the crystal lattice, e. g. the bonds of Si-O are much stronger and resistant to weathering than those of Al-O or the weaker bonds of Mg-O. The greater is the number of weak bonds of a crystal lattice of minerals, the greater is the susceptibility of the mineral to weathering. In the case of sedimentary rocks, weathering usually first attacks the cementing materials that combine the primary particles. The surface area of the solid phase exposed to chemical weathering is increased by physical weathering. Physical weathering disintegrates the bulk unity of rocks and sometimes even releases individual minerals. Hence, weathering is a non uniform process with its intensity depending upon parent material, climatic conditions, local water regime, chemical composition of the water and activity of microbiological organisms.

New materials form during weathering. Simple mineral salts typically dissolve in water and are often transported far from their origin. Hence, ground waters become more or less mineralized. Salts transported by surface and ground waters accumulating in oceans significantly contribute to the salinity of ocean waters. Or in dry climates, salt-affected soils originate by salts being brought permanently to the soil surface by evaporation. Some salts act as cement either directly or after transformation when sedimentary particles are formed. Here, with contributions from living organisms, grain sizes from sand to clay are the matrix of particles newly formed from the products of earlier weathering with contributions from living organisms. Residuals of weathering generally are not transported and usually remain and accumulate in place.

Dissolved inorganic and organic materials are an integral part of the water in each and every pore of any soil. Because soil water cannot be considered as "pure" water identical to distilled or deionized water available in a laboratory, it has become common practice to use 0.01 *N* $CaSO_4$ or $CaCl_2$ as water to measure soil water properties observed in the laboratory.

Because iron and aluminum oxides released by weathering are only conditionally dissolved under favorable pH and redox conditions, they are translocated only through small distances in the great majority of soils. In oxisols of humid tropical climates or in B horizons of spodosols (podsols) of cool, humid climates where favorable conditions exist, they accumulate in large amounts. Organic substances, especially those of lesser polymerization, contribute to the mobility of Fe and Al in the form of complexes.

As a result of weathering in the presence of biochemical reactions, restructuralization and crystallization, new secondary minerals are created. Among them, clay minerals are the most important inasmuch as they constitute the clay fraction of soils, frequently form complexes with organic substances and strongly influence the physical and chemical properties of soil [e. g. swelling, cation exchange capacity (CEC) and modulus of rupture].

Having mentioned the contribution of the biological activity of soil organisms toward the transformation of inorganic components of soils, we note that the residues of plants gradually transform into new organic substances. These substances such as humic or fulvic acids contribute and modify the CEC and buffer capacity of soils and are the dominant fixation agent of soil aggregation. The extent of their influence depends upon the quality and degree of polymerization of the soil organic matter and the environmental conditions which form the water and temperature regimes of soils.

All the above-mentioned as well as other substances occupy and partly fill in the porous space of soils in a characteristic way according to soil type and horizon. They uniquely influence the internal architecture and type of aggregation of each soil. Moreover, soil grains, residual minerals and newly formed clay minerals are usually covered by films of weathering products and continually transforming organic residues. We never find clean mineral surfaces within soils. Inasmuch as the internal architecture of soils is governed by prevailing conditions of soil evolution, its observation at any one time reveals heterogeneity at all spatial scales.

Many recent soils do not originate directly from freshly exposed rocks or solidified magma but are derived from substrates that are products of earlier weathering or of earlier soil genesis. Many of those materials were transported from the place of their origin to more or less distant positions. When we find traces of earlier evolution of soils, we speak of polygenesis. It contributes to the extreme heterogeneity of the internal arrangement of the solid phase of soils. Inasmuch as soil water is in direct contact with the internal architecture of soils, we can expect great variability of soil water behavior and its regimes. On the other hand, we assume that the dominant features of soil morphology which are relevant to soil classification allow us to group, classify and map soil water parameters and characteristics.

Owing to the action of climatic factors and soil organisms, horizons are formed within a soil profile according to topography and soil parent materials or rocks. Horizons are relatively thin layers having a thickness in the order of 10 cm and are typically characterized by their macro- and micro morphological features. Each horizon manifests its own processes of transformation and accumulation (or elutriation) of substances. Thus, along the vertical dimension of a soil we find a regular arrangement of heterogeneity. As a result of this arrangement, soil hydraulic theories applied to field situations must account for the fact that soil profiles are indeed layered. There are no soils without this deterministic vertical heterogeneity!

1.1.2 Soil Classification

The aim of soil classification systems is an objective definition of taxonomic units. Soils of each taxonomic unit should have identical evolutionary processes and identical basic soil properties. Taxonomic units should be defined such that a hierarchical arrangement is feasible from higher to lower levels. The influence of major soil forming factors (climate, parent rock, landscape and vegetation) are considered with various degrees of importance in each of the existing classification systems. In spite of the diversity of approaches of each system, fundamental features of a classification unit are the following:

1. Congruent soil water and thermal regimes,
2. Conformity in processes of accumulation and transformation of matter in each horizon,
3. Uniform character of a naturally occurring plant cover,
4. Identical sequences of horizons in the soil profile and
5. Conformity in micro morphological features in each soil horizon.

Preferably, characteristics which can be observed and measured in the field are used. However, they are frequently inadequate for accurate definition of characteristics and processes. In such instances, additional properties are determined in the laboratory.

The above criteria for classification of soil units are more complex than some hydrologists and engineers recognize owing to the fact that the majority of them classify soils according to particle size distribution and occasionally to parent rock. In soil classification systems, particle size distribution plays a relatively unimportant role and is usually related to one of the lowest level of classification units. However, some specific exceptions exist when particle size distribution affects the highest level of classification (e. g. arenosols) which we shall mention later in this chapter.

In the past the characteristic properties of soils and criteria for soil classification were not uniformly considered. And even now, there is no one unique classification system. Systems differ according to regions or countries because they are based on different concepts and degrees of appreciation of soil formation. Correspondingly, different criteria are used for hierarchical subdivisions. Recently, the most frequently used systems are FAO Classification (1985) and U. S. Soil Taxonomy (Soil Survey Staff, 1975 and 1990). For the earlier Soviet territory the Soviet System was used. In central Europe the German System and its modifications are prevalent, while the French System also includes successful soil classification in the tropics and develops a new reference system.

Because soil horizons are one of the main diagnostic features of soil classification, we briefly describe the more important horizons. With each soil classification system having its own nomenclature of horizons, it should be obvious that some deviations from the following descriptions must occur. The top horizon relatively rich in organic matter is the humus A horizon or also named the epipedon, and in forest soils we identify the O horizon. This humus horizon is further classified in greater detail according to the morphology and quality of the humus. The horizon of depletion, where either clay or sesquioxides have been leached out is the eluvial horizon E which is sometimes denoted as the A_2 horizon. The horizon enriched by clay translocated from upper horizons is the illuvial or argic B horizon. A spodic B horizon is illuviated by organic matter together with iron and aluminum. A ferralitic B horizon is rich in Fe and/or Al owing to strong weathering. A cambic B horizon is a young horizon of alteration and weathering. A natric B horizon has an exchangeable sodium percentage greater than 15%. Calcic, gypsic or salic horizons are characterized by accumulations of calcium carbonate, gypsum or salts, respectively. Redox-morphic horizons are also recognized. Partially weathered parent rock and unweathered parent rock are denoted as the Cr and R horizons, respectively.

1.1.2.1 FAO Soil Classification

The term *soil nomenclature* or *legend* may indeed be more appropriate than *soil classification* for the FAO Classification System inasmuch as it summarizes the classification systems of various schools and regions in order to elaborate the World Map of Soils. The FAO Classification System recognizes two classification levels: Major Soil Grouping at level I and Soil Unit at level II. At lower levels, soils are further classified according to substrate and soil climate. The FAO Classification System recognizes 28 Major Soil Groupings and 153 Soil Units.

Here we briefly characterize the Major Soil Groupings according to Driessen and Dudal (1989).

Organic Soils

HISTOSOLS (Greek *histos* = tissue) – Those soils have a large porosity and the hydraulic conductivity is small to nil depending upon the degree of decomposition of organic material. Those having more decomposed organic matter have smaller values of hydraulic conductivity as well as swelling and shrinking more than those with less decomposed organic matter.

Mineral soils conditioned by human influences

ANTHROSOLS (Greek *anthros* = man) – Soils whose properties depend not only upon the original native material but the nature and type of anthropogenic activity.

Mineral soils conditioned by the parent material

ANDOSOLS (Japanese *an* = dark and *do* = soil) – Soils formed from materials rich in volcanic glass, frequently on fresh deposits of ash. They have very large values of both porosity and hydraulic conductivity, excellent internal drainage and strong aggregate stability.

ARENOSOLS (Latin *arena* = sand) – Weakly developed soils derived from sand. They have very large values of hydraulic conductivity accompanied with extremely small water storage capacity.
VERTISOLS (Latin *verto* = turn) – "Heavy" clay soils that alternatively swell and shrink. Deep cracks which form in the dry season are partly filled with self-mulching topsoil material. After wetting slickenslides develop and undulating gilgai micro relief may occur owing to swelling pressures. Infiltration rates initially very large decrease to practically zero after a short time. Hydraulic conductivity and soil water storage depend upon soluble salts and the sodium content of the infiltrating water.

Mineral soils conditioned by topography and physiography

FLUVISOLS (Latin *fluvius* = river) – Young soils developed on alluvial deposits with their properties and eventual stratification depending upon the character of the alluvium. They are frequently influenced by hydromorphism.
GLEYSOLS (Russian *gley* = glue-like) – Hydromorphic soils with signs of excess wetness owing to shallow ground water. They originate on unconsolidated sediments. Chemical reduction occurring below the ground water level in the gleyic horizon destroys soil structure, increases compaction and decreases the hydraulic conductivity compared with soils of the same texture.
LEPTOSOLS (Greek *leptos* = shallow) – Weakly developed shallow soils on hard rock influenced frequently during their genesis by erosion. Their properties are determined primarily by the nature of the parent material and climate. At shallow depths, they are freely drained.
REGOSOLS (Greek *rhegos* = blanket) – Very weakly developed shallow soils with slight accumulation of organic matter originating from weathered unconsolidated material. The soil forming processes either lasted for a short time or were hindered by extreme climatic conditions (desert or permafrost). Their properties are related to the climate and the nature of the weathered rock.

Mineral soils conditioned by their limited age

CAMBISOLS (Latin *cambiare* = change) – Soils with a distinct B horizon where weathering is dominant and neither leaching nor accumulation occurs. Although their properties vary widely, the structural formation of the B horizon is always distinct.

Mineral soils conditioned by a wet (sub)tropical climate

PLINTHOSOLS (Greek *plinthos* = brick) – Iron rich soils formed in a tropical climate as a result of intense, long duration weathering of parent material. The plinthite layer rich in iron and aluminum is dense and not very permeable. Massive iron pans or discontinuous ironstone frequently develops at the soil surface owing to erosion and exposure of the plinthite layer. Inasmuch as Plinthosols develop in regions with annual precipitation exceeding evaporation, they frequently manifest features of waterlogging.

FERRALSOLS (Latin *ferrum* = iron and *al* = aluminum) – Tropical red and yellow soils rich in sesquioxides having a well-developed B horizon and a clay fraction dominated by kaolinite. With stable soil aggregates, the porosity, hydraulic conductivity and infiltration rates of ferralsols are large while soil water storage is rather limited. If the Al content exceeds that of Fe, the structural stability is lowered after deforestation and the soils are susceptible to surface sealing and compaction.

NITISOLS (Latin *nitidus* = shiny) – Deep, red tropical soils with illuviation of clay into a thick argic B horizon having a very stable structure of angular blocks which fall apart into shiny peds. The intensity of weathering is less than that in Ferralsols. Although kaolinite is the most abundant clay mineral, some swelling within soil pores persists. Moderate hydraulic conductivity and a large water storage capacity characterize the physical properties of Nitisols.

ACRISOLS (Latin *acris* = very acid) – Strongly weathered acid soils of the tropics having red and yellow subsurface colors. Plinthite and gleyic horizons are sometimes evident. After deforestation, the weak soil structure gives way to soil surface crusting and reduced infiltration rates. Waterlogging often occurs in local depressions.

ALISOLS (Latin *al* = aluminum) – Tropical acid soils somewhat less weathered than Acrisols containing large amounts of free aluminum which manifest a well-developed dense and slowly permeable illuvial B horizon. With swelling and shrinking clay minerals being present and microbiological activity constrained, the stability of the soil structure is fragile. Rain causes the soil surface to seal and crust with infiltration frequently being impaired.

LIXISOLS (Latin *lixivium* = washing) – Strongly weathered soils in which clays are washed down from the soil surface into the B horizon having a reddish or yellow color. With unstable aggregates, slaking at the soil surface frequently decreases infiltration and increases erosion hazards.

Mineral soils conditioned by a (semi-)arid climate

SOLONCHAKS (Russian *sol* = salt and *chak* = augmentation) – Soils with large accumulations of mineral soluble salts in areas where evaporation exceeds precipitation. Salts remain in the top part of the soil profile as a result of evaporation in the presence of highly mineralized shallow ground water. The structural stability and hydraulic conductivity of Solonchaks may fluctuate during the year owing to fluctuations of solute concentration and the chemical composition of the salts within the profile.

SOLONETZ (Russian *sol* = salt and *etz* = diminutive) – Relatively non saline soils having large proportions of sodium and a natric B horizon with an ESP (exchangeable sodium percentage) above 15. They swell and shrink, are without aggregation in the top A horizon and typically develop columnar structure in the B horizon. The hydraulic conductivity is strongly reduced by the large ESP. Many Solonetz soils develop by leaching salts from sodium rich Solonchaks. During the transition of a Solonetz to a Solonchak, the hydraulic conductivity increases as the concentration of soluble salts increases.

GYPSISOLS (Latin *gyps* = gypsum) – Soils with substantial accumulation of gypsum primarily in the B horizon often in sufficient quantities to form crusts at the soil surface. The gypsum originates from geological sediments. Depending on the local morphology, values of hydraulic conductivity range from very small to very large. The infiltration rate is close to zero in crusted soils. Water storage within the profile is good. Unfortunately, they are susceptible to water erosion.

CALCISOLS (Latin *calcis* = lime) – Soils with calcium carbonate accumulated at some depth below the surface. They originate and develop slowly on unconsolidated sediments and deposits and manifest a slight development of a cambic B horizon. They are well drained and have good water storage capacities. When crusted, infiltration is slow and water erosion occurs.

Mineral soils conditioned by a steppe climate

KASTANOZEMS (Latin *castaneo* = Russian *kashtan* = chestnut and Russian *zemlya* = earth) – Soils originating on loess. Sometimes a slightly developed cambic B horizon is above a lime or gypsum accumulation and at a depth of over 150 cm, salt accumulation may occur. Typically vegetated by sparsely occurring short grasses, they have smaller humus contents than those of chernozems. Although values of hydraulic conductivity are moderate and water storage capacity is good, soil surfaces without vegetation are prone to erosion.

CHERNOZEMS (Russian *chern* = black and *zemlya* = earth) – Black soils rich in organic matter originating on loess and under tall grass steppe vegetation. Infiltration during wet periods leaches carbonates below the humus horizon. They have favorable physical properties characterized by a crumb structure of water stable aggregates, good internal drainage and a good water storage capacity.

PHAEOZEMS (Greek *phaios* = dusky and Russian *zemlya* = earth) – Dark soils rich in organic matter similar to Chernozems but with a shallower humus horizon. Although both originate on loess, phaeozems develop in more humid climates. Their properties are similar to those of Chernozems.

GREYZEMS (Russian *zemlya* = earth) - Soils on unconsolidated materials where deciduous forests invaded the original grassland. Clay particles under a more humid climate are translocated into the argic (rich in clay) B horizon with a frequently existing polyhedral structure. They have good internal drainage and stable aggregation but their surfaces are sometimes susceptible to erosion.

Mineral soils conditioned by a (sub)humid temperate climate

LUVISOLS (Latin *luere* = wash) – Soils on unconsolidated sediments, frequently on loess, where clay has been washed down from the soil surface to accumulate in the argic B horizon. They have crumb structure within the topsoil and cubic to blocky structure in the B horizon. Most are well drained, but some B horizons obstruct percolation resulting in periodic waterlogging within the profile.

PODZOLUVISOLS (from podzols and luvisols) – Soils having the bleached E horizon of Podzols and the illuvial B horizon of Luvisols. They originate on unconsolidated materials including loess only under forests. They are compacted, have poor internal drainage and are prone to waterlogging.

PLANOSOLS (Latin *planus* = flat) – Soils with a leached eluvial E horizon and a compacted B horizon with a large amount of translocated clay. Owing to poor internal drainage accompanying an extremely small hydraulic conductivity in the B horizon, Planosols are seasonally waterlogged. Their humus horizon has poor structural stability. They harden in dry seasons and are muddy in wet seasons.

PODZOLS (Russian *pod* = under or Old Slavic *poda* = soil and *zola* = ash) – Soils with an eluvial E horizon and a spodic B horizon which is rich in sesquioxides and chelates translocated from the E horizon. They form on weathered siliceous rocks and sandy sediments, generally under coniferous forests which condition a strong acid reaction. Because internal drainage is impeded by the slowly permeable B horizon, waterlogging frequently develops. The water storage capacity of Podzols is small.

1.1.2.2 U. S. Soil Taxonomy

In U. S. Soil Taxonomy exactly defined horizons and their morphological features are the basis of the definition of all soil units in the U. S. Soil Taxonomy. Their definitions are completed by identifying the thermic and hydric regimes. The system recognizes sequential levels. The highest level describes Orders followed by Suborders, Great Groups, Subgroups and Families. Orders include soils of the same intensity and nature of soil forming processes which are characterized by a particular sequence of primary soil horizons. In Suborders soils are classified according to soil climate reflected by chemical, physical and morphological properties of horizons. Great Groups are related to the intensity of evolution of horizons and to their specific features. Subgroups are reserved for intergrades and modifiers. For Families we use adjectives and differentiate according to texture, mineralogy, cracks, pH etc. Greek and Latin terms are used for the nomenclature of soil units. From those terms syllables compose the final name of Great Groups in a prescribed manner. Hence, the suffix is for Order, the middle syllable is for Suborder and the first syllable is for Great Group. The following is a brief description of Orders.

ALFISOLS (The suffix *-alf* appears in Suborders and Great Groups) – Soils characterized by horizons of eluviation and illuviation. Udalfs are Alfisols with an udic (= humid) moisture regime and correspond to Luvisols (FAO). Boralfs (Eutroboralfs and Glossoboralfs) are Alfisols with a cold (cryic or frigid) temperature regime (boreal = northern). Eutroboralfs (eutrophic = fertile) have a base saturation greater than 60% while Glossoboralfs have smaller base saturations and are never dry. All of them correspond to Podzoluvisols (FAO). Aqualfs are Alfisols with an aquic water regime and are equivalent to Stagnogleyic Luvisols (FAO). There are 37 Great Groups of this Order.

ANDISOLS (*-ands*) – Soils corresponding to Andosols (FAO).

ARIDISOLS (*-id*) – Soils of arid regions with one or more distinct horizons of either enrichment or depletion of substances transported by water. Calciorthids (orthos = true) have a calcic horizon and correspond to Calcisols (FAO). Gypsiorthids are equivalent to Gypsisols (FAO) and Camorthids with a cambic horizon are identical to Cambic Arenosols (FAO). FAO equivalents to Great Groups include Solonchaks and Solonetz.

ENTISOLS (*-ents*) – Soils having no principal horizon below that of humus. Fluvents are related to Fluvisols (FAO) while Great Groups of Suborder Orthents correspond to Leptosols, Regosols and Arenosols of FAO.

HISTOSOLS (*-ist*) – Organic soils corresponding to Histosols (FAO).

INCEPTISOLS (*-ept*) – Soils with a diagnostic cambic horizon having no aridic hydric regime. They correspond to Cambisols (FAO) with some Great Groups equivalent to some Soil Units of Luvisols (FAO), e. g. Ustochrepts being equivalent to Calcic Luvisols.

MOLLISOLS (*-oll*) – Soils with a deep mollic humus horizon. Borolls with a frigid temperature regime and Ustolls with an ustic (nearly dry) moisture regime correspond to Chernozems (FAO) while Udolls with a udic (humid) moisture regime are equivalent to Phaeozems (FAO). Some Argiudolls and Argiborolls with an argic (clay) horizon correspond to Greyzems (FAO).

OXISOLS (*-ox*) – Soils with an oxic horizon characterized by kaolinites in the clay fraction and few weatherable minerals. They are intensively leached. Their Suborders and Great Groups are equivalent to Ferralsols and Plinthosols of FAO.

SPODOSOLS (*-ods*) – Soils with a diagnostic spodic horizon corresponding to Podzols (FAO).

ULTISOLS (*-ults*) – Soils with a diagnostic argic (illuvial, rich in clay) horizon having a small base saturation. Their Great Groups correspond to Acrisols and Nitisols (FAO).

VERTISOLS (*-ert*) – Soils equivalent to Vertisols of FAO.

1.1.3 Soil Mapping

The basic element of a soil and that of a soil map is the pedon. It is a three-dimensional body having a geographical surface area of 1 to 10 m^2. We usually describe its morphology on a two dimensional vertical cross section of a soil profile, i. e. on the wall of an excavated pit. Figure 1.1 illustrates the various

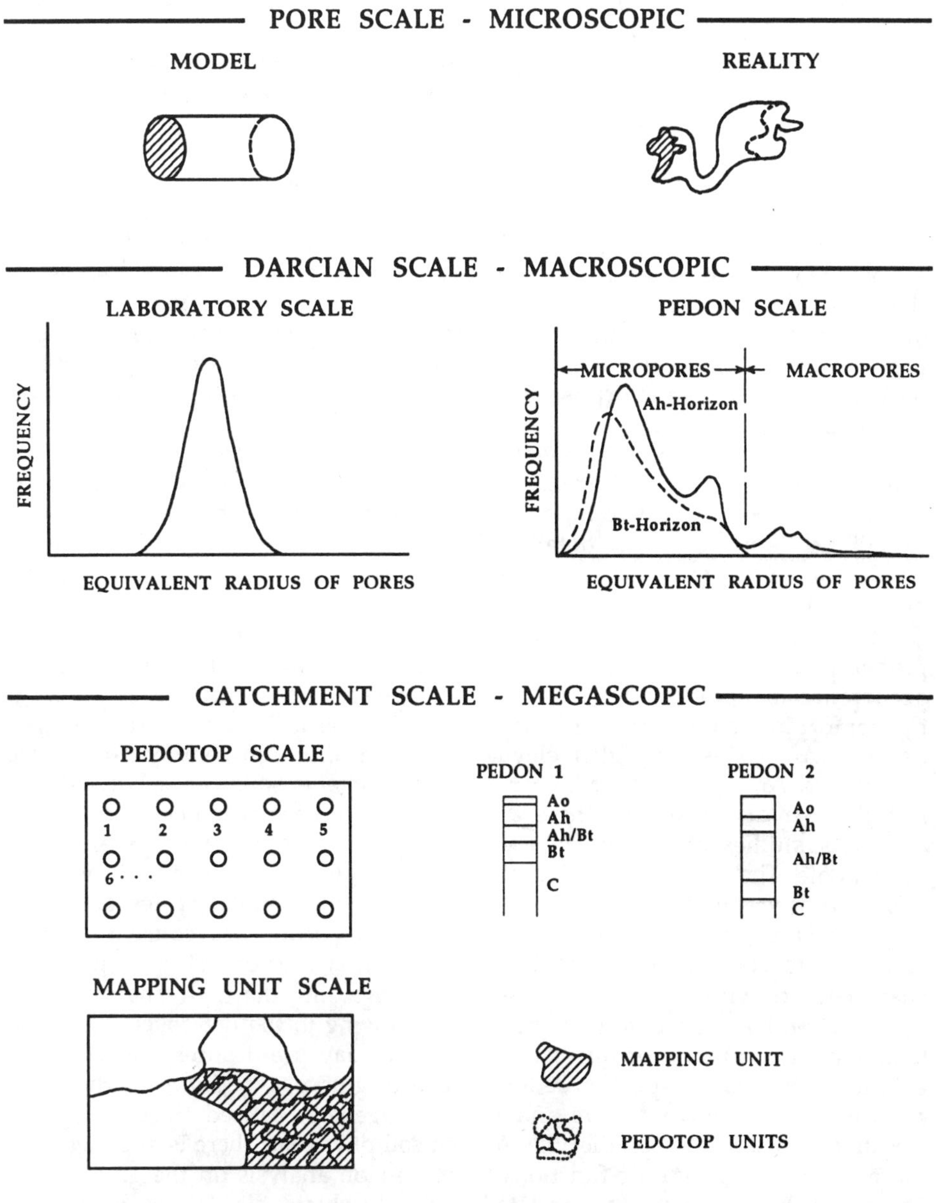

Figure 1.1. Scales in soil hydrology

scales of observation applicable to soil hydrology. We note that the pedon scale as well as the laboratory scale are macroscopic in size and are commonly denoted as the Darcian scale. At this scale, soil pores at the microscopic (or pore) scale are not individually described but considered as an ensemble defined in Chapter 2 as being a part of a representative elementary volume (REV) and through which water flows according to Darcy's equation defined in Chapter 5.

The positions of individual pedons are usually identified on maps of scale 1:2000 or 1:5000 with abbreviations of important properties. With the pedon classified according to the lowest level of the taxonomic system, we obtain the abstract form – a pedotop which is the lowest taxonomic unit of the soil mantle (soil cover). On geographically large scale maps (1:2000 is larger than 1:20000) we approximate the boundary of the domain of each pedotop either by simple interpolation or with the help of geostatistics. If the soil body consists of more than one contiguous pedon, it is termed a polypedon. The domain of one pedotop may contain inclusions of polypedons. In such cases a pedotop is polymorphic. A monomorphic pedotop has no inclusions of neighboring pedotops and polypedons.

When the area of a mapped region is orders of magnitude greater than that of an average district of one pedotop (i. e. if the scale of the map is decreased by orders of magnitude compared to that of pedotop maps), the soil cover cannot be depicted in all details of all districts of pedotops. The lowest taxonomic units, the pedotops are then clustered either according to the hierarchial structure of the classification system into higher taxonomic units. Or, the pedotops are grouped with regard to a close similarity of specified properties. Thus, we obtain higher mapping units. In such a mapping unit, the occurrence of polypedons frequently depends upon topography, e. g. hydromorphic soils occur in depressions of alluvia while around the depressions at slightly higher elevations of the alluvium semihydromorphic soils occur. And on the next higher terrace, lithogenic soils can occur without features of hydromorphism. These gradual transitions are called catena.

For studies of soil hydrology, the largest scales of pedotops are most appropriate. The random variation of soil hydrologic properties in each pedotop is statistically well-defined within this scale. Within smaller scales of higher mapping units, the variation of soil hydrologic properties consists of two components. One component is the random variation of soil physical characteristics within each of the pedotops composing the unit. The second is the variation having a deterministic character owing to the heterogeneity of the landscape, parent rock or micro climate which may pre-destine a specific soil evolution towards a specific pedotop. The origin of a pedotop is therefore determined by known factors and the heterogeneity caused by this type of variation of pedotops is predictable. Among soil physicists, there is a tendency to estimate the soil hydraulic functions by regression analysis on the given scale. Soil hydraulic functions are correlated to simply obtainable data (e. g. textural components, bulk density, organic matter content and measures of soil structure). However, a detailed preliminary study of the mutual relationships in the given region is indispensable and the knowledge gained is not simply transferable to other mapping units. Efforts to apply a general relationship (pedotransfer function) of small scale to soil hydrological studies into various pedoregions are futile.

1.2 CONCEPTS OF SOIL HYDROLOGY

All studies in soil hydrology eventually have a unique aim – a better understanding and description of hydrological processes. The individual elementary processes of infiltration, redistribution, drainage, evaporation and evapotranspiration are first analyzed and subsequently considered in combination during a particular sequence of events or season. Transport of solutes is also considered as an integral part of those processes. All such processes occur in soils and under actual meteorological situations. A proper physical understanding of them requires several levels of approximative studies.

As a first approximation we model the soil as a simple, homogeneous porous body temporarily forgetting the existence of horizons within its profile and the horizontal variation of its properties. In some instances a soil profile consisting of two horizons is modeled simply by considering a layer of a homogeneous soil overlain by a second having different hydraulic properties. For studying the behavior of soil water including flow and transport of matter, we use phenomenological (or macroscopic) descriptions. We describe what we can "see" with our apparatuses and we denote the scale where the phenomenological approach is applied as Darcian. Only when the physical interpretation of some phenomena requires a detailed discussion at the microscopic level will we temporarily abandon laws and equations based on a macroscopic scale of observation.

The elementary hydrologic processes for simply modeled soils and for trivial boundary conditions are described by analytical solutions of the basic macroscopic equations. The advantage of analytical solutions is a full understanding of the physical processes. Parallel to such mathematical analyses are carefully conducted experiments performed on repacked soil columns or on model porous materials under precise conditions in the laboratory.

The next level of approximation is the quantification of processes for real soils, i. e. field soils. Although the scale remains Darcian, we speak of it as the pedon scale. At this level the boundary conditions are usually less trivial than those used in the first level, and if they are sufficiently complex, numerical methods are applied to achieve particular solutions. These results, similar to an accurately performed field experiment, are regularly verified by field experimentation. The advantage of numerical simulation is the rapid production of a large number of "computer experiments" which partially substitute for tedious, time-consuming field experiments. Alternatively, numerical procedures allow us to study specific features of a process which are not accessible or readily observed by existing experimental techniques. We properly interpret the data physically by applying the knowledge we gained at the first approximation level.

From these pedon studies (often called "point scale" studies) we try to extend the results to the larger scale of a field or catchment. This megascopic scale, larger than Darcian, is usually denoted as catchment scale. See Fig. 1.1. Considering the principles of soil mapping, it is advisable to differentiate between two new categories within the catchment scale: (1) the pedotop scale

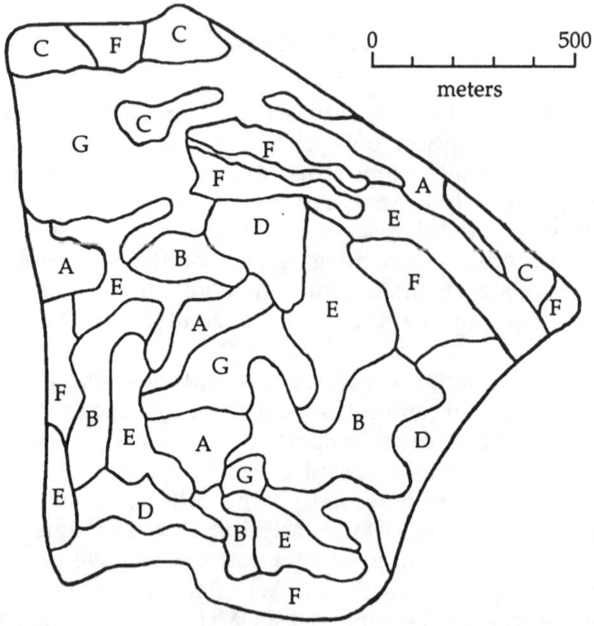

Figure 1.2. Pedologic map delineating seven pedotops (designated A through G) within a mapping unit associated with a 100-ha farm (Duffy et al., 1981).

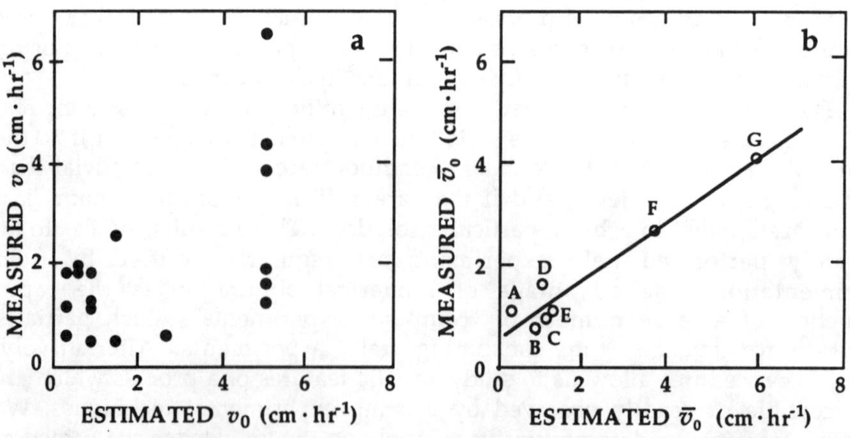

Figure 1.3. Measured quasi-steady state infiltration rates for the seven pedotops illustrated in Fig. 1.2 versus those estimated from soil texture: a. measured and estimated values v_0 at each location within the farm without considering pedotops, and b. measured mean and estimated mean values $\bar{v}_0$ within each pedotop.

and (2) the mapping unit scale. Both categories belong to megascopic observations in soil hydology but they differ in the structure of the soil mantle and therefore in the nature of the variability of soil physical properties. Within the pedotop scale, variability is strictly stochastic. Within the mapping unit scale, variability is both stochastic and deterministic. We illustrate this behavior of spatial variability in Fig. 1.2 which depicts seven pedotops located deterministically within a mapping unit across the landscape of a farm. The pedotops were mapped based on data from 293 bore holes taken approximately at 60-m intervals across the entire 100-ha farm. Measured values of surface water infiltration rates within each pedotop were found to be log-normally distributed. Estimates of infiltration rates were derived using soil survey interpretation methods (U. S. Dept. of Agr., 1951) based upon the texture of the surface soil where each infiltration rate was measured. If the seven pedotops are ignored, there is no relation between measured and estimated values of infiltration (Fig. 1.3a). On the other hand, if the infiltration rates are grouped together by pedotop, their measured and estimated geometric mean values are highly correlated with $r^2 = 0.936$ (Fig. 1.3b).

At the pedotop scale, methods used at the Darcian scale need to be modified with stochastic characteristics entering our equations and procedures. The stochastic structure of these hydraulic properties of field soils is studied by specific procedures. In some instances we obtain a set of deterministic pedon-scale observations spatially distributed across the field or catchment to define a newly formed stochastic or regionalized variable. In other instances, entirely new approaches are developed applicable only to the pedotop scale. Experience with the mapping unit scale remains sufficiently inadequate to preclude any generalizations.

Analogous to these briefly introduced concepts of soil hydrology, we proceed further into the content of the book.

2 SOIL POROUS SYSTEM

The term pore denotes that part of the soil space which is not filled by the soil solid phase. The shape, size and origin of pores play a role only in a detailed classification of the soil porous system. Inasmuch as all transport processes and the storage of water occur in the soil porous system, we must study it in detail.

2.1 SOIL POROSITY

The relative volume of pores is denoted as soil porosity

$$P = V_P / V_T \tag{2.1}$$

where V_p is the volume of pores and V_T the total bulk volume of soil. When (2.1) is multiplied by 100, we obtain P in percent. The volume of pores V_P is related to the volume of soil solid phase V_S by the void ratio

$$e = V_P / V_S \tag{2.2}$$

with mutual relations between P and e being

$$P = \frac{e}{1+e} \qquad \text{and} \qquad e = \frac{P}{1-P} \tag{2.3}$$

Void ratio is more appropriate than porosity if the reference term V_T is not constant as for example in clays when they swell or shrink.

Estimating the proper size of V_T is crucial. It may be obtained by the following theoretical treatment, see Fig. 2.1. Let us assume that when V_T is small enough ($V_T < V_P$) and centered in a pore, $P = 1$. When V_T is centered in the solid phase and small enough ($V_T < V_S$), $P = 0$. When V_T increases, we reach the situation when either $V_T = (V_P + \delta V_S)$ or $V_T = (V_S + \delta V_P)$ and porosity is neither 1 nor 0. Centering V_T in the solid phase, P is slightly above zero. With a gradual increase in V_T we obtain first a gradual decrease of P (center in the pore) or a gradual increase of P (center in the solid phase), until both values coincide. Before coincidence occurs, a damping oscillation of P may exist. At this stage of increasing V_T when porosity is independent upon the initial centering of V_T, we have reached the representative elementary volume REV (Bear, 1969 and 1972). This final volume V_T is called by Corey (1977) the minimum volume of a reference element. If V_T = REV, translating the center of V_T does not change the value of P. The concept of REV is important in methods of practical determination of porosity as well as the application of the theory of potential flow to the description of water flow in soils. Soils without structural development have an REV of about 100 cm^3 or less. Because the REV of aggregated soils depends upon the size and shape of their peds, its value may be one or more orders of magnitude greater.

Soil porosity manifests a broad range of values from less than 0.3 to more than 0.9. However, in the majority of mineral soils, P lies between 0.4 and 0.6. Some organic soils have values of P that exceed 0.9. Porosity depends upon the composition, texture and structure of soil. Soil structure is influenced by the content of organic matter, the quantity of inorganic cementing agents, the genesis of the soil and its horizons and more recently, by human activity.

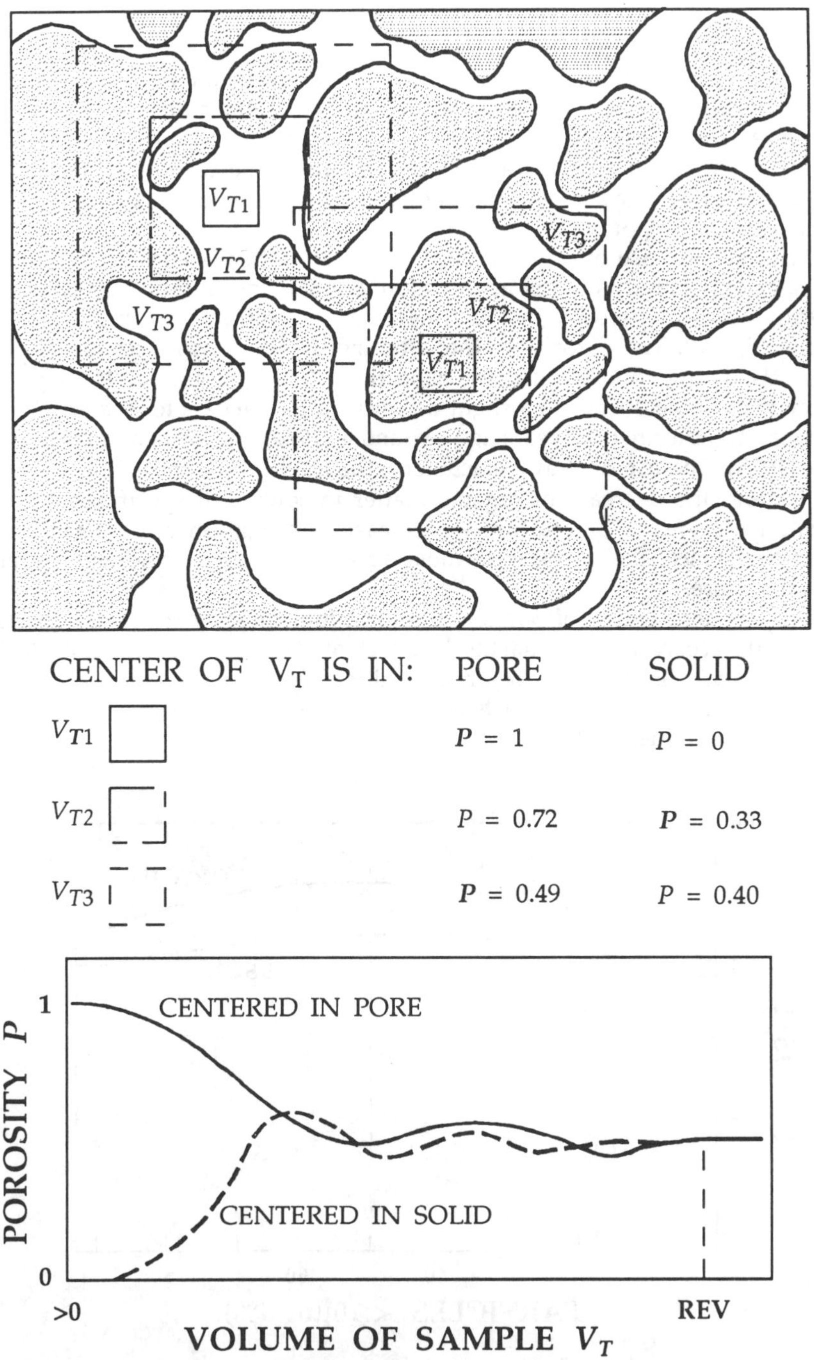

Figure 2.1. Estimati... Representative Elementary Volume REV by increasing the bulk v... il V_T to such an extent that porosity is independent of the posi... ter of V_T.

Theoretically, in the most dense arrangement of spherical particles of identical size, $P = 0.26$. Although this value is sometimes taken as the theoretical minimum porosity of sands, even sandy soils are mixtures of particles of various sizes and shapes. Hartge (1978) reviewed binary and ternary mixtures of particles having different sizes and shapes. Even in the case of sands, neither simple nor sophisticated geometric models of the arrangement of particles are applicable for use of porosity.

With an increase of clay content the porosity of soils rises, owing mainly to a greater opportunity for soil aggregation. For soils not intensively cultivated, the mean values of porosity in the A-horizon and sub soil are plotted versus texture in Fig. 2.2. The porosity of the A-horizon reaches its maximum value in loamy soils owing to the optimum conditions for aggregation. In clays, the tendency to slushiness leads to a slight decrease of *P*. The data for the graph stem from research performed 70 years ago by Janota (1924) and were summarized by Kutílek (1978). The use of heavy machinery, the application of fertilizers in high quantities and the decrease of organic matter in soils – all specific features of modern-day intensive agriculture – decrease aggregation and enhance soil compaction. Hence, the porosity of most agricultural soils no longer depends upon soil texture.

The range of values of porosity found in four major textural categories of mineral soils according to Schachtschabel et al. (1984) is given in Table 2.1. Except for clay soils, texture does not significantly influence porosity. In clay soils originating from volcanic ash which are classified as Andosols, porosity reaches extreme values of 0.7 to 0.8.

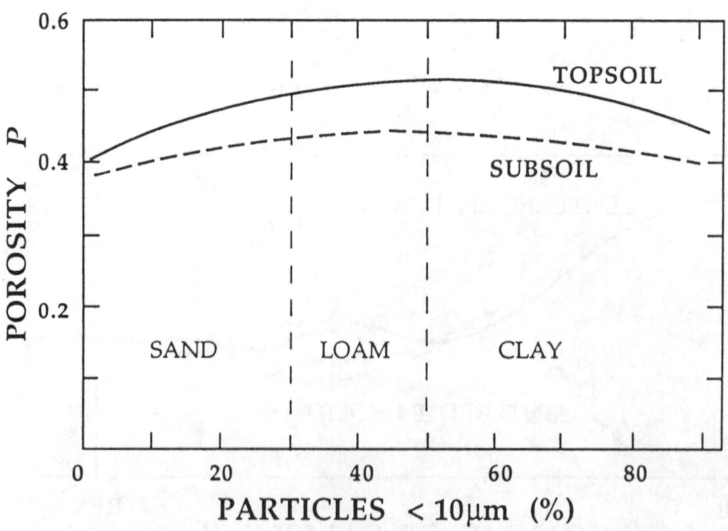

Figure 2.2. Dependence of porosity P upon the texture of the topsoil and subsoil of central European soils at the start of this century (Janota, 1924, and Kutílek, 1978).

Table 2.1. Range of the porosity and bulk density in mineral soils with $C_{ox} \leq 2\%$ (Schachtschabel et al., 1984).

Type of soil	Porosity P	Bulk density ρ_T
	%	$g \cdot cm^{-3}$
sandy soils	56 - 36	1.16 - 1.70
silty soils	56 - 39	1.26 - 1.61
loamy soils	55 - 30	1.20 - 1.85
clayey soils	70 - 35	0.88 - 1.72

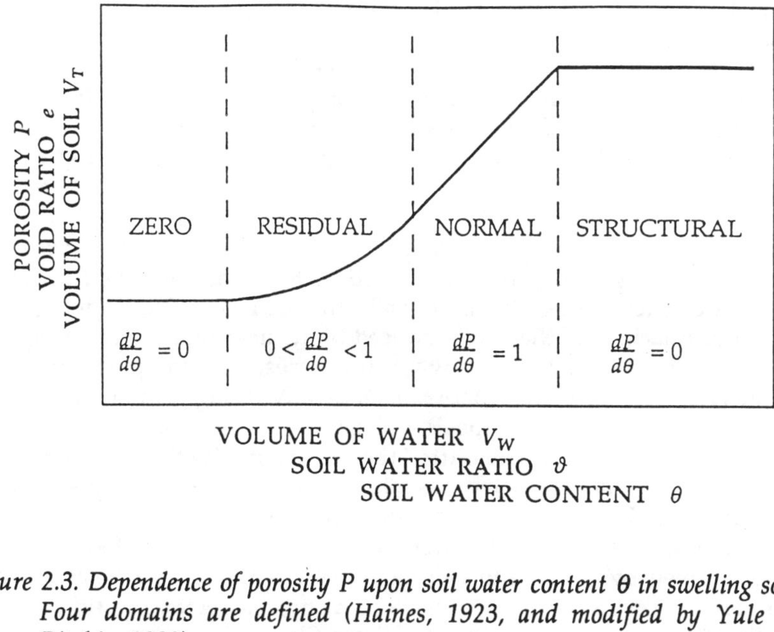

Figure 2.3. Dependence of porosity P upon soil water content θ in swelling soils. Four domains are defined (Haines, 1923, and modified by Yule and Ritchie, 1980).

In some clays and loams, P depends upon the soil water content θ inasmuch as soil swells when wetted and shrinks when dried. In such soils P increases with θ as shown in Fig. 2.3. This relationship is linear for large values of θ. At smaller values of θ, P changes non-linearly with θ. A detailed discussion is in section 3.1.

In the soil profile, the maximum porosity usually occurs in the top of the A_0 horizon. The porosity distribution with soil depth depends upon the development of each soil profile. For example, in the illuvial Bt horizon accumulated material causes a decrease in porosity. Or, the reduction of iron oxides and the destruction of the structure in hydromorphic gley G horizons lead to an abrupt decrease in porosity.

The porosity of organic soils is greater than that of mineral soils and reaches maximal values in highly decomposed low moor. As mineral substances increase in organic soils the porosity decreases.

2.2 CLASSIFICATION OF PORES

According to Corey (1977), the classification of soil pores should follow the laws of hydrostatics and hydrodynamics and leads to three categories:

1. Sub microscopic pores which are so small that they preclude clusters of water molecules to form fluid particles or continuous water flow paths. Inasmuch as convection does not exist in these pores, the laws of fluid mechanics are not applicable. Pores belonging to this category are oftentimes neglected.
2. Capillary pores where the shape of the interface between air and water is determined by the configuration of pores and by the forces on the interface. The resulting air-water interface is called the capillary meniscus. The flow of water in these pores is considered to be laminar and dominant in soils.
3. Macropores (non-capillary pores) are of sufficient size that capillary menisci are not formed. The shape of the interface between air and water is considered planar, and hence, capillary forces are nil. The flow of water inside of such pores can be either in the form of a film moving over all the irregularities of the walls induced by their roughness and shape, or in some cases, turbulent flow when the pores contain considerably more water. The shape of macropores is generally thought to be either tubular or planar. A tubular shape is caused by the action of soil fauna (e.g. earthworms) or originates after the decay of plant roots. Planar pores have the form of cracks and fissures owing mainly to cleavage planes and volume changes of the bulk soil when it dries.

A detailed description of systems of macropores as related to transport processes is given by Beven and German (1982). Although the relative volume of these pores is very small, it may have great impact relative to flows of water and its dissolved constituents within a soil profile.

In soil science and soil physics literature, capillary pores in the second category are frequently subdivided according to hydrostatic criteria as well as practical aspects of agriculture. Pores containing water below the permanent wilting point of plants are designated as fine pores, those with water between the wilting point and field capacity are medium pores and those with water above field capacity are coarse pores.

Further subdivisions of soil pores are usually based upon particular hydraulic pressures and equivalent pore radii. Details and the ambiguity of these approaches will be discussed in Chapter 4. The size of the pores can be defined by two approaches: i) The maximum radius of a sphere inscribed in the pore. The definition is applicable only to pores where the three main axes are of comparable size. Or, the pore has a tubular form and the radius of the inscribed circle is taken as the radius of the pore. For both, visualization of pores is required. ii) The equality of the behavior of a hydraulic parameter in a model consisting of capillaries with that of a real soil leads to the definition of equivalent radii. Hydrostatic methods based on the measurement of capillary pressure are described in Chapter 4.

In soil science, two terms are used to describe soil porosity – primary porosity and secondary porosity. Primary porosity denotes the relative volume of pores owing to the shape and size of primary particles. Secondary porosity denotes the action of other factors, e.g. tubular pores form after the decay of plant roots, or fissures develop as the result of drying of soils that swell and shrink. Hence, Katchinski (1965) defines aggregate porosity P_A and interaggregate porosity P_B in terms of the average porosity of the individual aggregates $\overline{P_I}$ as

$$P_A = \frac{\overline{P_I}\,(1-P)}{1-\overline{P_I}} \tag{2.4}$$

and

$$P_B = P - P_A, \tag{2.5}$$

respectively. The terms intrapedal pores and interpedal pores are sometimes used for P_A and P_B, respectively.

2.3 METHODS OF POROSITY MEASUREMENT

Methods of porosity measurement are based upon the determination of V_P and V_T as defined in (2.1). The *density method* relies on the expression

$$P = \frac{\rho_S - \rho_T}{\rho_S} \tag{2.6}$$

where ρ_S [$= m_S \cdot V_S^{-1}$(g·cm^{-3})] is the mean density of the soil solid particles, ρ_T [$= m_S \cdot V_T^{-1}$(g·cm^{-3})] the soil bulk density, m_S the mass of the soil solid phase [usually taken as the weight of the soil dried at 105°C for mineral soils, see also Table 3.1 in Chapter 3], V_S the volume of the soil solid phase measured pycnometrically and V_T the volume of the core sampler or the volume of water displaced when the aggregate (or the clod) coated with paraffin is submerged in water. In stony soils, V_T is measured by the volume of graded sand filling in the hole from which the soil sample was taken for drying and weighing.

Values of ρ_S for mineral soils range from 2.6 to 2.7 g·cm^{-3} which are close to 2.65 g·cm^{-3}, the density of quartz which is usually one of the main constituents of the majority of soils. In highly organic soils such as peats, bogs and moors, values of ρ_S occur in a rather wide range from 0.2 to well above 1.0 g·cm^{-3}. For these organic soils, the value of ρ_S generally decreases as the level of decomposition of organic matter increases.

Optical methods are used when the soil pores are able to be visualized.

This visualization is achieved by filling the pores with an initially fluid material that subsequently becomes a solid. The soil matrix is thus fixed into regions of soil particles and a contrasting material delineating regions of pore space. Thin sections of the fixed soil matrix subsequently produced reveal the area of pores A_P measured with microphotography is related to the total examined area A_T. Porosity is estimated with the assumption of equality of 2-dimensional and 3-dimensional arrangements (Delesse, 1848, and Rosiwal, 1898) which leads to

$$A_P / A_T = V_P / V_T \quad (2.7)$$

For measuring A_P, the method of secants or the Monte Carlo method is used. The principle of the method is nowadays applied to the description of the system of macropores. The impregnated macropores are counted by image analysis (Ringrose-Voase and Bullock, 1984).

Among the *volumetric methods*, procedures for measuring the volume of liquid filling the pores appear the simplest for obtaining a value of V_P. This simple concept is plagued with practical difficulties which include the entrapment of air, the altered hydrophility when the soil is dried and the volumetric change of the soil matrix accompanying the drying or wetting when water is used. If a non-wetting liquid (e.g. mercury) is used, the loose arrangement of soil does not permit an appropriate pressure to be defined. Another alternative of the volumetric method is the measurement of the gas pressure within the dried soil before and after contact with a known volume of gas. Boyle-Mariotte's law is used to calculate V_P. The volume changes accompanying the complete drying of a soil are disadvantageous features of the method.

Indirect methods are based upon the measurement of a particular physical property of soil dependent either upon the bulk density or porosity of the soil. They are advantageous for quickly estimating the change of P either in space or time. For example, the mechanical resistance of a soil measured by a penetrometer offers information about the variation of P if the soil water content is constant. The reference values of P should be known at a particular point or time. Gamma radiation attenuation and scattering methods offer much more exact data and will be discussed in Chapter 3.

2.4 SOIL POROUS SYSTEMS

Soil pores vary in size and shape, and manifest an interconnected lattice typical for each particular soil type, soil horizon, biological activity and human activity. The real size of pores, their continuity and forms are visualized microscopically on thin sections of soil (Kubiena, 1970, and Wilding and Hallmark, 1984). Computerized tomography CT (Crestana et al., 1986, and Steude et al., 1994), nuclear magnetic resonance NMR (Liu, 1994) and X-ray stereo radiographs (Tokunaga, 1988) offer new insight into 3-dimensional networks of pores.

X-ray computerized tomography CT and nuclear magnetic resonance NMR produce an image of a slice of finite thickness of a soil. If there is sufficient number of successive slices in the direction of the axis perpendicular to the slices, a 3-dimensional image can be obtained. Inasmuch as both techniques are non-invasive and do not disturb the soil system, they are suitable for repetitive measurements on the same soil body.

X-ray tomography utilizes a high energy X-ray source with the ray being detected after it passes through the plane of the soil. In order to obtain an image of a thin slice of soil, the source and detector (aligned at a fixed distance) rotate around the stationary soil sample, or vice versa. A series of attenuated X-rays are collected in order that all "points" on the slice are measured for a sufficient number of times. For each "point" on the matrix of the plane the data are analyzed with a computer to obtain values of soil bulk density or soil water content. Hence, the shape, size and continuity of the soil pores are obtained. The resolution of a typical medical CT which utilizes low energy X-rays is of tenths of a mm. The volume element (voxel) of such a medical CT is about 0.5-mm cube. Because the solid phase of a soil attenuates X-rays more strongly than the human body, higher energy X-rays and correspondingly smaller voxels are recommended for soils. Industrial CTs have voxels in the order of a 0.01-mm cube. A typical synchrotron CT having a resolution of a µm can be used to study thin water films and menisci in soil pores (Spanne et al., 1994).

NMR spectroscopic and imaging techniques utilize powerful magnets to reorient the nuclei of atoms with unpaired protons and neutrons. The atoms release energy upon removal of the magnetic field. In a practical sense, after nuclei in a static magnetic field are probed by a radio frequency magnetic pulse, they produce a measurable radio frequency signal. This signal is related to hydrogen concentration and thus for soils, it is related to the soil water content. Because NMR is also sensitive to many nuclei of non-zero spin angular momenta, an unwanted noise accompanies the detection of hydrogen. Ferromagnetic particles and films in soil are an annoyance. NMR is used primarily for the detection of preferential pores, or by-pass pores, their continuity and for measuring velocity fields in pores. The resolution of NMR is in tenths of a mm.

The form and distribution of soil pores can be quantified by the application of fractal theory and by methods of quantitative stereometry. However, neither of those progressive methods has been elaborated to a standard type of evaluation and for practical use in soil studies up to now. In soil hydraulics, the porous system is modeled and interpreted using soil water retention curves. Several different types of models have been used to describe experimentally measured soil water retention curves. Two types of models have received the greatest attention. The first is that derived from a bundle of parallel capillary tubes depicted in Fig. 4.19 in Chapter 4. This model has been used traditionally even though it has many disadvantages. The second is the lattice model of a 2- or 3-dimensional network shown in Fig. 4.21 of Chapter 4. This model is closer to reality especially for its applicability to percolation theory (Dullien, 1979, and Diaz et al., 1987). In both models, the size of pores is defined as a radius obtained from the equality of the hydrostatics in soil and that in the model. Hence, the size of soil pores is called the "equivalent radius" with no attempt made to quantify the real radius of the soil pores.

From the soil water retention curve, the equivalent pore size distribution curve is obtained from either of the models, see section 4.3.3. Examples are

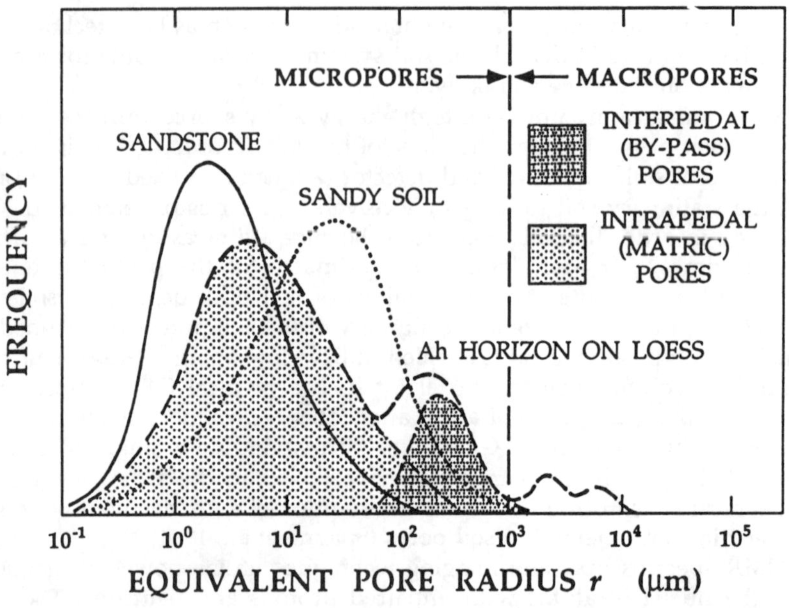

Figure 2.4. Schematic example of pore size distribution in parent rock and in soils.

given in Fig. 2.4. Inasmuch as the curves are smooth, they show a gradual change of the frequency of equivalent radii of pores. Because the nature of the curves is similar to probability distribution curves (probability density functions PDF), a quantitative description of the pore size distribution is achieved with the same parameters applied to a PDF. Hence, the mean equivalent radius is defined by the first moment and the second moment is the characteristic of the range of the equivalent radii of pores (Brutsaert, 1966).

Figure 2.4 provides pore size distributions of a sandstone and a sandy soil. The curve for the parent rock is different from that for the sandy soil. This difference manifests the alteration of the porous system caused by pedogenesis. The mean equivalent pore radius in the soil increases concomitantly with an increase of the range of equivalent radii. Similarly, comparisons of pore size distributions could be used to identify seasonal variations associated with swelling and shrinking, compaction and disaggregation.

Pore size distributions may have more than one modal pore size. The curve for the Bt horizon over a loess in Fig. 2.4 has two peaks. The principal peak represents the modal pore size within the soil aggregates (intrapedal pores) while the secondary peak represents the size of the inter-aggregate (interpedal) pores. The separate single modal curves can be derived from the bimodal pore size distribution curve from the Bt horizon of the loess. The value of the mean pore size from each curve can be related to transport processes in the soil. The interpedal pores are called "by-pass" or "short-cut" pores together with macropores because they both conduct an accelerated flux (Bouma,1984, and Othmer et al., 1991). Thus, the subcategory of interpedal micropores is grouped together with macropores into the class of pores preferentially conducting water and solutes. The intrapedal pores are frequently separated from the larger interpedal pores by various cutaneous film-like forms which cover the aggregates. Very small pores dominate within the cutaneously covered aggregates and the permeability within the aggregates is much smaller than that of the whole soil.

In addition to the above pore size distributions, a class of macropores can exist on the frequency curve illustrated at radii greater than 10^3 µm in Fig. 2.4. Macropores are developed frequently in the A horizon, partly caused by biological activity (e.g. earthworm holes and decayed roots), partly owing to shrinkage during drying of soils when abrupt fissures develop. Macropores that form as the result of plowing exist only temporarily between clods owing to internal instability. While biologically-induced macropores are usually stable, they are nevertheless destroyed primarily by cultivation.

Owing to the bimodal porosity and to macropores, the size of the REV increases substantially. According to Bouma (1984), a minimum of 20 elementary units of soil structure form a suitable REV.

2.5 SOIL SPECIFIC SURFACE

Soil specific surface A_m is the interface between the solid phase and the liquid and gaseous phases of soil related to the unit mass or volume of the soil. If A_m is expressed as the area related to the mass of soil, it has the dimension of L^2M^{-1}. If the area is related to the volume of soil, the dimension is $L^2L^{-3} = L^{-1}$. The magnitude of A_m ranges from 10^0 to 10^2 $m^2 \cdot g^{-1}$ or 10^6 m^{-1} to 10^8 m^{-1}.

Soil specific surface of sands can be estimated from a spherical soil particle model. This model fails for all other textural classes. The soil is assumed to be composed of spherical particles of known radii. The frequency distribution of the spherical particles is taken from the results of the textural analysis of the soil. The continuous function is discretized and in each i-th particle size group the specific area $(A_m)_i$ is calculated. The magnitude of A_m is simply

$$A_m = \sum_i (A_m)_i f_i \tag{2.8}$$

where f_i is the mass fraction of the i-th particle size group.

Soil specific area is usually measured by the amount of a substance required to adsorb a thin film over the surface of each and every soil particle. When dry soil is placed in an atmosphere of gas of a certain pressure, gas molecules attracted to the solid particles form a thin film on their surfaces. Adsorption theory (Brunauer et al., 1938) defines the gas pressure at which an average one-molecule thick film (monolayer) is formed. The amount of the adsorbed gas is measured either volumetrically or simply gravimetrically. Assuming the area occupied by one molecule, the soil specific surface is calculated.

The internal and external specific surfaces of clay minerals can be measured by using particular gaseous adsorbates. Because non polar adsorbates such as N, Ar and CCl_4 are adsorbed on only the external surface of clay minerals and other colloidal particles, A_m represents only the external surface. The value of A_m determined by these and other non polar molecules is therefore substantially smaller than that measured by the adsorption of polar molecules. On the other hand, polar adsorbates such as water or ethylene glycol cover both the internal and external surfaces of clay minerals. Only in allophanes, the organic polar molecules cannot enter the internal surface because this internal space is not expandable as it is e.g. in montmorillonites. Water vapor tends to form only one monolayer in the interlayer space of montmorillonites while ethylene glycol forms two layers, thus covering each of two neighboring surfaces by one monolayer. On the other hand, water forms thick hydration envelopes on exchangeable cations. Therefore, values of A_m as measured by these two different adsorbates are not equal.

Table 2.2. Specific surface of soils and soil components determined by adsorption of water vapor (Kutílek, 1962).

Soils and soil components	Specific surface
	$m^2 \cdot g^{-1}$
kaolinites	4 - 10
montmorillonites	150 - 500
illites	50 - 80
silt of quartz (<60 μm)	1.2
silt of quartz (<0.1 mm)	0.1
loamy cambisol	52
light clay rendzina	84
loamy gley	62
sandy loam podzol	24
silty loam chernozem	87
humus in podzol	37
humus in chernozem	940

Determination of the soil specific surface with water vapor is a very simple procedure. The dried soil is placed in a chamber where the relative pressure $p/p_o = 0.2$ (p is the vapor pressure and p_o the saturated vapor pressure at the given temperature). The pressure p is controlled either by a saturated solution of K-acetate or by 58% H_2SO_4 at $T = 20°C$.Water content w_m after reaching equilibrium is determined gravimetrically and $A_m = 3610w_m$ (Kutílek, 1962b). The range of A_m for soils and their constituents is given in Table 2.2. The method is based upon the evaluation of adsorption isotherms by the BET (Brunauer et al., 1938) modified to porous bodies where the size of pores may limit adsorption. When ethylene glycol is used as the polar adsorbate, the soil is initially over saturated with ethylene glycol and subsequently desorbed over a $CaCl_2$-glycol solvate. The evaluation is empirical (Mortland and Kemper, 1965).

PROBLEMS

1. Derive the identity of (2.1) and (2.6).
2. Estimate the soil specific surface of a sand having the following particle size distribution: particles < 2 mm ··· 100% by mass, < 0.63 mm ··· 88%, < 0.45 mm ··· 72%, < 0.10 mm ··· 16%, < 0.063 mm ··· 9% and< 0.045 mm ··· 0%.
3. Calculate the bulk density of a soil with volume fractions of solid 0.45 and of liquid 0.25 when the particle density is $\rho_S = 2.7\ g\cdot cm^{-3}$.
4. Calculate the weight of 0.5 m^3 of the soil from the previous problem.
5. What is the relationship between aggregate porosity P_A and the average porosity of aggregates $\overline{P}_1$? Is $P_A > \overline{P}_1$ or $P_A < \overline{P}_1$?
6. The theoretical value of the specific surface of montmorillonite is 800 $m^2\cdot g^{-1}$. What will be its estimate determined by water vapor adsorption, if its external surface is 50 $m^2\cdot g^{-1}$?
7. A well-aggregated A horizon is deteriorated by destroying its soil structure. What change, if any, will you expect in the value of the mode of the pore size distribution curve?

3 SOIL WATER

The liquid phase of a soil is never just pure water. There are always mineral salts and organic substances dissolved in the water. And, until recent times, they were usually thought of to be at such small concentrations that they were neglected. Saline soils were the exception where the concentration of salts was not neglected. Generally, in spite of the importance of its solutes, the liquid phase of soil is called soil water. Here, in this and other chapters the traditional approach of neglecting the solutes will be sustained unless their importance should be recognized in some parts of later chapters.

3.1 SOIL WATER CONTENT

Soil water content is defined in relation to either the volume of soil

$$\theta = V_W / V_T \tag{3.1}$$

or to the mass of the dry soil

$$w = m_W / m_S \tag{3.2}$$

where V_W is volume of water, V_T the bulk volume of soil, m_W the mass of water and m_S the mass of dry soil. All symbols related to soil water content are listed in Table 3.1. Soil air content is then $\theta_A = P - \theta$, or $\theta_A = V_A/V_T$.

Table 3.1. Symbols related to soil water content.

	solid phase	soil water	soil air	whole soil
volume	V_S	V_W	V_A	V_T
mass	m_S	m_W	m_A	m_T

particle density	$\rho_S = m_S/V_S$
bulk density	$\rho_T = m_S/V_T$
porosity	$P = (V_W + V_A)/V_T$
volumetric soil water content	$\theta = V_W/V_T$
mass soil water content (by weight)	$w = m_W/m_S$
soil water ratio	$\vartheta = V_W/V_S$
volume of pores	$V_P = V_W + V_A$

Expression (3.1) is preferred as it offers direct information regarding the ratio of pores filled with water. The storage of water W is simply obtained for the layer from z_1 to z_2 by

$$W = \int_{z_1}^{z_2} \theta \, dz \tag{3.3}$$

where W is in cm or mm, the same units as those for z. Because this dimension of length is compatible with that of rainfall or evaporation, W is a very useful tool for hydrologic balancing. For example, if the difference in θ measured before and after a rain $\Delta\theta = 0.2$ in the top 12 cm, the amount of water that infiltrated is $\Delta W = 0.2 \cdot 12$ cm = 2.4 cm. When we measure $\overline{\theta}$ as the mean value of each layer in a soil profile, the water stored in the profile is just the sum of W for each of the individual layers.

In loams and clays, the volume of the soil is usually dependent upon θ. If, after we take an undisturbed moist soil sample, the soil dries, the soil shrinks with its volume being reduced. On the other hand, when a soil wets, it swells with its volume increasing, see Fig. 2.3 (Chapter 2). Inasmuch as V_T is not constant in (3.1), we replace it by a constant reference parameter V_S, the volume of the solid phase. We obtain the soil water ratio ϑ (Smiles and Rosenthal, 1968)

$$\vartheta = V_W / V_S \tag{3.4}$$

where ϑ is dimensionless and is analogous to the void ratio e [see (2.2)] as θ is analogous to P.

We distinguish three domains in $V_T(V_W)$ or in $e(\vartheta)$ based upon the magnitude of change of V_T as a function of the content of water V_W, see Fig. 2.3 (Haines, 1923):

1. In the *normal domain*, the change of the soil volume V_T equals the change of the water volume V_W. Hence, $de / d\vartheta = 1$ and $e = \vartheta$.
2. In the *residual domain*, the change in the volume of soil V_T is smaller than that of the water V_W. Hence, $0 < de / d\vartheta < 1$ and $e > \vartheta$. When the soil is drained or dried, only a portion of the volume initially occupied by water is replaced by air. And, when the soil wets, the soil volume increases, but a portion of the pores remains filled with air.
3. In the *zero domain* characterized by $de / d\vartheta = 0$, e = constant and $e > \vartheta$, V_T has reached its minimum value and remains constant provided that the water content is below that threshold value between the residual and zero domains. The replacement of air by water corresponds to an increase of θ or ϑ, respectively.

The normal domain is sometimes preceded by the structural domain (Yule and Ritchie, 1980) where water is lost from a very wet soil without a macroscopic change of the volume of soil, $de / d\vartheta = 0$. Water is released from the stable macropores. When we deal with clods or with remolded soil material, this phase may not be detected. Threshold values of θ between individual domains depend upon soil texture, soil structure, mineralogy of the clay fraction, organic matter content, exchangeable sodium percentage and the electrical conductivity of the soil solution.

Mutual relationships between the various expressions of soil water content are simply

$$\theta = w\rho_T / \rho_W = w\rho_T \tag{3.5}$$

$$\vartheta = (1+e)\,\theta \tag{3.6}$$

and

$$\vartheta = w\rho_S \tag{3.7}$$

where ρ_S is the mean density of the soil particles, ρ_T the soil bulk density and ρ_W the density of water. The relative filling of pores by water is described by the relative soil water content θ_R

$$\theta_R = \theta / P \tag{3.8}$$

where $0 \le \theta_R \le 1$ or by the effective soil water content θ_E

$$\theta_E = (\theta - \theta_r)/(P - \theta_r) \tag{3.9}$$

where θ_r is the residual soil water content (when θ is so small that liquid water continuity practically ceases to exist). In the field when $\theta_S < P$ owing to entrapped air, P in (3.9) is usually replaced by θ_S.

3.2 MEASUREMENT OF SOIL WATER CONTENT

There are destructive and non-destructive procedures for measuring soil water content. Destructive procedures entail a soil sample being taken from the field each time a water content value is desired. The water content of the soil sample, usually obtained with a soil auger or soil core sampler, is subsequently determined in the laboratory. The field soil profile is thus disturbed each time a soil water content sample is removed. The two main disadvantages of the method are obvious. First, the soil profile is continually disturbed with an ever-increasing number of holes that form a system of macropores that influence the water regime of the original soil profile. Second, the repetition of soil water observations at a particular location is not feasible. The difference between soil water content values at two consecutive times of sampling is caused by a real change of soil water content as well as soil heterogeneity. Values of soil water content at the two different locations at the same time are not identical owing to spatial variability. Nonetheless, the method is appropriate when the soil water content is measured infrequently. It is also well adapted for measuring the soil water content near the soil surface horizon, especially when frequent cultivation is practiced. And, the method is regularly used for the construction of calibration curves in non-destructive methods.

Non-destructive methods rely upon a sensor permanently placed in the soil with an evaluation unit connected to its cables at the time of measurement or the sensor being inserted each time an observation is desired into the soil by means of a vertical hole lined with an access tube. Repetitive measurements throughout a long term period does not disturb the soil and values of soil water content are known immediately after reading the sensor.

According to the nature of the methods, the procedures are classified as either indirect or direct. Indirect methods are based upon the measurement of a physical property of the soil dependent upon the soil water content. Direct methods allow the mass of water in the soil itself to be measured.

The dependence of a soil physical parameter upon the soil water content is theoretically related to the physical parameter of each constituent in the mixture according to the relation (Mushkin, 1971)

$$\sum_i \left(\frac{b_i - b}{b_i + 2b} \right) C_i = 0 \qquad (3.10)$$

where b_i is the physical parameter of the i-th constituent or phase, respectively, in the mixture, b the parameter for the entire mixture (i. e. the bulk soil) and C_i the volumetric concentration of the i-th constituent or phase, respectively. Equation (3.10) is valid for a spatially independent distribution of each of the constituents. For various forms of clustering and for spatially dependent geometrical arrangements of the constituents, (3.10) must be modified. Considering all of the difficulties in the application of (3.10) to real soils, it is simpler and more accurate to prepare an empirical calibration curve for the particular method and soil.

If the sensor is at a remote distance from the soil, the methods are known as remote sensing.

A great number of methods for measuring soil water content are reported in the literature. The most common methods as well as those commercially available or distinctively attractive are presented here.

3.2.1 Gravimetric Method

This method, a direct, destructive procedure, is a standard technique commonly used to obtain reference data on soil water content. It is used as a standard for the construction of calibration curves of indirect methods.

The soil sample is taken in the field either as an "undisturbed" sample of known volume V_T (core sample) or as a disturbed sample where the volume V_T is irrelevant and not measured. In the laboratory the sample is weighed, oven-dried at the temperature of 105°C and weighed again. The change of weight is assumed to be owing to the removal of water. When a high accuracy is required, the drying process is controlled by weighing and drying until a constant soil weight is reached. In routine procedures, 24 hours is taken as adequate for drying 100 cm^3 of soil. For organic soils, drying is achieved at a lower temperature, usually at 60°C in partial vacuum with the use of desiccants. Drying organic soils at the lesser temperature diminishes substantially the loss of organic matter by volatilization.

The accuracy of the method depends upon proper sampling – water should not drain out of the sample if the water content is large, and water lost by evaporation during the period between sampling and weighing should be prevented. The degree of perfection achieved with the drying process also contributes to the accuracy of the method. The advantage of the method is the direct determination of the mass of water without a special calibration. Additionally, the procedure is technically simple and easily carried out with routine laboratory equipment. On the other hand, the procedure is laborious and time consuming, repetition of measurements at the same location is not feasible and the difference in soil water content for two consecutive readings includes the impact of spatial heterogeneity. Moreover, repetitive measurements on a small area permanently alter the original nature of the soil profile. Modifications of the method include the replacement of drying with an extraction of the soil water or with a measure of a chemical reaction between the soil water and a specific chemical compound. These modifications speed up the procedure but usually reduce the accuracy of the method.

3.2.2 Electrical Resistance Method

This indirect, non-destructive method is based upon the change of the electrical resistance R_E of a soil with a change of soil water content θ. The relationship $R_E(\theta)$ can be either hyperbolic or exponential. A hyperbolic relationship is

$$R_E = \frac{a}{\theta^b}(1+\alpha\Delta T) \qquad (3.11)$$

where a, b and α are empirical constants depending upon the concentration and nature of the dissolved matter in the soil water, the soil bulk density, the nature and geometry of the solid phase and ΔT the temperature difference from the reference temperature. An exponential relationship for $R_E(\theta)$ can also be formulated.

The electrical resistance is measured between two metallic electrodes inserted into a porous body. In order to keep a constant geometrical arrangement of the electrodes and their good contact with the porous body, the electrodes are placed in a small porous block made of gypsum, nylon or other material which does not readily deteriorate when buried in soil. Although the water content of the block is in quasi-equilibrium with that of the soil, their values are not identical. True equilibrium is reached through the equality of the total potential head H in the block and that in the neighboring soil, see in detail Chapters 4 and 5. The relationship between H and the water content of either the soil or the block is strongly hysteretic. That is, the relation depends upon the magnitude and direction of the previous water content. Did the water content increase or decrease, and at what rate of change. As the nature of hysteresis is not identical for both the soil and the porous block, an error is inherent in the measurement of soil water content using the electrical resistance method.

Electrical resistances of gypsum blocks are relatively insensitive to the concentration of solutes in soil water compared with those of blocks produced of inert materials. Because of the slight water solubility of gypsum, the electrical resistance between the electrodes is constantly buffered by the dissolved gypsum even when variations of soil salinity exist owing to fertilization or salinization within soil profiles. However, the slow dissolution of gypsum results in deterioration of the blocks in relatively short time periods of less than 1 to 3 years.

Calibration curves needed for the determination of θ from a resistance R_E are carried out either in the field or in the laboratory. The field procedure is more laborious but the calibration obtained is more site specific. Blocks are installed in the field and when the electrical resistance is measured, the soil water content is determined gravimetrically. In the laboratory, the blocks are placed together with the surrounding soil into the pressure plate apparatus. After an initial wetting the blocks and soil are drained by incremental increases of applied pressure. In the laboratory procedure, the soil is usually packed artificially, and therefore, additional recalibration in the field is recommended. An example of some calibration curves is given in Fig. 3.1.

The relationship $R_E(\theta)$ is very steep in dry soil for θ_R ranging between 0 and about 0.35. Near water saturation for $\theta_R > 0.9$, the resistance R_E is not sensitive to water content changes. Hence, the method is not fully applicable to the entire range of θ, and as a result, the method is only infrequently used.

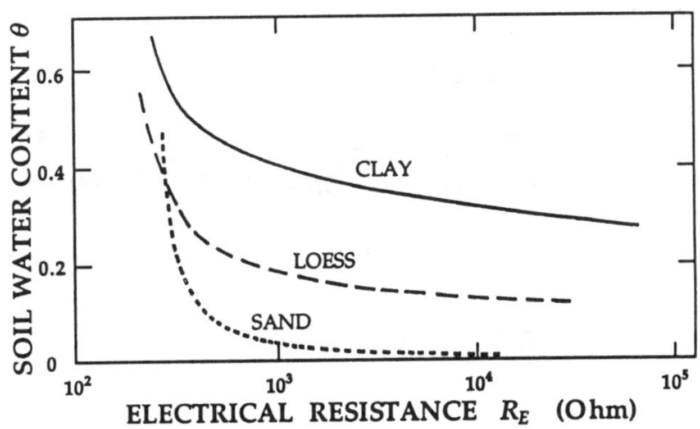

Figure 3.1. Examples of calibration curves for the electrical resistance method carried out in the laboratory with gypsum blocks for three soil samples.

3.2.3 Capacitance Methods

The dielectric constant of a substance depends upon the polarization of its molecules in an electric field. If the molecules have a permanent dipole, this permanent orientation of poles contributes to their induced orientation in the electric field. Inasmuch as water molecules have a large permanent dipole moment, the dielectric constant ε of water is 81 for frequencies less than 1000 Mhz. Values of ε for the soil solid phase are about 4 to 8 while that for air is 1. Because the soil is a mixture of solid, air and water, with a constant content of solid phase, the value of ε for the mixture depends strongly upon the water content of the soil. The dependence of ε upon soil water content θ is illustrated in Fig. 3.2.

When a capacitor contains soil as the dielectricum between the two electrodes, measurements of ε for the soil can be used for estimating the value of θ. Capacitance methods are therefore based upon the measurement of either ε or of an electrical parameter dependent upon ε. The methods belong to the indirect, non-destructive procedures.

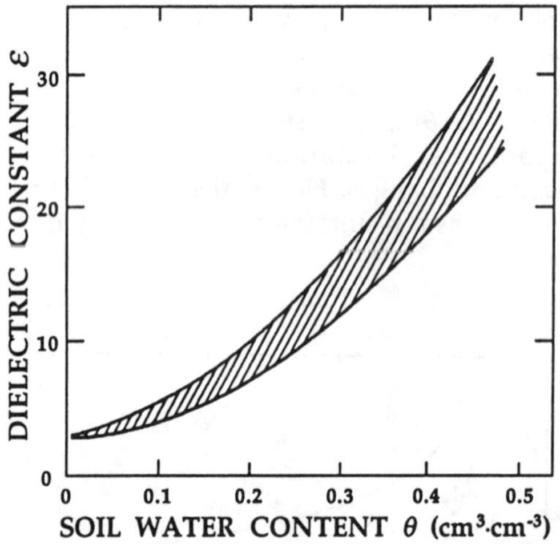

Figure 3.2. Range of dependence of the dielectric constant ε of the soil upon the soil water content θ. The relationship depends upon soil type and frequency range of the apparatus.

A simple, cheap instrument is the resonance meter utilizing a very high frequency of 30 to 60 Mhz to measure the ε of the soil (Kuráž et al., 1970). A choice of lower limit of 30 Mhz prevents the macroscopic electric dipoles associated with interfacial polarization from responding to the applied electric field. More sophisticated instruments are available (Kuráž and Matoušek, 1977, and Dean et al., 1987). The capacitor usually has a cylindrical form housing the electronic circuitry with two metallic bands placed on the periphery of the cylinder at a vertical distance of about 5 to 8 cm. The electrical field is formed in the neighboring soil around the electrodes to a radial distance of influence approximately 50% greater than the distance between the electrodes, see Fig. 3.3. Soil in this ring of influence is the dielectricum of the capacitor. The radial distance should be maximized while the depth of resolution should be minimized. A reasonable compromise for the sensor given in Fig. 3.3 is a distance of 5 to 8 cm between the electrodes.

For the purposes of taking measurements in the field, the sensors are permanently buried at desired soil depths or are inserted into a plastic access tube. Values relative to θ are read on the evaluating instrument linked by a cable to the sensor.

A calibration curve of the measured electrical parameter versus θ unique for each soil is established. The majority of the calibration curves is nearly a straight line, especially when the restricted range of θ in practical field situations is considered. A linear calibration curve provides a constant accuracy of the estimation of the soil water content and for practical use, only two data readings are required for field calibration. First, θ is determined gravimetrically on soil samples taken when the vertical holes are drilled for the installation of the

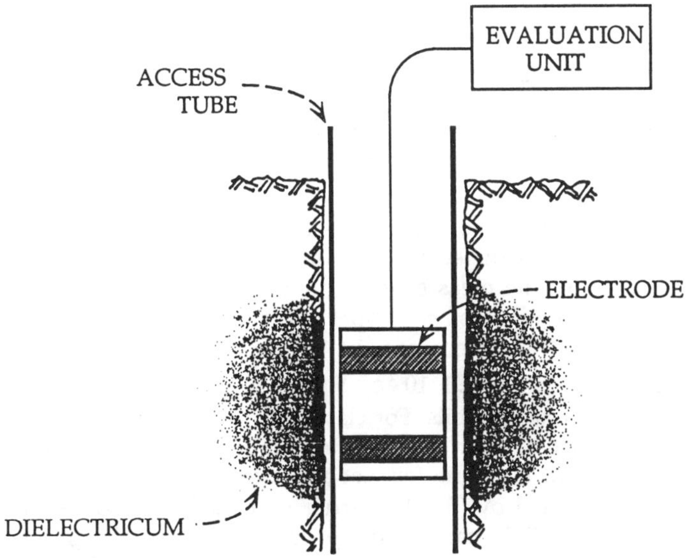

Figure 3.3. Schematic arrangement of the instrumentation of the capacitance method (Kuráž and Matoušek, 1977) for field measurements of soil water content.

access tubes or for burying the sensors. The sensors are immediately read at the same depths as the samples were taken. Inasmuch as the installation is easier in wet soil, the first reading is for wet soil. A second reading accompanied with soil sampling is taken when the soil is relatively dry. Soil is sampled at distances of about 1 m from the access tube. A straight line drawn through the two points serves as a calibration.

Advantages of the capacitance method are its theoretical accuracy, fine resolution with depth and relatively low price. The method is able to detect steep gradients in q as well as a freezing front in the soil. Mainly because the water in close vicinity of the soil solid particle surfaces does not have the same value of e as that of free water, calibration is required for each soil. A second disadvantage of the method is its sensitivity to small irregularities and cavities along the access tube. In swelling-shrinking soils fissures frequently develop between the material of the access tube and the soil. Inasmuch as these fissures behave as macropores and contain only air except during flooding, they contribute substantially to the dielectric constant of the soil and render readings of q inaccurate.

Time domain reflectometry (TDR) belongs to the methods based upon the variation of e of soil with soil water content. The propagation velocity of an electromagnetic wave at a frequency above 50 MHz along parallel metallic rods inserted into soil depends upon e as

$$v = c / \sqrt{\varepsilon} \tag{3.12}$$

where c is the velocity of light in vacuum ($3 \cdot 10^8$ m·s^{-1}). Although the propagation velocity of the pulse is not directly measured, it is deduced from the length of the rods L along which the wave travels. The transit time t is the time interval between the entrance of the wave into the soil and the reflection of the signal after it reached the end of the rod. Substituting the propagation velocity calculated from $v = 2L/t$ into (3.12), we obtain

$$\varepsilon = (ct/2L)^2. \tag{3.13}$$

We note for the majority of soils that it is acceptable to neglect the imaginary component of the complex dielectric constant in (3.13).

The main aim of the measuring technique is to read the transit time t of the pulse and obtain a value of ε. Topp et al. (1980) established the empirical relationship for the soil water content

$$\theta = -5.3 \cdot 10^{-2} + 2.91 \cdot 10^{-2} \varepsilon - 5.5 \cdot 10^{-4} \varepsilon^2 + 4.3 \cdot 10^{-6} \varepsilon^3 \tag{3.14}$$

which is acceptable for sandy soils. For clays and organosols, a unique, empirical calibration $\varepsilon(\theta)$ for each soil series or soil type is recommended. If we have only two measured values of $\theta(\varepsilon)$, one for wet and one for dry soil, we translate and rotate the curve expressed by (3.14) in order that the curve passes through our two measured points. This calibration procedure is simple and sufficiently accurate for any soil. On the other hand, (3.14) will suffice for most all soils (Zegelin, 1992) if our field problem requires knowing only relative changes of θ.

Because the transit time in (3.13) is extremely small (10^{-9} s), the measuring technique must be highly sophisticated. The TDR unit must have the capability of measuring the time elapsed from the time the propagated wave enters the soil until it arrives along the rod wave guide to the end of the rod. The electronic equipment required for this method is the most expensive of soil water content techniques.

The main components of a TDR unit are a signal generator which produces the voltage pulse, a voltage sampler which measures the voltage on the transmission line at selected time intervals and a timer which controls the action of both components. An internal computer produces and analyzes the measured voltage with the data displayed as a voltage versus time relationship. Coaxial cables connect the TDR unit to the transmission lines (rods) which are inserted in the soil. If we work with two rods as wave guides, an impedance-matching transformer is placed between the TDR and the transmission line. By using three rods, the necessity of a transformer is eliminated. In addition to this advantage, 3 or 4 rods offer a more clear signal (Zegelin et al., 1992). The minimum length of rods is dictated by the resolution of travel time, and therefore, L = 10 cm is usually taken as the minimal limit although Malicki et al. (1992) recently succeeded in using lengths of rods substantially below this value.

When a pulse passes the interface of two materials having different values of ε, a portion of the signal is reflected and a portion passes the interface. With more than one peak appearing on the voltage versus time graph in order to read the reference and resolution times, the accuracy of the method is improved (Malicki et al., 1992). It is important to keep the rods strictly parallel within the soil. After entering soil, the pulse travels along the rods and the voltage of the signal gradually decreases. At the end of the rods, because the

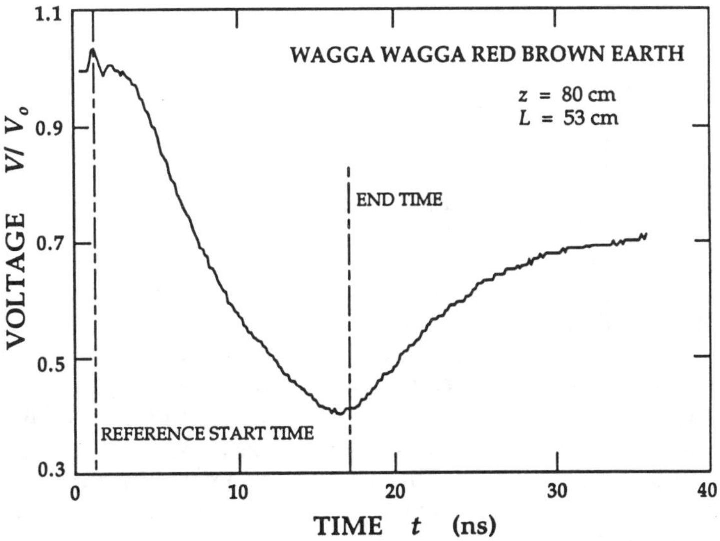

Figure 3.4. Normalized voltage versus transit time for a 3-rod probe used to measure soil water content with the time domain reflectometry method (Zegelin et al., 1992).

wave is reflected back along the wave guides and increases the voltage, the travel time is relatively distinct, see Fig. 3.4. The metallic rods may be installed vertically, horizontally or at any angle within the soil profile. All orientations give ε and θ in the soil averaged over the length of the rods. When steep changes of soil water content with depth are expected, the rods should be installed horizontally. When the rods are permanently buried in the soil, they are permanently attached to cables which are terminated in a connector panel above the soil surface where the TDR unit is attached during data reading.

Salts dissolved in the soil water influence the height of the transmitted pulse. A larger concentration of salts causes a greater electrical conductivity and a larger portion of the electrical signal to be lost into the surrounding soil. Using this principle, Dalton (1992) simultaneously measured the contents of both solute and water in soil.

3.2.4 Gamma Radiation Methods

Gamma radiation offers indirect, non-destructive methods for measuring soil water content. When gamma radiation, a highly penetrating electromagnetic form of energy, penetrates matter, it is absorbed and scattered. Photoelectric absorption occurs owing to the interaction of γ photons with electrons in the K or L shell of the absorbing atom. Having absorbed all of the energy of the photon, the electron is ejected from its shell. Gamma radiation is absorbed and scattered during the Compton effect when the γ photon interacts with the outer electron of the absorber atom. After only part of its energy is absorbed by the electron, the photon with lower energy is scattered in a new direction. Eventually, after multiple scattering, the lower energy photons are absorbed photo electrically. Pair production is a nuclear reaction that reduces the energy of a γ photon by causing ionization. It does not play a role when the initial energy of the γ radiation is less than 1.02 MeV.

Absorption and scattering, the effects upon which the γ radiation methods are based, are each dependent upon the density of the soil ρ_v [$= (m_W + m_S)/V_T$]. If the bulk density ρ_T ($= m_S/V_T$) is kept constant, the effects depend upon only the soil water content m_W/V_T ($= V_W/V_T = \theta$).

The gamma radiation attenuation method is based upon absorption of monoenergetic radiation. The most commonly used sources of monoenergetic radiation are ^{241}Am with a radiation energy of 60 keV and ^{137}Cs with a radiation energy of 662 keV. Radiation from the source is collimated by a narrow window made in lead to reduce the scattering of radiation before it reaches the detector. The beam of photons that pass through the soil is detected by the scintillator. Its pulses are counted in an appropriate time interval, usually 30 s or 1 min. Attenuation of the γ rays is described by

$$\frac{dI}{dx} = -\nu \rho I \qquad (3.15)$$

where I is the intensity of the γ beam in the direction of x and ν and ρ are the mass attenuation coefficient and density of the matter, respectively. After integration over the distance L, the length of beam path through the matter, and with the attenuation coefficients of the soil, water and the walls of the container or the access tube (3.15) becomes (Middleboe et al., 1976)

$$\theta = \frac{\ln\left(I_d / I_w\right)}{\nu_W L} \qquad (3.16)$$

where I_d and I_w are the beam intensities for dry and wet soil, respectively, ν_W is the attenuation coefficient of water and L the distance between the source and the detector. The method is excellent in the laboratory where the geometry is accurate, a window for collimation at the detector is available and a pulse height analyzer discards all energies below that of the original monoenergetic radiation. For a field soil surface reading of θ, the source is placed at a small depth and the detector is just on the surface, see Fig. 3.5. In the field it is difficult to install the two access tubes strictly parallel at a given distance L.

When the masses of soil and water both change within the volume being measured, additional observations are necessary. In the laboratory, two sources of different monoenergetic radiation are used (Nofziger and Swartzendruber, 1974). In the field, two independent methods of measurement of θ and ρ_T have to be used. In addition to γ radiation which is sensitive to both, one of the methods sensitive only to θ should be applied, as e.g. the capacitance method.

Gamma scattering, based upon the Compton effect, provides two techniques for field measurement of soil water content or soil bulk density. The methodology is nearly identical to that of attenuation (see Fig. 3.5), except that the calibrations are empirical and have to be established for each situation.

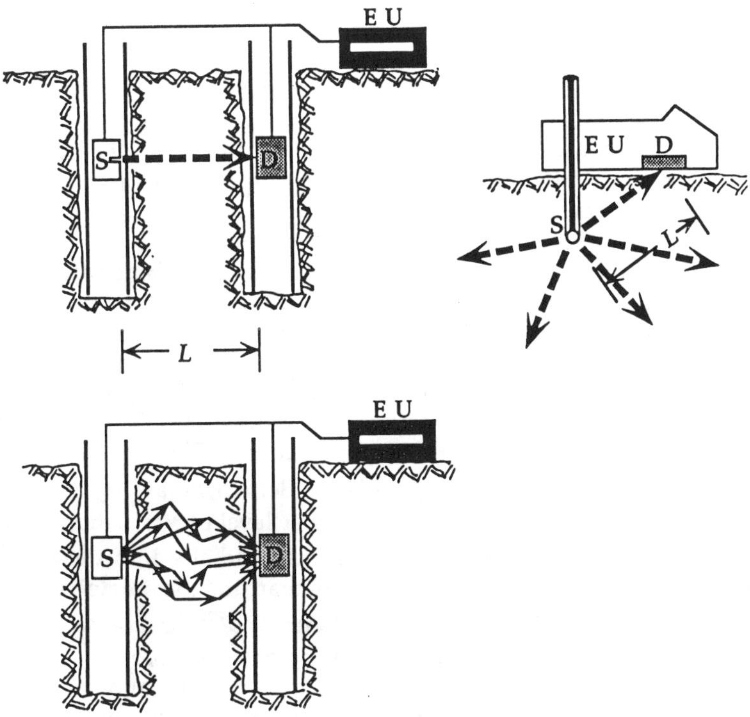

Figure 3.5. Schematic arrangement for γ-attenuation (top part) and γ-scattering methods (bottom part) in field measurements of soil water content. S is the source of γ-rays, D the detector and EU the evaluation unit.

3.2.5 Neutron Method

The neutron method, an indirect, non-destructive procedure, utilizes a source of fast neutrons (kinetic energy of about 2 MeV) that are slowed down when they interact with the medium surrounding the source. Through collisions with the nuclei of elements in the soil, the originally fast neutrons lose their energy down to that of thermal neutrons (about 0.025 eV), i. e. to the energy related to the temperature of the medium. The lighter is the nucleus, the more energy is lost for one elastic collision. Because small atoms are the best moderators of fast neutrons, the most efficient atom for neutron moderation is ^{1}H. Only 18 collisions are needed to reduce the original 2 MeV energy to 0.025 eV. For ^{7}Li, the number of collisions is 68, for ^{12}C it is 115 and for ^{16}O it is 152. Owing to the two hydrogen atoms in its molecule, water is an excellent moderator of fast neutrons. The wetter the soil, the greater is the number of slow neutrons. Thermalized neutrons form a cloud surrounding a source of fast neutrons. Although the size of the cloud in a dry soil is greater than that in a wet soil, the concentration of slow neutrons is smaller in the dry soil. This behavior is employed in the neutron method.

Fast neutrons disintegrate and also are absorbed by other nuclei. Free neutrons are not stable and disintegrate with a half-life of 13 s. Neutron absorption depends upon the energy of the neutron and the nature of the surrounding nuclei. With absorption a new nucleus is formed. In some instances when the new nucleus is unstable, it disintegrates emitting radiation. Nuclei of this type of reaction are rarely present in soil and their concentration is so small that this induced radioactivity cannot arise. Generally, there is no hazard of artificially inducing radioactivity in either the soil or the access tubes by using a source of fast neutrons.

The equipment of the neutron method consists of a protective shield that houses the probe and the electronic counting system. The probe is a metallic cylinder containing the radioactive source of fast neutrons. The cylinder is about 3 to 4 cm diameter with a length of about 20 cm. The neutron source is a mixture of an α emitter and a fine powder of Be. Nowadays, ^{241}Am is usually preferred as the α emitter because it has a lower level of γ radiation than that of the originally used ^{226}Ra. Fast neutrons, emitted radially into the surrounding medium, are the product of α particles bombarding Be nuclei and have an energy of about 2 MeV. The half-lives of ^{241}Am and ^{226}Ra are both sufficiently long to ensure a constant activity during the use of the probe.

Because the source emits α particles, fast neutrons and γ radiation, the user is protected from the radiation by a shield. During the measurement, the depth probe is inserted into the soil access tube and separated from the shield. The shield is an integral component of the neutron surface gauge and is not separated from the neutron probe, see Fig. 3.6. Shielding of the fast neutrons is realized by their moderation in a material containing large amounts of hydrogen (e.g. paraffin or polyethylene). The walls of the shield are made of lead for protection against γ radiation. When the depth probe is released from the shield and lowered into the soil, the user is protected by the soil.

Access tubes are made of aluminum or steel. The material of the tubes has little influence on the thermalization of fast neutrons. The variation of OH^- groups in soil minerals does not play a practical role in the thermalization of fast neutrons. Although in organic soils, the large content of H and C in organic molecules contributes to the thermalization, its effect on the calibration can be considered constant and part of the background observation of a dry soil.

In close proximity to the source is the detector which detects only slow neutrons – not fast ones. Slow neutron pulses are preamplified in the probe and after that, sent through a cable to the counting system. Counting rates are always indicated in counts per minute (cpm). Modern neutron probes have a microprocessor which keeps in memory the calibration curve so that soil water content is directly read or recorded.

The calibration curve (Fig. 3.7) is the relationship between the relative pulse count rate and the soil water content. In order to avoid electronic drifts, the influence of temperature etc., the pulse count rate measured for a soil location is divided by that of a reference standard material inside the protective

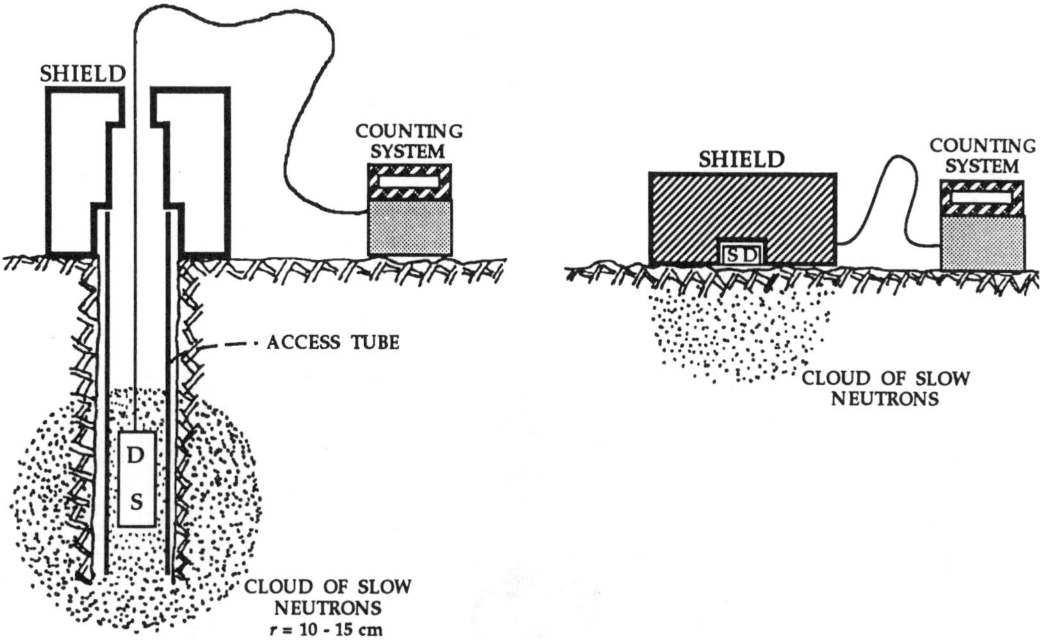

Figure 3.6. Neutron probe used for measurement of soil water content in the profile (left) and on the soil surface (right).

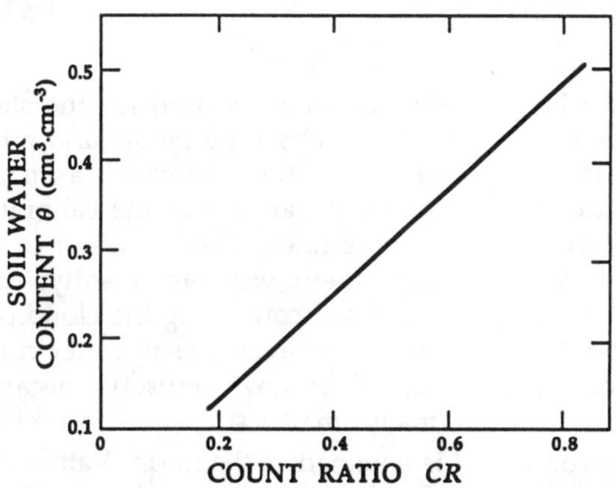

Figure 3.7. Example of the calibration curve of soil water content θ versus count ratio CR for a neutron depth probe.

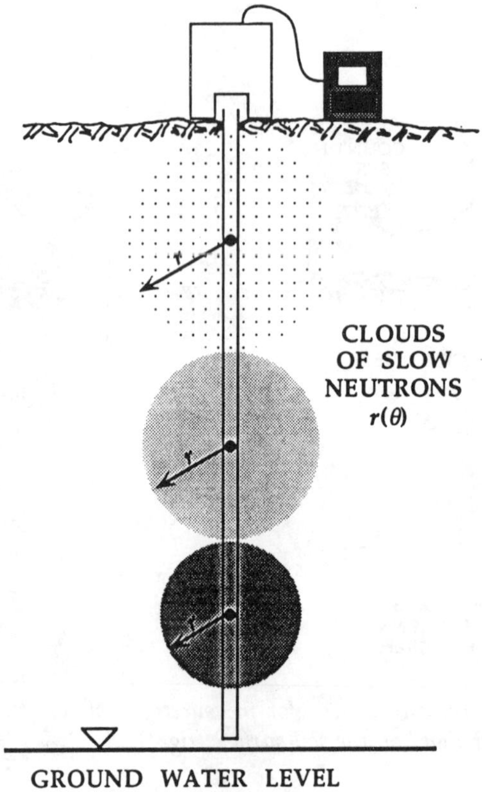

Figure 3.8. Spheres of influence in the field measurement of soil water content with a neutron probe.

shield. Sometimes a large volume of water is used for the standard reference material. The value of the reference is observed before and after each series of soil measurements, or at whatever time interval assures precision of measurement. Hence, the independent variable of the calibration curve is the count ratio CR = [(soil rate)/(reference rate)]. The soil water content obtained by gravimetric sampling is the dependent variable. Ideally, the same sample should be taken as is the portion of soil containing the cloud of slow neutrons. The CR reading should correspond to an average soil water content responsible for the 20 to 30 cm diameter cloud of slow neutrons. Because such a large sample cannot easily be taken, many smaller soil samples are taken for each CR reading and the appropriate $\overline{\theta}$ is obtained as the mean. Values of CR and θ have to be taken at various stages of soil wetness to obtain a calibration curve over the entire soil water content range. A linear regression initially suffices although a polynomial can be evaluated with perhaps some slight curvature may eventually result using an abundance of observations.

When a great accuracy in absolute values of θ is not required, the factory-supplied calibration curve can be adjusted to fit a limited set of field data. A few readings suffice to simply shift the factory curve, and with still more readings the curve may be slightly rotated to pass through the experimental data for a particular location. In layered soils, calibration curves for each layer need not be parallel.

Detailed instructions on the measuring procedures are given by the producers of neutron probes. Theoretical reviews together with detailed instructions on calibration appear in the literature (Gardner, 1965, Middleboe et al., 1976, and Greacen, 1981).

When the neutron method is used to evaluate the soil water regime, readings taken at constant depth intervals do not necessarily relate to the spheres of influence. Indeed, they overlap, see Fig. 3.8. Hence, sharp steps in θ with depth cannot be detected. For an analysis of such errors and those owing to spatial variability, refer to Haverkamp et al. (1984) and Vauclin et al. (1984).

3.2.6 Remote Sensing

Remote sensing occurs when the sensor is not in direct contact with the soil. The spectral band of electromagnetic radiation most sensitive to soil water content is utilized – wavelengths equal to and longer than those of visible radiation. Detection is performed usually from an airplane or satellite.

The detection in visible and infrared spectra has these advantages – geometric accuracy, good resolution, easy interpretation and relatively simple procedures. With an increase of wavelength, those advantages disappear. The main disadvantage of the visible spectrum is that observations can be made only in the day-time without cloudiness, and the detected soil water content is limited to a surface layer of about 1 cm or less. Increasing the wavelength to the infrared reduces some of the disadvantages and the depth of soil layer detected is increased to a few centimeters.

Working passively or actively with the microwave spectrum holds great potential opportunity. The signal of passive radiometers is influenced primarily by the soil temperature which reflects among others the soil water content. The resolution depth is dependent proportionally to the wavelength and it is inversely proportional to the dielectric constant of the soil water system. Practically, the resolution depth is in centimeters.

The radiation of the soil surface is composed of emitted and reflected portions. Generally, the surface appears as a "colored picture" in the microwave spectrum. This "colorization" averages the properties of the soil layer. The "colorization" is detected by radiometers, filtered and identified with regard to the acting factors. Finally, an estimate of the soil water content in the surface layer is obtained. The roughness of the surface and the vegetative canopy decrease the sensitivity of the method.

Detailed information is abundantly available in the literature (e.g. Jackson, 1993; Jackson et al., 1993; Njoku and Kong, 1977; Ulaby and Elachi, 1990; and Wang et al., 1983).

PROBLEMS

1. The weight of a 100 cm^3 undisturbed soil sample was 175.5 g. After drying at 105°C the weight was reduced to 144.8 g. What was the water content by weight w and by volume θ? What was the bulk density and porosity if the particle density $\rho_S = 2.7$ $g \cdot cm^{-3}$?
2. What is the storage of water W (mm) in a soil when we have determined the water content w by weight and the bulk density ρ_T at the following depths?

z (cm)	w ($g \cdot g^{-1}$)	ρ_T ($g \cdot cm^{-3}$)
0 - 20	0.21	1.15
20 - 30	0.28	1.37
30 - 50	0.30	1.41

3. What was the mean evaporation rate ($mm \cdot d^{-1}$) from a soil column initially saturated to $\theta_S = 0.46$ on the first day and θ was measured on the fifth day at the following depths?

z (cm)	θ ($cm^3 \cdot cm^{-3}$)
0 - 10	0.35
10 - 20	0.40
20 - 30	0.43
30 - 40	0.45
> 40	0.46

Drainage at the bottom of the column was prevented.

4. List the advantages and disadvantages of the various methods for measuring soil water content. Which method would you select for continuously measuring the soil water content θ in a solonchak developed on a layered alluvium?
5. What was the amount of water ($m^3 \cdot ha^{-1}$) that drained down from the top 60 cm of soil profile when the surface was protected from evaporation by sheets of plastic and by a mulch? The first and last day readings of count ratio CR using a neutron probe at depths z were:

z (cm)	first day CR	last day CR
20	0.731	0.559
40	0.668	0.583
60	0.675	0.621

For all soil layers use the following equation for the calibration of the neutron meter: $\theta = 0.712\ CR - 0.025$.

4 SOIL WATER HYDROSTATICS

In this chapter we are studying the system of soil and water in an equilibrium state, characterized by zero fluxes in the soil. The soil water content does not change in time and it is in equilibrium with internal and external forces acting on the system. Under such equilibrium conditions values of soil water content at two different points in the soil need not be identical. Indeed, the distribution of soil water content in space does not pertain to the existence of a water flux if an equilibrium of forces exists.

Although we do not pay attention to the rate at which the equilibrium state was reached, we must know how the equilibrium was obtained. It is usually important to know the recent "history" of the system, e.g. was the equilibrium reached by wetting or drying of the soil?

First we shall study the role of the forces grouped according to their nature into separate categories. Then we shall show that the sum of the action of all force fields can be usefully studied within the concept of the soil water potential. Generally, the difference of the potential at two separate points relates to the driving force of the flow of soil water similar to the difference of temperature at the distant ends of a metallic rod that produces heat flow. Soil water potential is the pivotal concept of soil hydrology which treats the theory of soil water flow and retention.

4.1 INTERFACE PHENOMENA

In the soil, the coexistence of all three phases – solid, liquid and gas – is the most frequent. Force fields existing at the interfaces of solid-liquid, liquid-gas and solid-gas influence the behavior of soil water. Dissolved matter in the soil water contributes to and modifies the action of these force fields. The longer term chemical action of water upon the solid phase that causes alteration of both the soil minerals and the soil solution owing to weathering will not be considered. Here, we deal only with relatively short time effects. Emphasis will be given to macroscopic concepts with individual forces not being discussed in detail unless we need to explain some particular phenomenon.

The action between the soil solid surface including its sorbed ions and water takes place in close proximity of the surface. This portion of soil water is commonly called adsorption water. It can be most simply studied in the absence of other force fields and is usually observed when limited amounts of water vapor are adsorbed on the solid surface.

Water filling only parts of the smaller spaces in an unsaturated soil forms curved interfaces between air and water. Menisci or capillary forces arise from these interfaces. Capillarity plays an important role in retention of water in unsaturated soils. Most equations describing the flow of water in unsaturated soils are based upon this fundamental phenomenon of capillarity.

When soil solution penetrates between two parallel clay surfaces, the surfaces tend to shift apart – a phenomenon called swelling.

These three phenomena – adsorption, capillarity and swelling – will next be discussed to the extent needed for an understanding of their participation in soil hydrology. Theoretical, detailed discussion of soil water interactions can be found in Iwata et al. (1988) where the nature of individual force fields is more completely presented.

4.1.1 Adsorption of Water Vapor

When dry soil is allowed to come into contact with air containing water vapor of a certain partial pressure p (i.e. with air of a certain humidity), the weight of the soil increases. This gain in weight is caused by sorption of water vapor in the soil. Adsorption continues until an equilibrium between the soil water content w and the partial pressure p of water vapor is reached. The greater is p, the greater will be the corresponding value of w at equilibrium. Because the ability of a soil to adsorb water vapor from the air is called hygroscopicity, soil water originating from adsorption is sometimes called hygroscopic water.

The nature of this soil water vapor adsorption process is completely different from the simple process of vapor condensation to its liquid phase. When we describe water adsorption in soils, several considerations are necessary. First, water molecules have a polar character given by the space configuration of two hydrogen atoms and one oxygen atom. The positive pole is oriented and attracted to a negative charge of the soil solid surface. The polar character of the water molecules associates with the polarization of the solid surface. Exchangeable cations interact with the electric field of the solid surface and with the first layers of the adsorbate as they tend to form envelopes of water molecules. The hydrogen bond of water molecules on one side and the same of the OH groups of the crystal lattice of minerals or of COOH reaction groups in molecules of the soil organic matter contribute to the adsorption process. The arrangement of water molecules in crystallite clusters of liquid water is comparable to the hexagonal arrangement of silicate tetrahedral of soil minerals. Owing to this similarity, the slightly modified and stabilized arrangement of water molecules caused by the adsorption process has been described as quasi-crystalline water (Low, 1962).

The adsorptive force field decreases with the distance from the solid surface. The free energy of the solid surface decreases with the thickness of the adsorption film. Because the sum of the forces reduces the energy of the water molecules, adsorption is accompanied by adsorption heat. With some physically defined corrections, the adsorption heat can be studied from the heat of immersion. This heat is generated and can be measured when dry soil is immersed in water. Inasmuch as the heat of immersion decreases as the thickness of adsorption films on the solid surface increases, the greater is the water content of clay before immersion, the smaller is the heat. Fig. 4.1 shows heat of immersion as the energy related to the surface ($J{\cdot}m^2$) decreasing nearly exponentially with the water content of the clay before immersion expressed as the relative partial pressure p/p_o where p_o is the partial pressure of water vapor at saturation.

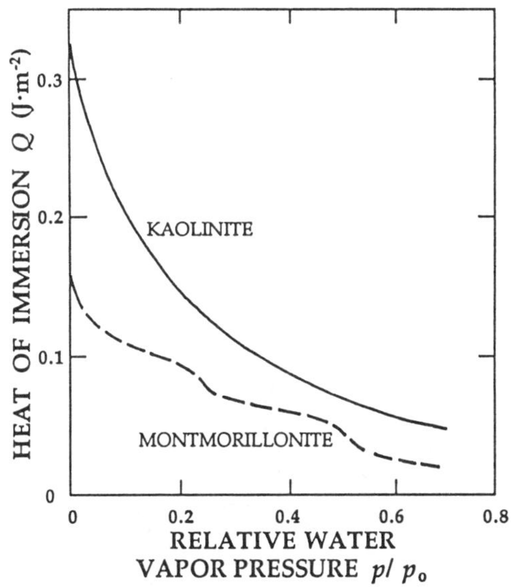

Figure 4.1. Heat of immersion related to unit surface area (J·m^{-2}) plotted against relative water vapor pressure p/p_o which was in equilibrium with the soil water content before immersion (Kutílek, 1969).

Adsorbed water vapor expressed as soil water content w increases with relative partial pressure p/p_o and soil specific surface A_m and decreases with temperature T. The influence of the quality of the solid surface is expressed by the functional relationship of the adsorption heat. Inasmuch as the porosity of the soil does not play a role in vapor adsorption, the mass soil water content w is used exclusively for both theoretical and experimental studies.

The relationship $w(p/p_o)$ determined at a constant temperature is called an adsorption isotherm. It is constructed as a smooth curve passing through equilibrium points of w at corresponding values of p/p_o. For soils and for most soil constituents the isotherm has a sigmoidal shape as illustrated in Fig. 4.2. The exact nature of the isotherm depends upon soil texture, mineralogical composition of the clay fraction, exchangeable cations and contents of organic matter and free oxides. Curves obtained from initially dry samples subjected to monotonic increases of p/p_o are called adsorption isotherms while those obtained as a result of drying are called desorption isotherms. Such curves which are not identical as shown in Fig. 4.2 exhibit hysteresis. Hysteresis of adsorption-desorption isotherms is related to the difference of energy of adsorption and desorption at a certain range of p/p_o. Moreover, hysteresis is also caused by the different spatial connectivity of ultra pores during this process. At much higher values of p/p_o, hysteresis of the contact angle plays a role in capillarity as will be discussed in the next section. Some authors quoted by Iwata et al. (1988) did not obtain smooth adsorption isotherms but detected multiple step-like curves with about five inflection points.

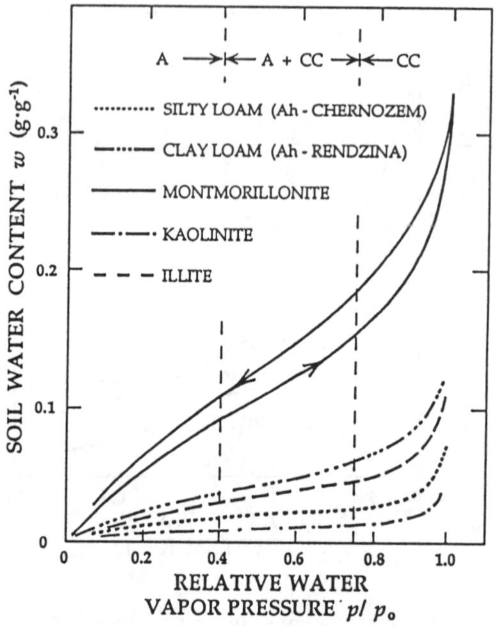

Figure 4.2. Adsorption isotherms of water vapor on clay minerals and soils. The three domains of the processes are indicated by A for adsorption only, A + CC for adsorption plus capillary condensation and CC for capillary condensation being dominant. The hysteretic nature of the isotherms is shown for montmorillonite (Kutílek, 1962).

Various types of equations have been proposed for the theoretical analysis of the adsorption isotherm. The most frequently used is the BET equation derived for multimolecular adsorption by Brunauer, Emmet and Teller (1938). Its derivation is based upon some assumptions which are not fulfilled in soils. For example, assuming unlimited adsorption is not valid owing to the size of the soil ultra pores as well as exchangeable cations were not considered. However, a comparative study has shown that the BET equation in its simplest form for unlimited adsorption is completely appropriate (Kutílek, 1962a):

$$\frac{p}{w\,(p_o - p)} = \frac{1}{w_m C} - \frac{C-1}{w_m C}\,\frac{p}{p_o} \tag{4.1}$$

$$C = \exp\left(-\,\frac{q_1 - q_L}{RT}\right) \tag{4.2}$$

where q_1 is the adsorption heat, q_L the heat of condensation, R the Universal gas constant, T the temperature, C the adsorption constant which characterizes hydrophility of the solid surface according to (4.2) and w_m the water content (by mass) for a monolayer. From the measured adsorption isotherm we can compute the adsorption heat of the monolayer and its water content w_m. After approximately two molecular layers are formed, the capillary condensation starts from $p/p_0 \approx 0.4$ to 0.5 and it dominates the process for $p/p_0 > 0.7$ to 0.8. Capillary condensation is related to the alteration of p_0 above the curved surface, the greater is the curvature, the lower is p_0 above this curved water level (see the details in sections 4.1.2 and 4.3.1). When we study the adsorption process, we are restricted to the appropriate part of the adsorption isotherm at $p/p_0 < 0.4$ in order to avoid the influence of capillary phenomena. In early studies of soil physics, a characteristic water content at $p/p_0 = 0.95$ was introduced and named the hygroscopic coefficient, θ_H (Mitscherlich, 1901). Inasmuch as the term hygroscopic coefficient should truly characterize only the hygroscopic properties of a soil, the term defined for such a large value of p/p_0 which includes capillary condensation has become obsolete.

A soil water content in equilibrium with $p/p_0 \approx 0.7$ to 0.8 appears as the smallest physically admissible value of the residual water content θ_r because here is the boundary between water with bulk liquid flow characteristics and adsorption water. Adsorption water transport is usually formulated differently than bulk liquid water flow.

4.1.2 Capillarity

When water meets a solid surface, a contact angle γ is formed, see Fig. 4.3. This contact angle is observed when a drop of water is placed on a plane solid surface, or when the solid plate is submerged into water. The magnitude of γ is used to distinguish three classes of wetting for the solid surface. For $\gamma = 0°$, the surface is completely wet and the solid is fully hydrophilic. A non-complete wetting of the surface occurs for $0 < \gamma < 90°$ and the solid is partly hydrophilic. A non-wetting surface exhibits $\gamma \geq 90°$ and the solid is hydrophobic.

At a point of contact between a solid, liquid and gas, three interface boundaries exist: (i) interface between the solid S and the liquid L, (ii) between the solid S and gas G and (iii) between the liquid L and gas G. Macroscopically, we can define at that point the three interface tensions: σ_{SL}, σ_{SG} and σ_{LG}. Accordingly, equilibrium exists when

$$\sigma_{LG} \cos\gamma + \sigma_{SL} - \sigma_{SG} = 0 \tag{4.3}$$

or

$$\cos\gamma = \frac{\sigma_{SG} - \sigma_{SL}}{\sigma_{LG}} . \tag{4.4}$$

On a molecular scale we must admit that an adsorption layer of the liquid first exists on the solid surface. A drop of the same liquid formed on this adsorption layer allows interface tensions to exist between the adsorption layer and gas or liquid, respectively. In the derivation, the surface energy of the adsorption layer is defined with regard to the arrangement of molecules or bonding sites of the solid phase. Under certain conditions γ is not definable and the adsorption layer is influenced by the disjoining pressure (Derjaguin and Churaev, 1972).

When the drop of liquid moves along a solid surface we observe different values of the contact angle. When the liquid proceeds to the dry surface, the contact angle is γ_1. When the liquid recedes from the wetted surface, we observe

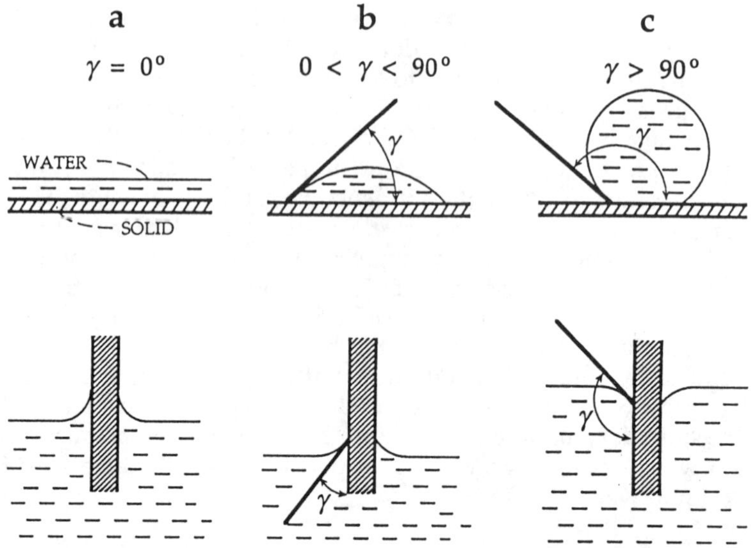

Figure 4.3. Manifestation of the contact angle γ for a liquid on a solid: a. $\gamma = 0°$, complete wetting, b. $0° < \gamma < 90°$, incomplete wetting and c. $\gamma > 90°$, non-wetting (hydrophobicity).

the contact angle γ_2. Generally, $\gamma_1 > \gamma_2$, see Fig. 4.4. Similarly, different contact angles are observed when we first insert a capillary tube into liquid compared to that when we pull it out. The phenomenon is called hysteresis of the contact angle and it is one of the reasons for hysteretic behavior of water in soils.

Soil solid particles are covered irregularly by free iron oxides, macromolecules of soil humus, and thin layers of clay minerals. Each of the constituents of the soil solid phase shows different values of γ. Rarely is a silica surface "clean", and only then, $\gamma = 0$. What we observe macroscopically is a composite value for γ. The actual value of γ is microscopically distinct but spatially heterogeneous owing to the distribution and nature of the covering of the solid particles. Generally, the highest and next highest values of γ are those of the soil humus and free iron oxides, respectively. Kaolinite is among the clay minerals having the smallest value of γ. Apparently, exchangeable cations do not influence γ significantly (Kutílek, 1963). Burning organic residues substantially increases γ with the soil surface especially being hydrophobicized (Savage et al., 1972). Surfactants tend to decrease γ. And in general, all agricultural chemicals as well as other anthropogenic pollutants act steadily to change the contact angle (Letey et al., 1962, and Pelishek et al., 1962).

If the solid plate is inserted into water, the air-water interface close to the plate is curved, and its curvature depends upon the contact angle γ, see Fig. 4.3. When two parallel plates close together are partly submerged in water, the curvatures join and form a portion of a nearly cylindrical surface. If we insert a thin capillary tube into water, a nearly spherical surface is produced inside the

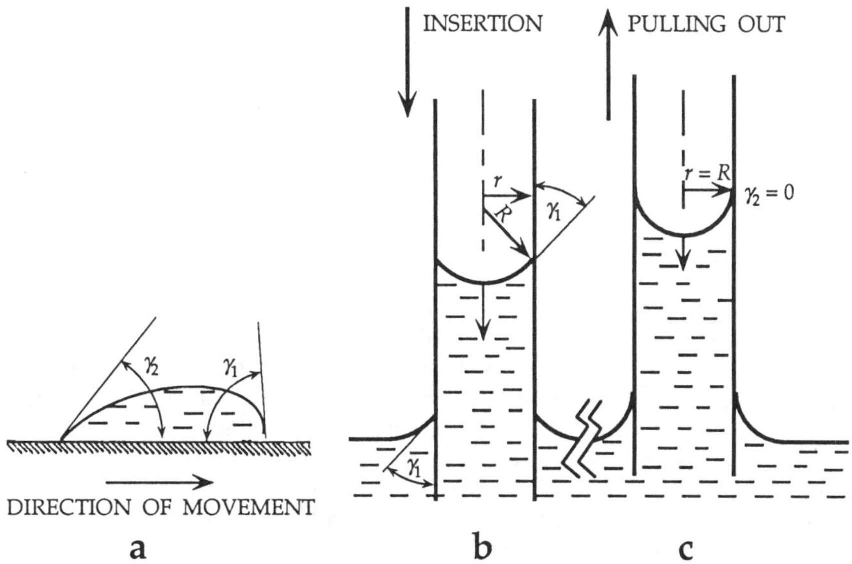

Figure 4.4. Manifestation of the hysteretic nature of the contact angle γ. a. water drop moves along the solid surface, b. a capillary tube is inserted into water and c. a capillary tube was first fully immersed in water and subsequently pulled partially out.

tube by the air-water interface. Owing to the curvature of the surface of the water, the surface tension is changed from that of a plane water surface.

The surface tension of a plane water surface is the result of attractive forces acting from the surrounding molecules upon the molecules that compose the surface. Very simply, on a molecular scale, there is a force of attraction between the bottom "hemisphere" of water molecules composing the air-water interface and those immediately surrounding the "hemisphere". The resulting sum of attractive forces is a force normal to the water surface and directed into the water. Integrating these forces over the area L^2 yields the surface pressure p_r (Pa) where the subscript r relates to the reference level. Macroscopically, the term surface tension σ is usually employed as the force per unit length needed for an increase of the surface area with its value for water at 20°C being $7.28 \cdot 10^{-2}$ $N{\cdot}m^{-1}$ (earlier, 72.8 $dynes{\cdot}cm^{-1}$). Or, if the work required for the extension of the water surface is considered, we have $7.28 \cdot 10^{-2}$ $J{\cdot}m^{-2}$ (earlier, 72.8 $ergs{\cdot}cm^{-2}$).

When the air-water interface is curved, the pressure below this curved surface is different from that of a plane surface. On a molecular scale, we can schematically demonstrate that for a concave surface, a greater portion of the attractive forces is canceled than that for a plane surface. Hence, for the concave surface, its surface pressure is decreased. In capillaries, the surface pressure is decreased when $0° < \gamma < 90°$, i.e. when the walls are hydrophilic, see Fig. 4.4. If the water curvature is convex, the surface pressure is increased when compared with that of a plane water surface. In a capillary this hydrophobicity of the solid surface is linked to $\gamma > 90°$. Hence, the surface pressure in a capillary is

$$p_s = p_r + p_\sigma \tag{4.5}$$

where p_r is the reference pressure at the plane water level. Note for a concave curvature, $p_\sigma < 0$, and for a convex curvature $p_\sigma > 0$. This additional pressure p_σ is formulated according to Laplace's expression

$$|p_\sigma| = \sigma \left(\frac{1}{R_1} + \frac{1}{R_2} \right). \tag{4.6}$$

Here, σ is the surface tension of the plane water level (its values for various temperatures are given in physical tables) and R_1 and R_2 are the main radii of curvature. Equation (4.6) and relations derived from it are valid if the disjoining pressure of very fine films can be neglected. For very fine capillaries of about $r < 100$ nm, the disjoining pressure starts to play a significant role and (4.6) together with derived equations have to be modified according to Derjaguin and Churaev (1972). However, the details of scientific discussions of all phenomena related to very thin films of water (i.e. in the dry region of soils) are far beyond the scope of this book.

In a cylindrical capillary for a spherical meniscus

$$|p_\sigma| = \frac{2\sigma}{R} \tag{4.7}$$

where R is the radius of the meniscus, see Fig. 4.4. Similarly, for the curved air-water interface between two parallel plates

$$|p_\sigma| = \frac{\sigma}{R}. \tag{4.8}$$

Let us assume for the moment that the walls of the capillary are hydrophilic. Because the surface pressure of the plane water level around the capillary is greater than that of the meniscus, a non-equilibrium condition is generated when the bottom end of the capillary is placed in bulk water. Water rises into the capillary because of this inequality of pressure until the difference of pressures is equilibrated by the hydrostatic pressure at depth h

$$|p_\sigma| = \rho_W g h \tag{4.9}$$

where g is the gravitational acceleration at the earth's surface, h the height of the water column in the capillary, sometimes denoted as capillary rise, and ρ_W the density of water. Note in Fig. 4.5 the magnitude of h for the concave meniscus is less than zero. If we installed instruments for measuring the hydrostatic pressure, a straight line passing through zero at the plane water level would be observed. Below the water level, the hydrostatic pressure is positive, and above it in the water column of the capillary, the pressure is negative, see Fig. 4.5. The atmospheric pressure is taken as the reference. From this discussion it follows that the soil water pressure is negative above the free (ground) water level.

Equating values of p_σ from both (4.7) and (4.9) we obtain

$$|h| = \frac{2\sigma}{R \rho_W g}. \tag{4.10}$$

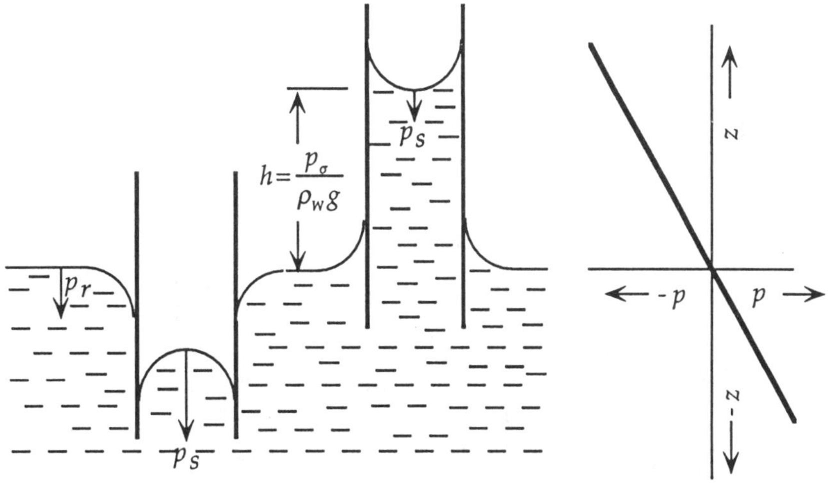

Figure 4.5. Capillary depression (left) for $\gamma > 90°$ and capillary rise (right) for $0° < \gamma < 90°$. For capillary rise, the water below the meniscus shows negative pressure (relative to the pressure on the plane of the water table).

Recognizing the radius r of the capillary is related to the contact angle γ by $R = r/\cos\gamma$, we obtain

$$|h| = \frac{2\sigma\cos\gamma}{r\,\rho_W g}. \tag{4.11}$$

A more accurate expression is obtained if ρ_W in the denominator is replaced by $(\rho_W - \rho_G)$. For water at 20°C and with $\gamma = 0°$, (4.11) is simply

$$|h| \approx 0.15/r \tag{4.12}$$

for length units in cm.

Earlier in Fig. 4.4, we demonstrated the hysteresis of the contact angle γ. This hysteretic behavior can also be demonstrated with capillary tubes. Let us assume that we have two tubes of the same material and of the same radius. We insert the bottom end of one capillary into water and allow the water to rise. We totally submerge the second capillary in water, and subsequently, pull it partly out while still keeping the bottom end in contact with the free water level. In the first capillary, the wetting contact angle γ_1 causes a smaller curvature than that of the draining contact angle γ_2. Inasmuch as $\gamma_2 < \gamma_1$, the height in the first capillary is smaller than that in the second one, see Fig. 4.4.

When a capillary is pulled out of the free bulk water, the column of water remains in the capillary. It is kept inside by the difference of surface pressures between the top and bottom menisci (Fig. 4.6) with the height of the hanging water column equal to

$$h = \frac{p_2 - p_1}{\rho_W g} \tag{4.13}$$

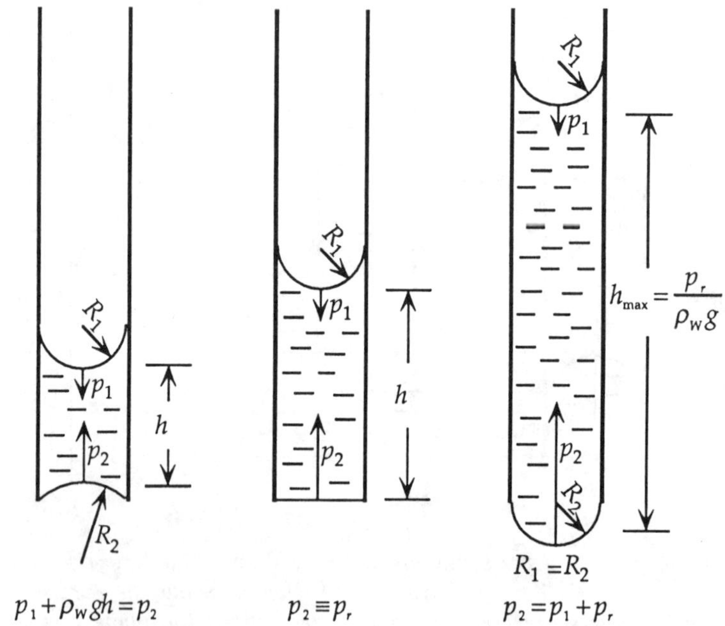

Figure 4.6. Equilibrium conditions in capillary tubes for "hanging water". The radius of curvature R_1 is constant.

where the subscripts 1 and 2 indicate the top and bottom menisci, respectively. Or, with (4.10), we obtain

$$h = \frac{2\sigma}{\rho_w g}\left(\frac{1}{R_2} - \frac{1}{R_1}\right). \tag{4.14}$$

For complete wetting of the tube, (4.14) becomes

$$h = \frac{2\sigma}{\rho_w g}\left(\frac{1}{R_2} - \frac{1}{r}\right). \tag{4.15}$$

The change of the height of the water hanging in the capillary is accompanied by a change of R_2. Because $(p_2)_{max} = p_r + p_1$, we obtain $h_{max} = p_r/\rho_w g$ for complete wetting.

When the capillarity of water lenses between spherical particles is studied, we have to recognize that the air-water interface is convex-concave (see Fig. 4.7) where R_1 is positive and R_2 is negative in (4.6). The height of capillary rise h_s into an idealized porous medium formed by spheres depends then upon forces acting on the contact points of the gas-liquid interface. During capillary rise the total force not only changes its magnitude but also its direction. It is an attenuated wave function which oscillates between positive and negative values (Lu et al., 1994). Geometry of packing plays an important role. When h_s is related to the height of capillary rise h_t into a tube, Lu et al. derived

$$\frac{h_s}{h_t} = C\left(\frac{1+\cos\gamma}{\cos\gamma}\right) \tag{4.16}$$

where C is a constant which has values ranging from 0.82 for cubic packing to 1.5 for rhombohedral packing with the radius of the tube identical to that of the spheres forming the column.

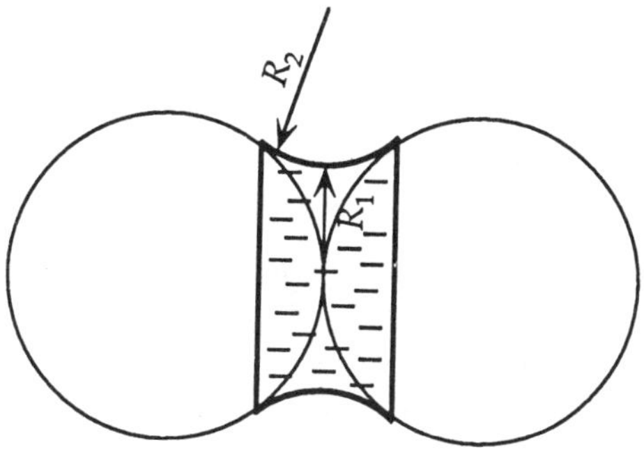

Figure 4.7. A convex-concave (anticlastic) air-water interface at the contact of two spherical particles. The water lens has two radii of curvature R_1 and R_2.

4.1.3 Swelling

The majority of soils change their volume when their water content θ is altered, especially when θ is not far from saturation. The increase in water content accompanied by an increase of volume of such soils is called swelling. When soils lose water either by draining or drying, their volume decreases until a certain threshold value of θ is reached. Below this value, shrinkage is accompanied by the formation of fissures (cracks). The extent of swelling has already been presented in Fig. 2.3. The swelling index is the ratio of the volume change for a soil saturated with water to that saturated with a non-polar liquid.

It can be seen in Fig. 2.3 that swelling can continue even after a soil is fully saturated with water provided that further water is available. Hence, θ_S is not a reliable reference value in swelling soils. The pressure needed to prevent a soil from swelling is called the swelling pressure. Swelling depends upon the clay content, mineralogical composition of clay and upon the exchangeable cations. Monovalent exchangeable cations induce greater swelling than divalent cations. When the clay fraction is dominated by montmorillonite, soils swell more than those dominated by chlorite or illite. The swelling of kaolinitic soils is negligible.

An explanation of swelling starts with simple model of two parallel plates, which can demonstrate the behavior of, e.g. montmorillonitic sheets of clay. Iwata et al. (1988) describe in detail the repulsive and attractive forces acting upon two parallel plates. Attraction energy is caused by van der Waals short distance forces and by adhesion forces which depend upon the orientation of the plates and upon the pH of the soil solution. When the pH decreases, the charges at the edges of the plates become positive and edge-bonding develops. This type of bonding avoids the parallel arrangement of plates and induces a spatial arrangement of plate-like montmorillonites. Repulsive forces between two parallel plates with negatively charged surfaces consist of two classes: (i) short distance (up to about 8 nm) forces owing to double layer forces and (ii) long distance forces caused by the difference in electrolyte concentration between the plates and that of the outer solution, see Fig. 4.8. Because the chemical potential of the solution of the interlayer is greater than that of the outer solution, an addition of water causes the plates to shift apart until an equality of the two potentials is reached. If this shift is prevented, the necessary applied pressure is the swelling pressure. The total potential energy for the two charged plates is the sum of the repulsive and attractive potential energies – a sum that is not necessarily a monotonic function with distance.

Norrish (1972), using montmorillonite as the model material of the swelling of plate-like clay particles, speaks about three stages of swelling:

First stage: The initial distance between the two sheets is less than 2 nm. As water is added, the distance increases in discrete steps associated with each molecular water layer formed between the sheets. With divalent and trivalent cations present, swelling ends at this first stage. In their presence, even when immersed in excess water, the bundles of sheets persist forming a structural arrangement of tactoids, see Fig. 4.9. On the other hand, when monovalent cations dominate, the second stage of swelling follows. We note that allophane has no first stage of swelling.

Second stage: At the start of the second stage, the bonding of water molecules to the solid surface continues to contribute to the swelling process. The distance between neighboring sheets rises smoothly up to tens of nm. Because the sheets are separated so widely, the edge-to-face forces begin to be important for holding the sheets together.

Third stage: The sheets are completely separated and form an arrangement caused by edge-to-face and edge-to-edge forces. The bonds influence the thixotropic behavior of the suspensions. Thixotropy is the isothermal reversible process, when after mixing, a suspension of the soil without a definable arrangement of sheets is transformed into the cluster arrangement mentioned above.

In unsaturated clay, the capillarity owing to the cylindrical menisci of water placed between two sheets contributes to the attractive forces, and hence, reduces the resulting repulsive force. In soils, through the arrangement of soil particles of various sizes in their micro aggregates, capillarity reduces the swelling of bulk soil. The cementing agents of aggregation (free iron oxides, humus etc.) alter the solid surface as they form thin irregular films on clay particles. For the first stage of swelling, the 2:1 crystal lattice minerals are more loose and the interlayer is more accessible for water. But the second and third stages of swelling are usually sharply reduced owing to the action of cementing agents. Inasmuch as the cementing is competitive with edge-to-face and edge-

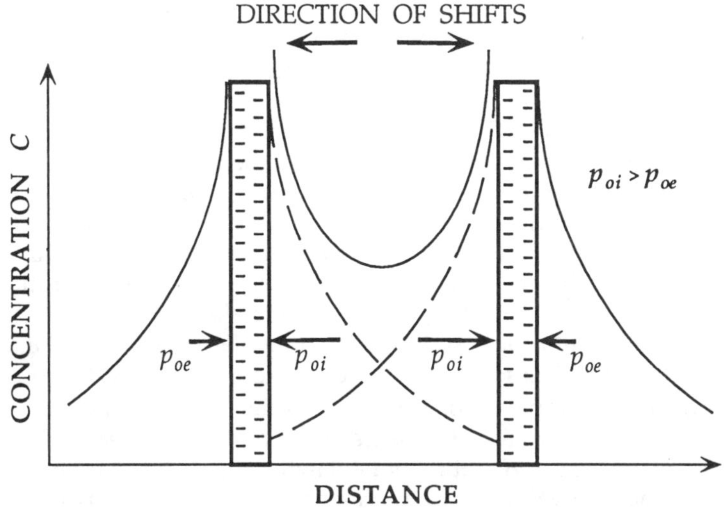

Figure 4.8. A model of swelling in plate-like montmorillonite.

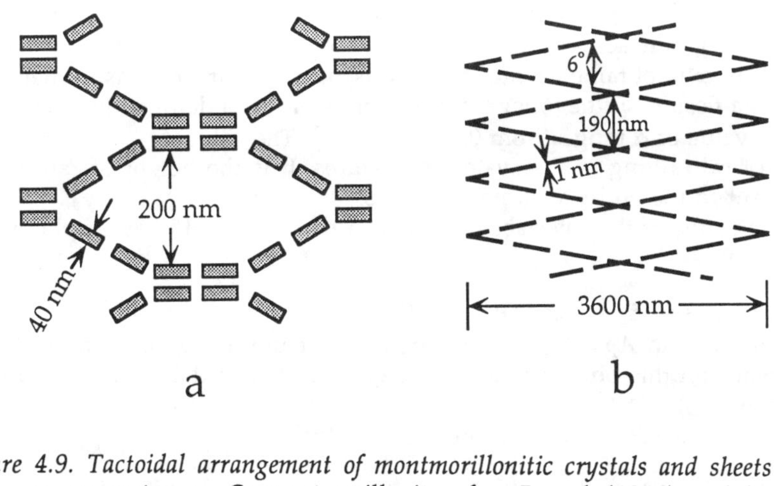

Figure 4.9. Tactoidal arrangement of montmorillonitic crystals and sheets in excess water: a. Ca-montmorillonite after Bernal (1946) and b. Na-montmorillonite after M'Even and Pratt (1957).

to-edge forces, the tactoidal arrangement of sheets is greatly deformed and never completed in natural soils. However, in spite of cementation, the thixotropic behavior of the soil water system is detectable in natural soils. For example, the mechanical properties as well as soil water retention depend upon the length of time elapsed from a mechanical disturbance of soil. Immediately after mixing or plowing a soil, its shear strength is small and, subsequently increases with time.

Similarly, the soil water ratio ϑ increases with time after mixing the soil providing water has access to the system and the negative pressure head h (see 4.2.1) is kept constant. Swelling and the thixotropic behavior of θ or ϑ are time dependent functions that are empirically described as exponential functions (Kutílek, 1978). Swelling and the thixotropic behavior of the subsoil are strongly reduced because of the overburden pressure of the topsoil.

The opposite process to swelling is shrinkage when the soil water ratio ϑ or the water content θ decreases. The system of soil fissures plays an important role in soil hydrology. The beginning of surface runoff when rainfall occurs is drastically delayed by the existence of fissures. Rates of evaporation are also influenced by fissures but to a lesser extent than infiltration rates. For a quantitative description of a system of fissures, let us assume that the REV is large enough to fully describe the system. The threshold value of ϑ_F for the initiation of the development of fissures is in the normal range of swelling, see Fig. 2.3. To obtain the volume of fissures we measure the area of fissures on the topographical surface A_{CO} and the total length of the sum of individual fissures L_{CO}. Both values are valid for the representative elementary area REA which is equivalent to REV for a 2-dimensional system. If we assume the width of fissures decrease exponentially with depth (Kutílek and Novák, 1976), the volume of fissures V_F becomes

$$V_F = \int_0^\infty A_{CO} \exp(-\alpha z)\, dz \qquad (4.17)$$

which reduces to $V_F = A_{CO}/\alpha$. The vertical coordinate z is zero at the topographical surface and increases positively with soil depth. The empirical coefficient α is obtained from repetitious pairs of direct measurement of the width of a fissure on the topographical surface and at a defined depth, e.g. $z = 20$ cm. The value of α ranges from 0.01 to 0.1 cm^{-1}. The value A_{CO}/α represents the height of rain filling in the system of fissures. It is the height of rain which is equivalent to the entire volume of fissures.

The area of the interface A_F between the soil and the atmosphere within the fissures is

$$A_F = \int_0^\infty 2L_{CO} \exp(-\alpha z)\, dz \qquad (4.18)$$

which reduces to $A_F = 2L_{CO}/\alpha$. Evaporation or infiltration into the soil matrix takes place on this boundary A_F together with that of the REA (discounted by the value A_{CO}).

4.2 SOIL WATER POTENTIAL

4.2.1 Concepts of Soil Water Potential

Let us assume that we have packed a fine sandy soil into a long cylinder provided with a porous plate or a fine sieve at its bottom. The bottom end of the soil column is submerged into a vessel containing free water. As water rises into the soil the water level in the vessel is kept at a constant elevation, see Fig. 4.10. By measuring the soil water content at various heights, we find after some

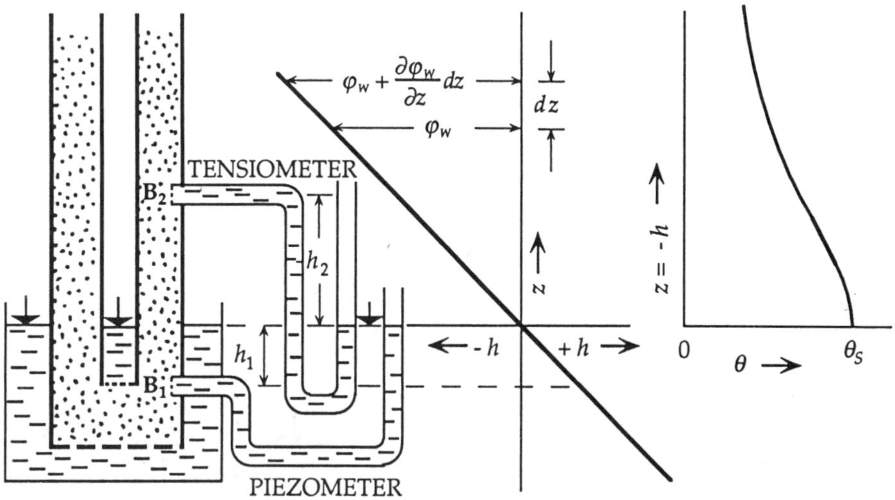

Figure 4.10. Derivation of soil water potential head h for a homogeneous soil column when equilibrium is reached above the defined ground water level GWL. Positive pressure heads below GWL are measured with piezometers and negative pressure heads above GWL are measured with tensiometers. The distribution of soil water content above GWL at equilibrium is identical to the soil water retention curve SWRC.

time that the values of soil water content begin to be constant at each elevation. Immediately above the free water level in the vessel the value of the soil water content is at near saturation and it decreases with increasing height above the free water level. If evaporation from the system is prevented, all conditions necessary for an equilibrium state will eventually be established. The equilibrium state is characterized by no flow of water and by the soil water content remaining constant in time at every elevation within the soil.

We shall discuss in detail the energy status of a hypothetical water particle. Starting at the free water level, the particle has some energy which will be considered as referential or at the reference level. The energy related to the unit of water is its potential. According to convention, the total potential Φ of water at the free water level is zero. When the water particle is lifted in the gravitational field to elevation z, its energy associated with the gravitational field is increased proportionally to z, and this component of the total potential Φ will be called the gravitational potential ϕ_g. At elevation z the particle is attracted by the force fields of the soil as e.g. adsorption and capillarity, but the force fields need not be defined separately – only a composite force is necessary. This composite soil force field being component of the total potential Φ is called the soil water potential ϕ_w. It is sometimes called the wetness, matric or capillary potential. (On the other hand, we could have considered the water particle initially at elevation z and allow it to fall and decrease its gravitational energy to that of the reference level. The particle loses some of its energy owing to the "fall" in the force fields of the soil at z.) Inasmuch as the water in our soil column is in equilibrium, the water particle, being moved from one elevation to another, is obliged to neither gain or lose total energy. Hence, any gain in energy ϕ_g must equal the loss of energy ϕ_w. According to our experiment, $\phi_g > 0$ and $\phi_w < 0$. Therefore, $\phi_g = -\phi_w$. Let us now check if the system is in equilibrium, i.e. if the values of the total potential Φ are in all points of the soil column and of the water in the vessel in equilibrium. Because we have assumed that a reference value of $\Phi = 0$ exists at the free water level, the same zero value of Φ $(= \phi_g + \phi_w)$ should exist everywhere in the system. At the elevation z we find $\Phi = 0$ if $\phi_g = -\phi_w$.

After the above experiment we now mathematically describe the total potential (Buckingham, 1907). We assume that the soil water potential at elevation z above the free water level in Fig. 4.10 is ϕ_w. The infinitesimal mass of water dm will be transferred from elevation z to $(z + dz)$. The work done by the force applied for the removal of dm of water from the force fields of the soil is $\phi_w dm$. The potential at the elevation $(z + dz)$ is $[\phi_w + (d\phi_w/dz)]$ and the work done by the attractive forces of the soil at $(z + dz)$ is

$$\left[\phi_w + \left(\frac{\partial \phi_w}{\partial z}dz\right)\right]dm.$$

The total work W_1 done by the removal and attraction of the water mass dm is

$$W_1 = \phi_w\, dm - \left[\phi_w + \left(\frac{\partial \phi_w}{\partial z}dz\right)\right]dm. \qquad (4.19)$$

Because the water particle dm was transported along dz in the gravitational field, the work done in this field is

$$W_2 = -g\, dm\, dz. \qquad (4.20)$$

For the system in equilibrium $(W_1 + W_2) = 0$, and

$$\frac{\partial \phi_w}{\partial z} = -g. \qquad (4.21)$$

When (4.21) is integrated with respect to z from $z = 0$ $(\phi_w = 0)$ to z, we obtain

$$\phi_w = -g\,z. \qquad (4.22)$$

Hence, the soil water potential is proportionally related to the height above the free water level, or ground water level, and is identical except for its sign to that of the gravitational potential.

Consider the system of soil pores modeled as an ensemble of parallel capillaries decreasing gradually in diameter as shown in Fig. 4.19 for equilibrium conditions. The position of the water height $|h_i|$ in each i-th capillary of radius r_i is given by (4.11). At equilibrium the gravitational potential

$$\phi_g = g\,z \qquad (4.23)$$

equals $-\phi_w$ of each i-th capillary designated here as $-\phi_i$. The value of ϕ_i is obtained by substituting $-h_i$ from (4.11) for z in (4.23). Hence,

$$\phi_i = -\frac{2\sigma\cos\gamma}{r_i\rho_W}. \qquad (4.24)$$

Inasmuch as only capillarity was considered, the component ϕ_i of the total potential derived here is referred to as the capillary potential. The relationship between the capillary potential and the radius of the capillary is restricted physically to capillary water columns less than 10 m.

Because the mechanical approach used thus far in this chapter does not necessarily allow us to consider separately all existing force fields, a thermodynamic approach offers another alternative. Thermodynamics describes the transformation of heat and work in the equilibrium state of a system. The work, formulated as the change of the Gibbs free energy of the system, can be expressed as the algebraic summation of the change of the free energy of all components of the system with the term chemical potential being used. Instead of absolute values, the difference related to a reference system is used. In the soil water system, the reference is bulk water. Consequently, the total potential is the summation of the component potentials analogous to the chemical potential of composed systems (ISSS Bull., 1976)

$$\Phi = \phi_g + \phi_w + \phi_o + \phi_a + \phi_e \tag{4.25}$$

where ϕ_g is the gravitational potential and ϕ_w is the soil water potential – both of which have the meaning already explained in this chapter. Osmotic potential ϕ_o is owing to the difference in the chemical composition of the soil solution related to free, pure, bulk water at the same elevation. The pneumatic potential ϕ_a accounts for air pressure inside the soil pores being different from the outside atmospheric air pressure acting upon the reference water. When an external mechanical pressure such as the overburden pressure of the topsoil layers acts upon the soil, the magnitude of change of the total potential is expressed by the envelope potential ϕ_e. The contribution of ϕ_e is usually negligible for sandy soils and becomes more important for soils having greater clay contents. In sands, because the external pressure is transmitted almost completely by the direct contacts of the grains, especially in pre consolidated sands, the potential of the pore water is not altered by the external pressure. When the external pressure is transmitted from particle to particle via the water film envelopes around the particles, the pore water is greatly influenced and the value of ϕ_e is significant. We note here that the pressure potential measured with a tensiometer is $\phi_p = (\phi_w + \phi_a + \phi_e)$ provided that the solution within the tensiometer cup is at chemical equilibrium with that in the soil. The provision is not fully met in saline soils or even in non saline soils whenever the concentration of the soil solution changes abruptly in the vicinity of the tensiometer cup.

The formulation of the total potential as accepted by the Terminology Committee of ISSS (1976) has been critically reviewed by e.g. Corey and Klute (1985) and Iwata et al. (1988). In the theoretical development of the soil water potential analogous to the chemical potential, the thermodynamic quantities (partial internal energy and entropy) are considered as constant at all points of the soil water as is the case in homogeneous systems. Because the soil solution is a heterogeneous system, the above assumption is not strictly true (Iwata et al., 1988). A strict interpretation of the soil water potential should relate to the heterogeneity of the system. Such a strict approach may be helpful when the transport of chemicals in the soil solution or when soil water relationships in the dry region of the retention curve are considered. In most soil hydrological equations, the macroscopic definition of Φ in (4.25) or that equal to $(\phi_g + \phi_p)$ is completely adequate.

4.2.2 Units of the Potential

The potential of soil water is expressed as the energy related to a unit of water and its dimension, therefore, depends upon the choice of the unit of water – *mass, volume* or *weight*. When energy J is related to a unit *mass* of water, a unit of potential is $J \cdot kg^{-1}$ and has dimensions $[L^2T^{-2}]$. Although this definition is exact, it is not used in practical computations because it leads to complicated expressions for some terms. When energy J is related to a unit *volume* of water, a unit of potential is $J \cdot m^{-3}$. However, because J is a N·m, a unit of potential is that of pressure Pa having dimensions $[ML^{-1}T^{-2}]$. Potential defined in this manner has sometimes been designated as suction. If we assume that water is incompressible and its density is independent of the total potential, the mass in the first definition can be replaced by volume. The assumption is not fully exact, but is acceptable for the majority of situations. The assumption becomes less acceptable when the soil is nearly air-dry (i.e. θ in equilibrium with $p/p_o \leq 0.8$) inasmuch as the density of the first several molecular layers of water adsorbed on the solid soil particle surfaces may differ from that of free water. When energy J is related to a unit weight of water, a unit of potential is $J \cdot N^{-1}$ or simply m, and has the dimension [L]. For this definition, we use the symbol h which is approximately equivalent to the term pressure head in hydraulics. The degree of this approximation is the same as that used for the second definition. This third definition is often preferred owing to the simplicity of the units of other terms depending upon the potential unit.

Mutual conversion of units of potential, including obsolete ones, is provided in Table 4.1. The term pF is also used in some texts for potential. Its definition, $pF = \log_{10}|h|$, utilizes the pressure head h (cm) from the above definition of potential based upon a unit *weight* of water.

In the majority of situations the simplest definition of the total potential is

$$\Phi = \phi_w + \phi_g \tag{4.26}$$

and with the potential expressed as energy per unit weight of water

$$H = h + z \tag{4.27}$$

where z is the distance from the reference level, positive upwards. For soils not fully saturated with water, $h < 0$. In the remainder of this text we shall clearly declare when we deviate from this convention.

4.2.3 Measuring Techniques

4.2.3.1 Piezometers and Tensiometers

If in our experiment with the sand column, Fig. 4.10, a vertical cylindrical hole is drilled in the sand and lined by a tube having a sieve across its bottom, water will gradually fill this tube to the elevation equal to that of the free water level outside the soil. We have thus simulated the drilling of a well and the insertion of a piezometer in a soil. A piezometer is used for measuring the pressure of

Table 4.1. Conversion of units of the soil water potential

	$J \cdot kg^{-1}$	$N \cdot m^{-2}$ (Pa)	$dyn \cdot cm^{-2}$	bar	$kp \cdot cm^{-2}$	psi	cm H_2O	torr
Energy per mass								
1 $J \cdot kg^{-1}$	1	10^3	10^4	10^{-2}	$1.02 \cdot 10^{-2}$	$1.45 \cdot 10^{-1}$	$1.02 \cdot 10$	7.50
1 $erg \cdot g^{-1}$	10^4	10	10^8	10^2	$1.02 \cdot 10^2$	$1.45 \cdot 10^3$	$1.02 \cdot 10^5$	$7.50 \cdot 10^4$
1 $cal \cdot g^{-1}$	$2.39 \cdot 10^{-4}$	$2.39 \cdot 10^{-1}$	2.39	$2.39 \cdot 10^{-6}$	$2.44 \cdot 10^{-6}$	$3.46 \cdot 10^{-5}$	$2.44 \cdot 10^{-3}$	$1.78 \cdot 10^{-3}$
1 $cal \cdot mol^{-1}$	$4.31 \cdot 10^{-3}$	4.31	$4.31 \cdot 10$	$4.31 \cdot 10^{-5}$	$4.40 \cdot 10^{-5}$	$6.25 \cdot 10^{-4}$	$4.40 \cdot 10^{-2}$	$3.23 \cdot 10^{-2}$
Pressure								
1 $N \cdot m^{-2}$ (1 Pa)	10^{-3}	1	10	10^{-5}	$1.02 \cdot 10^{-5}$	$1.45 \cdot 10^{-4}$	$1.02 \cdot 10^{-2}$	$7.50 \cdot 10^{-3}$
1 $dyn \cdot cm^{-2}$	10^{-4}	10^{-1}	1	10^{-6}	$1.02 \cdot 10^{-6}$	$1.45 \cdot 10^{-5}$	$1.02 \cdot 10^{-3}$	$7.50 \cdot 10^{-4}$
1 bar	10^2	10^5	10^6	1	1.02	$1.45 \cdot 10$	$1.02 \cdot 10^3$	$7.50 \cdot 10^2$
1 $kp \cdot cm^{-2}$	$9.81 \cdot 10$	$9.81 \cdot 10^4$	$9.81 \cdot 10^5$	$9.81 \cdot 10^{-1}$	1	$1.42 \cdot 10$	10^3	$7.35 \cdot 10^2$
1 psi	6.89	$6.89 \cdot 10^3$	$6.89 \cdot 10^{-4}$	$6.89 \cdot 10^{-2}$	$7.02 \cdot 10^{-2}$	1	$7.02 \cdot 10$	$5.17 \cdot 10$
Pressure head								
1 cm H_2O	$9.81 \cdot 10^{-2}$	$9.81 \cdot 10$	$9.81 \cdot 10^2$	$9.81 \cdot 10^{-4}$	10^{-3}	$1.42 \cdot 10^{-2}$	1	$7.35 \cdot 10^{-1}$
1 torr	$1.33 \cdot 10^{-1}$	$1.33 \cdot 10^2$	$1.33 \cdot 10^3$	$1.33 \cdot 10^{-3}$	$1.36 \cdot 10^{-3}$	$1.93 \cdot 10^{-2}$	1.36	1

pore water at the point B_1 where it is attached to the soil. The height of the water level above B_1 is h_1, and this column of water acts upon the bottom B_1 with pressure

$$p_1 = \rho_W g h_1 \tag{4.28}$$

where ρ_W is the density of water [ML^{-3}], g the acceleration of gravity [LT^{-2}] and h_1 the height of the water column [L]. The same quantity would be measured if the tube of the piezometer were inserted from the side and bent in the form of a manometer. The water level in the free arm above elevation B_1 is a measure of the pressure of the soil water at point B_1 with reference to the atmospheric pressure. Moreover, if equilibrium exists, the pressure acting upon the soil water at any depth h below the ground water level is the same as the pressure at depth h below the free water level in the vessel.

If a porous cup connected to a manometer initially filled with water is inserted into the sand at point B_2 at an elevation h_2 above the free water level, at equilibrium the water level in the free arm of the manometer will eventually equal that of the free water level in the vessel. The walls of the porous cup should be semipermeable, i.e. permeable to water but not to air in the range of measurements. An instrument composed of a porous cup and a manometer is called a tensiometer. The pressure p_2 at point B_2 measured with the tensiometer is calculated from

$$p_2 = -\rho_W g h_2 . \tag{4.29}$$

The water level in the manometer being below the elevation of point B_2 indicates negative pressure at point B_2. Additional tensiometric observations would reveal, for the equilibrium condition, a linear function $p(z)$ with negative values above the ground water level and positive values below, see Fig. 4.10. Both positive and negative pressures are equivalent to soil water potential ϕ_w mentioned earlier.

A tensiometer as drawn schematically in Fig. 4.10 is not normally used in soil hydrology because it would require a hole of great depth for the manometer and access for its reading. A frequently used construction is sketched in Fig. 4.11. Although the details in construction may differ, a porous cup at the bottom of an impervious tube of specified length enables hydraulic contact between the soil water and the tensiometer. The negative pressure of the water is measured either by a mercury manometer, or by a pressure transducer connected to electronic instrumentation. A mercury manometer is shown in Fig. 4.11. Before a measurement can be taken, a zero value of the tensiometer could be determined, i.e. the height of mercury in the manometric tube with reference to the mercury level in the vessel when the pressure at the porous cup is zero. We use the analogy to the hydrostatics of a siphon with one arm submerged into mercury and the other arm into water (see Fig. 4.11) to calculate the value of h_o. Hence,

$$h_0 = h_1 / (\rho_{Hg} - 1). \tag{4.30}$$

The value of the pressure head h at the porous cup is calculated from the reading h_{oo} of the tensiometer using $h = [(1 - \rho_{Hg})h_{oo}]$. An alternative for the value of h at the porous cup located at depth L ($L > 0$), avoiding the use of the zero reading, is $h = [(1 - \rho_{Hg})h_R + h_2 + L]$ where h_R is simply the height of mercury in the manometer and $h_2 = (h_1 - L)$. Because a change of the mercury

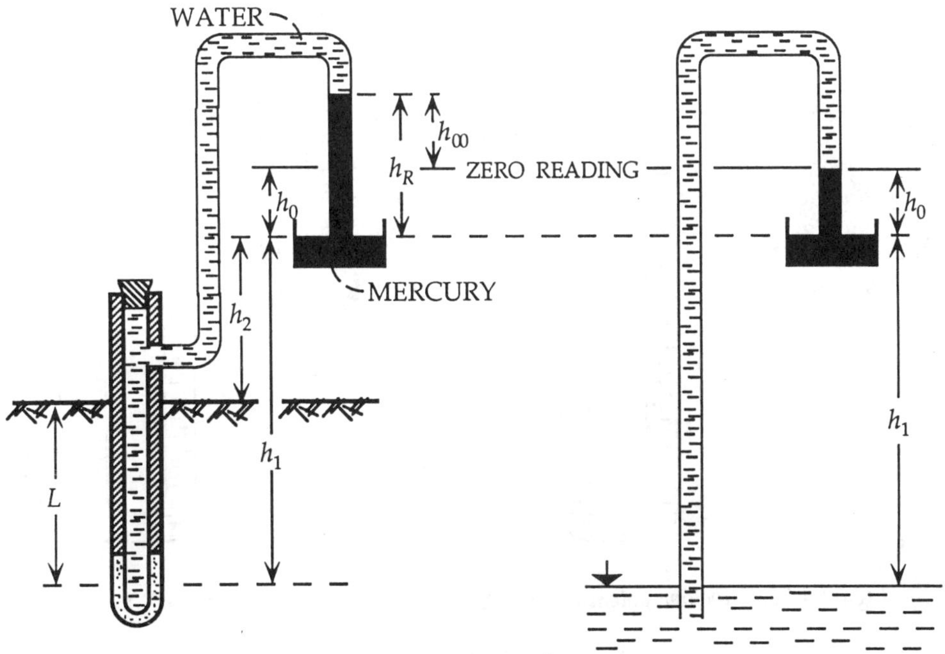

Figure 4.11. Scheme of a tensiometer in the field (left). The zero reading is analogous to the rise of mercury in the siphon having its bottom arm submerged in the water and its top arm submerged in mercury (right).

level in the manometer is accompanied with a volume of water moving between the tensiometer and the surrounding soil, the response time of the tensiometer to the change of water pressure in the soil may be excessive. Hence, a more rapid response time with a reading from a transducer may be preferred. Automatic data logging and greater accuracy are easily achieved with a transducer. Tensiometers are applicable to a pressure range of 0 to -80 kPa (0 to -800 cm) and a small range of positive values. For smaller values of pressure (i.e. to greater absolute values) the use of psychrometers is described in section 4.3.1.

With the tensiometer measuring all mechanical components of the potential, the measured pressure head is equivalent to $\phi_p = (\phi_w + \phi_a + \phi_e)$.

The envelope pressure ϕ_e can be demonstrated if the soil is placed in a plastic container which transfers the external forces acting upon the soil, see Fig. 4.12. When the soil is unsaturated, the tensiometer shows a pressure head equivalent to the soil water potential ϕ_w. When an outside pressure is applied, the water level in the manometric tube changes its position with the difference being the envelope potential ϕ_e. Here, the tensiometer measures the sum of both components $(\phi_w + \phi_e)$. Similarly, a change of air pressure in the soil pores related to a reference atmospheric pressure results in a change in the pneumatic potential ϕ_a. Such a change also influences the reading of a tensiometer.

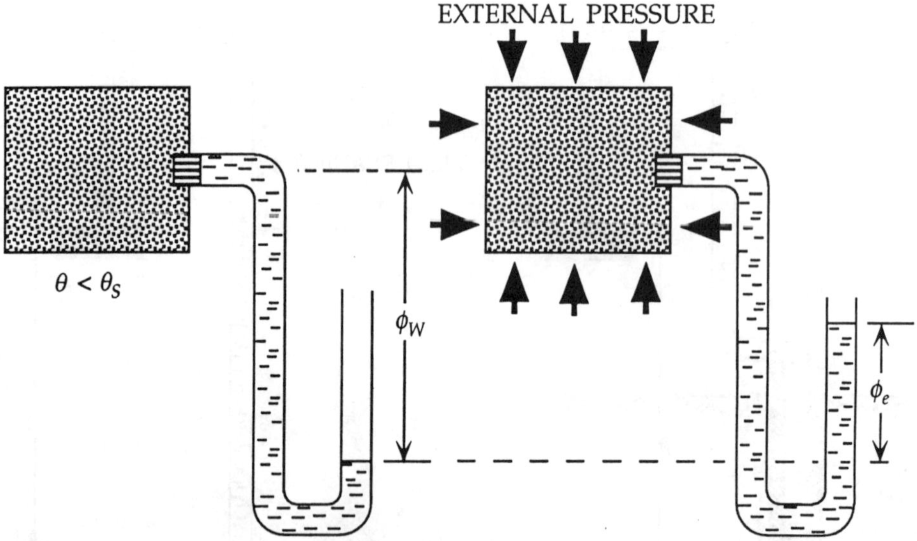

Figure 4.12. Envelope potential head ϕ_e in soil with soil water potential head ϕ_W.

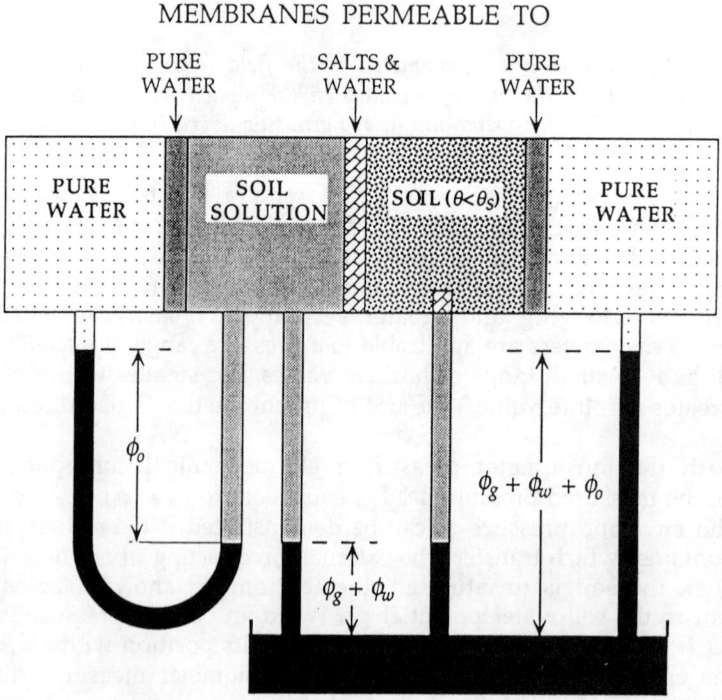

Figure 4.13. Demonstration of gravitational ϕ_g, osmotic ϕ_o and soil water potential ϕ_w heads after Richards (1965).

In a similar way Richards (1965) explained the meaning of the osmotic potential with a set of pools, and a manometer or tensiometer attached to each of the pools as illustrated in Fig. 4.13. The pools are separated by semipermeable membranes permeable only to specified substances. The central pool contains an unsaturated soil with water actually being a solution. The soil is separated by a membrane permeable only to water from the pool on the right containing pure water and by a membrane permeable only to soil solution from the pool on the left containing soil solution. This latter pool containing the solution of the same concentration as the soil water is separated by a membrane permeable only to water from the next pool to its left containing pure water. The differential manometer on the left shows a pressure height identical to the osmotic potential ϕ_o, while the manometer attached to the pool with the soil solution shows the sum of the gravitational and the soil water potential (ϕ_g + ϕ_w). The same value of ($\phi_g + \phi_w$) is read if a tensiometer is inserted into the soil inasmuch as its porous cup is permeable to both water and its dissolved substances. The manometer attached to the pool with pure water shows the sum of the potentials ($\phi_g + \phi_w + \phi_o$).

Diurnal variations of tensiometer readings p_T in the field are frequently reported. They usually occur owing to the action of two factors: (1) a change of θ owing to redistribution, drainage and evaporation processes, and (2) thermal effects which influence readings in a complex manner. The latter are more pronounced when we use mercury tensiometers. Volumetric changes of water, mercury and air with temperature partially transform the function of the tensiometer to that of a thermometer. If the instrument is not well deaerated, expansions and contractions of air bubbles in the tubes are manifested by changes in the length of the column of mercury in the arm of the manometer. The instrument performs also as a shortcut, heating the tensiometer cup and its surrounding in the daytime and cooling them at night. Hence, local fluxes of water vapor and eventual condensation of water vapor on the cool cup can influence the reading. In the construction of tensiometers all tubes should be made as small as possible and heat conducting materials should be avoided. Instead of mercury manometers, pressure transducers are preferred. The effect of diurnal thermal fluctuations upon the readings should be minimized by taking readings at the same time of day, preferably in the morning (Richards, 1949).

4.2.3.2 Psychrometers

Thermodynamics shows that the potential of a water solution at its reference water level equals the potential of water vapor immediately above the water level. For a plane water level of pure water we take conventionally $\Phi = 0$ and $\phi_w = 0$. The water vapor pressure above that water level is p_o. When the water level is curved as in a capillary tube, the surface pressure is decreased and the water vapor pressure also decreases. Let us now consider two vertical capillary tubes with their bottom ends submerged into free water. The diameters of the tubes differ slightly so that the height of rise of water differs only by dh. We have

water level height	water vapor pressure	surface pressure
h	p	p_s
$h + dh$	$p - dp$	$p_s - dp_s$
dh	$dp = \rho_G g dh$	$dp_s = \rho_W g dh$

We also have

$$\frac{dp}{dp_s} = \frac{\rho_G}{\rho_W} \tag{4.31}$$

where ρ_G and ρ_W are the densities of water vapor and liquid water, respectively. For $\rho_G = n\,\overline{M}/V$, the ideal gas law $pV = nRT$ becomes

$$p = \frac{RT}{\overline{M}}\rho_G \tag{4.32}$$

where R is the Universal molar gas constant [J·mol^{-1}·K^{-1}], T temperature [K], $\overline{M}$ molar mass [kg·mol^{-1}], V volume [m^3] and n number of moles. Combining (4.32) with (4.31) and remembering from (4.5) that if $p = p_o$, $p_s = p_r$ we have

$$\frac{RT}{\overline{M}}\int_{p_0}^{p}\frac{dp}{p} = g\int_0^h dh. \tag{4.33}$$

Integration provides the pressure head of the soil water

$$h = \frac{RT}{\overline{M}g}\ln(p/p_0). \tag{4.34}$$

For $T = 20°C$, $h = 1.46 \cdot 10^4 \ln(p/p_o)$ [m]. If (4.34) is combined with (4.10) we obtain the expression for capillary condensation in the wet part of the adsorption isotherm in Fig. 4.2. The same equation as (4.34) in units of energy by weight (J·kg^{-1}) has been developed in physical chemistry for the calculation of the chemical potential of a solution, if the water vapor pressure is measured above the plane level of the solution. Inasmuch as water in the soil is never chemically pure water, but a solution of a particular concentration, the sum of both potential components (the osmotic ϕ_o and the soil water ϕ_w) is determined from the relative water vapor pressure above an unsaturated soil by

$$\phi_o + \phi_w = \frac{RT}{\overline{M}}\ln(p/p_0). \tag{4.35}$$

Their mutual separation would be very difficult. The soil water content at ordinary field conditions corresponds to values $1 \geq p/p_o > 0.95$. Psychrometers used for measuring this combined potential ($\phi_o + \phi_w$) have to be extremely accurate. They are constructed on the principle of a thermocouple enclosed in a small cell provided with porous walls. Inasmuch as the potential of water in the walls of the cell is in equilibrium with that of the surrounding soil water, the water vapor pressure inside the cell is identical with that in the soil pores. Inside the cell we measure the temperature of the "dry bulb" of the thermometer, and using the Peltier effect, the temperature of the junction in the cell is reduced by an application of electromotive force. Water vapor condenses on this cool junction and the "wet bulb" temperature is read. The accuracy of the procedure is usually in the order of h = -100 cm.

4.2.3.3 Heat Dissipation Method

The heat dissipation method was developed for a relatively simple estimate of soil water potential within the entire range of most practical problems in soil hydrology, i.e. from 0 to -1500 kPa. The method is based upon the assumption that a thermal pulse in a porous body is deformed in relation to the thermal conductivity of the medium. The extent of pulse deformation is represented by the temperature difference ΔT before and just after the pulse is applied. For a large value of ΔT, little energy is conducted away from the source. This conduction of energy depends upon the soil thermal conductivity K_T which increases sigmoidally with soil water content. Fortunately, when K_T is plotted against soil water potential, the relation is nearly linear. The procedure is sometimes called the thermal conductivity sensor method.

For practical reasons, the heater and thermal sensor are both incorporated into a porous ceramic block. With the block inserted into the soil, a quasi-equilibrium of potentials is assumed between the block and the soil. Two temperature measurements are usually taken 1 minute apart. The first measurement represents the initial temperature T_i of the sensor and surrounding soil prior to a heat pulse that has a duration of 1 minute. When the heating ends, the final temperature T_f is read. For the temperature change $\Delta T = T_f - T_i$ the soil water potential is determined from a calibration curve. Instead of measuring the temperature difference, recordings of the voltage output (mV) are used. A nearly linear relationship between mV and the value of soil water potential exists in the range of -20 to -350 kPa. At smaller values of potential, the relationship is curved but applicable to -1500 kPa (Phene et al., 1992) Although a delay in reaching equilibrium and hysteresis are possible sources of errors, the method has the advantage of measuring potentials far beyond the range of tensiometers. Similarly, electrical resistance described earlier in section 3.2.2 as a soil water content method can be calibrated to soil water potential (Marshall and Holmes, 1979).

4.2.4 Potential in Swelling Soils

When a swelling soil is loaded with a pressure equal to the swelling pressure, the soil is incapable of swelling and cannot accept water from a free water pool. If the loading pressure is reduced, the soil swells and increases its water content. It is therefore necessary to introduce the component of the potential dependent upon the envelope pressure and the consolidation characteristics of the soil. We assume for the purpose of this introduction that all other conditions remain unaltered and all other components of the potential remain invariant. In field situations, the loading is the sum of the weight of the soil column of height z in the upper layers plus the pressure p_S acting upon the soil topographic surface

$$\int_0^z \rho_v \, g \, dz + p_S \qquad (4.36)$$

with $\rho_v = (m_S + m_W)/V_T$ where m_S is the mass of the dry soil, m_W the mass of soil water and V_T the volume of the soil. When the consolidation characteristic α is considered, the envelope potential is defined as

$$\phi_e = \alpha \left[\int_0^z \rho_v \, g \, dz + p_S \right] \tag{4.37}$$

where ϕ_e is in units of pressure ($N \cdot m^{-2}$). It is equivalent to the work done owing to a unit of the soil at depth z accepting water, causing the upper soil of thickness z to be lifted. The subsoil is assumed to be intact and not in any way involved with the change in swelling. If the soil is inert (sand), all the load is transferred on the contacts of the solid particles and the change of the soil (sand) water content does not influence its volume, $\alpha = 0$. In an unconsolidated, fully swelling soil comprising the topographical surface, if the change in its water content fully equals its change of volume (i.e. an equivalent uplift of the topographical surface), $\alpha = 1$. Between these two extremes the transitional situations exist with (Groenevelt and Bolt, 1972)

$$\alpha = \frac{1}{p_e} \int_0^{p_e} \left(\frac{\partial e}{\partial \vartheta} \right)_{p_e} dp_e \tag{4.38}$$

where e is the void ratio, ϑ the soil water ratio and p_e the total pressure loaded on the element. If the influence of p_e upon α is neglected, we obtain an approximation of the envelope pressure head h_e in units of energy per weight (Philip, 1969b)

$$h_e = \frac{1}{\rho_W \, g} \frac{de}{d\vartheta} \left[\int_0^z \rho_v \, g \, dz + p_S \right]. \tag{4.39}$$

In the majority of practical instances, pre consolidated loamy and clayey soils manifest values of α in the range from 0.2 to 0.3.

4.3 SOIL WATER RETENTION CURVE

We have demonstrated that a particular soil water potential ϕ_w is related to the soil water content θ. The potential ϕ_w is usually expressed as pressure head h for the convenience of computation and measuring technique. The functional relationship $h(\theta)$ is typical for the given soil having its particular status of consolidation, geometrical arrangement of particles and aggregation and other chemical and biological features. The function $h(\theta)$ when plotted is called a soil water retention curve SWRC, see Fig. 4.14. In the literature, the terms capillary pressure curve (Dullien, 1979) or soil moisture characteristic curve (Childs, 1969, and Hillel, 1971) has also been used. Because h extends over three to four orders of magnitude for fluctuations of θ commonly occurring in the field, h is frequently plotted with a logarithmic scale, see Fig. 4.14. Most often, it is tacitly assumed that $h(\theta)$ is determined at a constant temperature because properties and behavior of the soil water system are temperature dependent, e.g. the surface tension of water. However, in practical studies, the temperature dependence of $h(\theta)$ has been neglected until now without fully demonstrating the range of errors owing to such negligence. We shall say more about it at the end of this chapter.

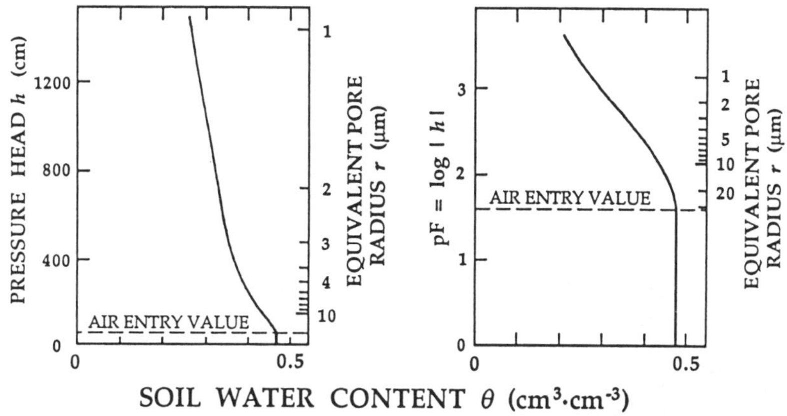

Figure 4.14. Soil water retention curve SWRC using a linear scale for pressure head h (left) and a logarithmic scale for h (cm) where $pF = \log_{10}h$ *(right). The right-hand curve is commonly called a pF curve. Equivalent pore radii are plotted on the vertical axes assuming a parallel capillary tube model.*

Similar to the hysteresis of adsorption-desorption isotherms, the soil water retention curve manifests hysteresis. That is, θ in the drying (or drainage) branch of $h(\theta)$ is larger than θ in the wetting branch for the same value of h.

In Fig. 4.15 hysteresis loops of the SWRC are schematically demonstrated. The soil was assumed to be initially fully saturated with water, completely without an air phase, with a saturated water content θ_S equal to the porosity P. Subsequently, the soil was gradually drained and dried, respectively at each h_j, the equilibrium θ_j was determined at sufficient values of j to ascertain the primary drying curve PDC continuously to a soil water content equal to zero. Next, the soil was gradually wetted along several values of θ_j until the primary wetting curve PWC was measured and at $h = 0$, $\theta = \theta_W$. The wetting curve does not converge to θ_S, and hence, $\theta_W \neq \theta_S$. The volumetric domain $\theta_{Ar} = (\theta_S - \theta_W)$ is filled by an incoherently distributed air phase. This condition is sometimes called entrapped air which we can approximately demonstrate as mutually unconnected micro bubbles of air. Having no chance to escape as a continuous flux, these incoherently distributed micro bubbles of air can only disappear either by a long lasting process of dissolution and diffusion in soil water, or by a drastic reduction of the pressure of the external atmosphere. When we start now with drainage having $\theta = \theta_W$ at $h = 0$, we obtain a main drainage curve MDC which is not identical to the PDC. If instead of drying the soil to a defined value of $p/p_o < 0.98$ using a controlled humidity chamber, we use a pressure plate apparatus, the MDC will asymptotically approach the "real" residual water content θ_{Wr} which characterizes the boundary between the coherent and incoherent water phase distributions. Luckner et al. (1989) found the approximate relationship $\theta_{Wr} \approx 2\theta_{Ar}$. The wetting curve which starts at θ_{Wr} is the main wetting curve MWC, and it is not identical with the PWC. At $h = 0$, the

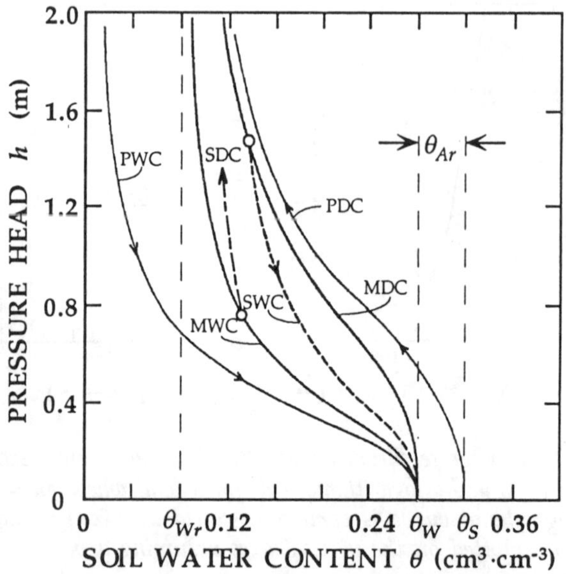

Figure 4.15. Hysteresis of the SWRC for a coarse-textured soil (Luckner et al., 1989). PDC is the primary drainage curve, PWC the primary wetting curve, MDC the main drainage curve, MWC the main wetting curve, SWC a scanning wetting curve and SDC a scanning drainage curve.

MWC reaches again θ_W. Thus, we obtain a closed hysteretic loop of main curves with an exclusion of the incoherent distribution of air close to $\theta = \theta_S$ and the incoherent distribution of water close to $\theta = 0$ (Luckner et al., 1989). These two incoherent domains of isolated air or water, respectively, are transported by different mechanisms than those for the bulk flux of either of the coherent phases. When drainage on the MDC is interrupted at point D_1 by the soil gradually wetting, the scanning wetting curve SWC is "above" the main wetting curve and the values of θ corresponding to a given h are larger than those on the main wetting curve MWC but smaller than those on the main drainage curve MDC. Similarly, when the wetting along the main wetting curve MWC is interrupted by the soil draining before reaching θ_W, the scanning drainage curve is "below" the MDC. When the hysteretic loop is gradually shortened as in Fig. 4.15, the scanning curves never intersect each other. Hysteresis is attributed to the action of several factors – the enclosed air in "blind pores", the ink-bottle effect of the influence of the rosary shaped pores where the radius of the meniscus depends upon the direction of reaching a given level and, moreover, the difference in wetting angle occurring for an advancing versus receding liquid front over the solid surface. In our description of hysteresis we could have proceeded vice versa, i.e. we could have started with a completely dry soil at $\theta = 0$.

A treatise on the role of individual factors controlling hysteresis is rather speculative. The theory of independent domains was first applied (Poulovassilis, 1962) and further on, the theory of dependent domains was discussed (Mualem and Miller, 1979, and Mualem, 1984). Both approaches were based upon the theories of Everett (1954, 1967). The independent domain theory assumes that all pore domains are free to drain independently. In reality, only those pores which have free access to the outside air can drain. This access is dependent upon the state of the surrounding pores – be they water or air filled. To account for this dependence, a domain dependence factor $P_d < 1$ is applied and its relationship to θ is searched. If $P_d = 1$, the dependent domain model is, in this case, equal to the independent domain model. For the wetting process, the blockage to the entering water is usually assumed to be of negligible importance and the independent domain theory is applied. Direct measurement (CT and NMR) and evaluation of soil porous systems in appropriate models can offer parameters needed for conceptual description of hysteresis.

Generally, hysteresis decreases the rate of change in θ when wetting is interrupted and replaced by drainage, and vice versa.

4.3.1 Construction of the Soil Water Retention Curve

A soil water retention curve can be determined either directly in the field, or in the laboratory on "undisturbed" core samples. Field methods usually require the installation of tensiometers (or other means of ascertaining the soil water potential) and access tubes (or borings) into the soil profile for measuring θ by an appropriate technique. Measuring under specified conditions such as for drainage or infiltration is preferred. If such conditions are not met, the experimental data are scattered about a mean of scanning hysteresis curves which are *a priori* unknown and, hence, the soil water retention curve is poorly defined inasmuch as the data for both wetting and drying are confounded. Field methods have the advantage of observing the soil water property at several soil depths simultaneously under natural field conditions that include overburden pressure, swelling and shrinkage and the effects of local water quality. Their disadvantages include a) only a small part of the SWRC is determined, usually in the range of $0 \geq h > -750$ cm, b) the method can be labor intensive during times that are inconvenient to the observer, c) non-isothermal conditions prevail and d) the main hysteresis branches cannot be determined with most measurements being limited to one or two particular scanning curves.

For a laboratory determination of the SWRC on "undisturbed" core samples, the principle of hydrostatic equilibrium is applied. The soil core is placed upon a layer of fine sand saturated with water in a sand tank (Fig. 4.16) connected hydraulically to an outflow vessel. First, the soil is water saturated to the extent practically possible by positioning the outflow vessel at the same level as the soil core. After the soil is fully wet, the first value of negative

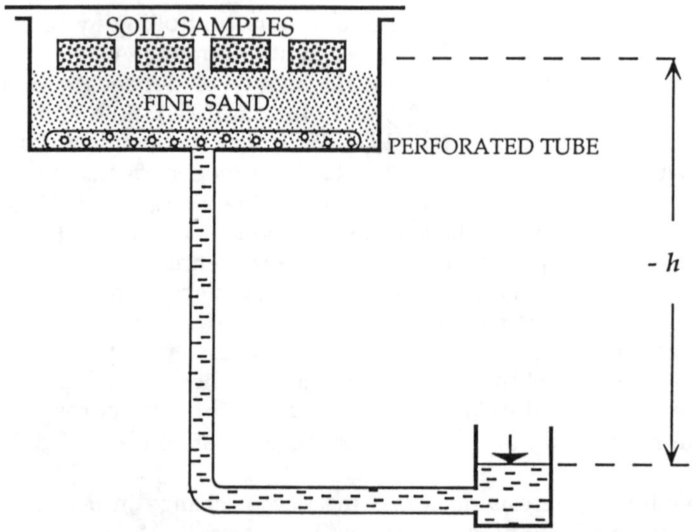

Figure 4.16. A sand tank with h being the applied negative pressure head.

pressure head h is induced by lowering the outflow vessel. This new position of the water level produces an outflow of water. The soil water content after reaching equilibrium with the imposed pressure head h is determined either gravimetrically, or from the volume of the outflow if only a single soil core is treated. We proceed to increase gradually the values of $|h|$ in equilibrium with each water content up to the air entry value of the sand, i.e. until the sand starts to be drained at a maximum value of $|h|$ equal to about 100 cm. In the range $0 > h = h_A$ the wet sand represents a semipermeable membrane – it is permeable to water and impermeable to air. When $h < h_A$, air enters the sand disrupting the continuity of water in its large pores that renders further drainage of the soil to be impeded and prohibitively slow.

When a porous plate replaces the sand, the apparatus is called a "tension" plate (Fig. 4.17). The principle of measurement remains the same, but the critical value of $|h|$ can be increased by selecting plates having smaller pore sizes than those of a sand. If each core is placed individually on the "tension" plate, the change of the soil water content resulting from the change in h is directly read on the outflow burette.

For smaller values of the soil water potential (or pressure head), usually in the range -20 kPa > ϕ_w > -1.5 MPa (-200 > h > -15,000 cm), a pressure plate apparatus is used. Instead of lowering the pressure hydraulically below the plate, the air pressure is increased above the porous plate (Fig. 4.17) which is hydraulically connected to the outflow vessel, usually a burette. Because other components of the total potential are assumed constant, $\phi_w = - \phi_a$. In both the "tension" plate and the pressure plate methods, the porous plate must be impermeable to air within the range of applied pressure.

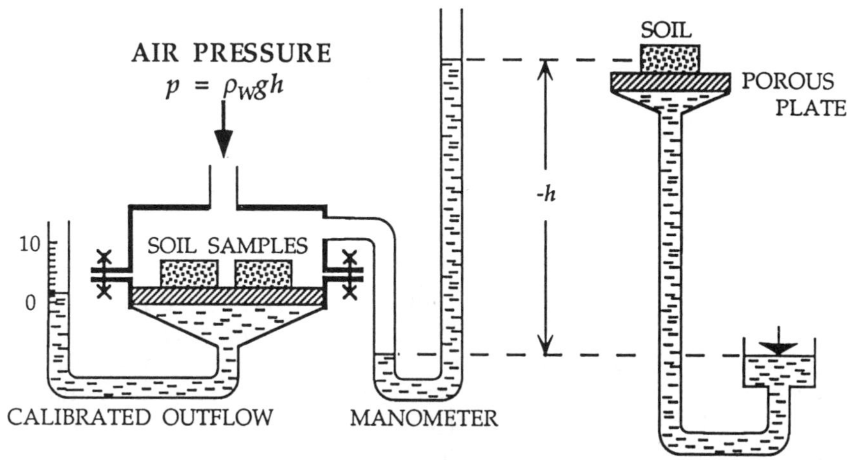

Figure 4.17. Pressure plate apparatus (left) provided with a liquid manometer for the sake of instuctiveness. Routinely, a pressure gauge is normally used in the laboratory. The tension plate apparatus (right) is shown with the same value of h which yields identical values of soil water content for identical soil samples in each of the apparatuses.

For extremely small values of soil water potential, commonly below -1.5 MPa, the desiccator method is sometimes used whereby the water vapor pressure is controlled with an aqueous solution of specified concentration within the desiccator. Here, however, the osmotic component of the potential is included, see (4.35). Hence, the SWRC obtained with the desiccator method cannot be connected to that obtained with "tension" or pressure plates.

In the laboratory procedure using soil core samples, the main drainage branch of the soil water retention curve is usually obtained. The opposite procedure for incrementally wetting an initially dry sample for different equilibrium values of h meets the very practical difficulty of extremely long equilibrium times. Hence, only drainage curves are measured in routine procedures.

The advantages of the laboratory method are a) a large number of soil cores can be simultaneously analyzed, b) a labor-intensive procedure is easily avoided and c) increments of imposed values of h can be so chosen to provide a smooth curve passing through the experimental data. The primary disadvantages include a) the volume of the soil core may not be equal or larger than the REV, especially in structured soils, b) in order to avoid extremely long time periods for equilibrium, the height of the core should not exceed 3 to 4 cm, and yet with such short samples, the vertical continuity of pores is restricted, c) large macropores can be deformed during sampling and d) owing to the manner in which h is imposed, soil water contents near saturation can be very inaccurate. Except perhaps for sandy soils, Bork and Diekkrüger (1990) have shown that laboratory determined SWRC are systematically and significantly shifted to larger water contents compared to those obtained in the field. Therefore, if SWRC determined in the laboratory are to be used for the solution of field problems, they should be objectively corrected, e.g. by the solution of a simple inverse problem.

4.3.2 Analysis of Soil Water Retention Curves

Some characteristic values are significant for each SWRC, see Fig. 4.15. After subjecting a soil for a long period with a pressure head $h = 0$, complete water saturation θ_S is eventually reached when $\theta_S = P$ the porosity. When the soil is quickly wetted, as is usually the case, 3 to 8% of the soil is occupied by entrapped air. This smallest maximum extreme of θ is denoted in Fig. 4.15 as θ_W with $\theta_W = P(1 - n)$ where $n = 0.05$ to 0.15 even though $h = 0$. On the MDC the air entry value h_A is sometimes found at the minimum value of h when $d\theta/dh$ remains equal to zero. When the same criterion is applied to the main wetting branch, the water entry value h_W is also similarly defined. However, in some soils, $h_A = h_W = 0$, or more frequently, $h_W = 0$. The air entry value h_A is well defined and well measurable in coarse textured soils and in modeling graded sands in the laboratory. In medium and fine textured field soils the estimation of h_A is inaccurate and some authors pose the question on the existence of h_A in the majority of topsoils (Nielsen et al., 1986). However, in subsoils and in deep layers of sediments, the saturated "capillary fringe" above the zero pressure ground water level is traditionally accepted, thus indicating the existence of both h_A and h_W. If macropores as fissures and cavities produced by the edaphon exist in a soil, the value of θ changes abruptly from θ_S to θ_{SM} when h is decreased by an infinitesimally small increment dh starting from $h = 0$. Under such behavior, the term air entry value is referred to the soil matrix with the value of θ_{SM} being constant from $h = 0 - dh$ to $h = h_A$ where θ_{SM} denotes saturation of the soil matrix. At a very small value of h (and of θ as well) we meet again the condition of $d\theta/dh \rightarrow 0$ on the MDC. The corresponding value of θ is called the residual (or minimal or intrinsic) soil water content. When it is the limit of the continuity of liquid water in soil, we denote it as θ_{Wr} (Luckner et al., 1989). When it is just a fitting parameter obtained when experimental data are fitted to an empirical equation of a SWRC, we denote it as θ_r.

An analytic expression of the SWRC is advantageous for the solution of practical problems. Except for values of θ close to θ_S the SWRC resembles the hyperbola

$$|h| = a\,\theta^{-b} \tag{4.40}$$

where a and b are empirical constants. Because of the large variations of θ_S that occur in the field, the utility of (4.40) is increased substantially if the soil water content θ in (4.40) is replaced by the relative soil water content $\theta_R = \theta/P$ or by the relative soil water saturation $\theta_R^* = \theta/\theta_S$ (Simmons et al., 1979). An analysis based upon capillary and adsorption theories obtained by Mitchurin (1975) led to the more complex expression

$$|h| = a\,A_m\left(w - 2w_m\right)^{-b} \tag{4.41}$$

where a and b are physically based constants defined in a model where the soil is represented by individual particles, A_m the specific surface of the particles, w the mass water content and w_m the mass water content when a monomolecular layer of water exists on the particles. The advantage of (4.41) is its physical interpretation of the coefficients in (4.40) that applies in the region $-\infty < h < -0.3$ MPa.

After an exhaustive study of experimentally determined SWRC on many soils, Brooks and Corey (1964) rewrote (4.40) to be

$$\theta_E = \left(\frac{h_A}{h}\right)^{\lambda} \tag{4.42}$$

where the effective water content $\theta_E = (\theta - \theta_r)/(\theta_S - \theta_r)$ and λ, the pore size distribution index, is a characteristic of the soil with values approximately equal to 2 to 5. The value of λ is large for soils having a uniform pore size distribution and small for soils with a wide range of pore sizes.

Among the less frequently used expressions of $\theta(h)$ are hyperbolic, error function or exponential equations of other authors. For the convenience of analytic or approximate solutions of some elementary hydrologic processes, the SWRC can be formulated by still another equation which is well suited to the mathematical development of the solution (Broadbridge and White, 1988), however its practical applicability for experimental data has not yet been proved.

Inasmuch as (4.42) does not offer a satisfactory description of the SWRC in the wet region, especially for soils not exhibiting a distinct value of h_A or h_W, van Genuchten (1980) proposed the equation

$$\theta_E = \left[1 + \left(\alpha|h|\right)^n\right]^{-m} \tag{4.43}$$

where α, n and m are fitting parameters with their limitations being $\alpha > 0$, $n > 1$, $|h| \geq 0$ and $0 < m < 1$. Values of n occur between 1.2 and 4 and those of α between 10^{-3} and 10^{-2} cm^{-1}. Because of computational expediency, values m have arbitrarily been taken equal to $(1 - 1/n)$. van Genuchten and Nielsen (1985) have proposed for pragmatic reasons to merely consider θ_r and θ_S as empirical fitting parameters. Note that the physically real residual water content on the SWRC in Fig. 4.15 was denoted by θ_{Wr}. Equation (4.43) can be adopted to describe each of the branches of the hysteretic loop. The detailed procedure for expressing MDC, MWC and the scanning curves by a modified (4.43) is described by Luckner et al. (1989). Usually, $\alpha_w \approx 2\alpha_d$ where w denotes wetting and d is for drainage. Equation (4.43) does not allow the existence of h_A and (4.42) is not suitable if an inflection point exists on the SWRC. Hence, a compromise description of the SWRC is achieved when θ_E is replaced by θ_e, especially if the SWRC is further used for the determination of the unsaturated hydraulic conductivity (Šír et al., 1985). The definition of θ_e requires that θ be replaced by $\theta_e = (\theta - \theta_b)/(\theta_a - \theta_b)$, and hence, (4.43) becomes

$$\theta_e = \left[1 + \left(\alpha|h|\right)^n\right]^{-m} \tag{4.44}$$

where θ_a and θ_b are considered only as fitting parameters without physical significance and having values $\theta_a > \theta_b$ and $\theta_a \geq \theta_S$. Note however, in the range $h_A \leq h \leq 0$ that this SWRC described by (4.44) is valid only in the range $\theta_b \leq \theta \leq \theta_S$ for $\theta_b > 0$, or $0 \leq \theta \leq \theta_S$ if $\theta_b < 0$. All three equations (4.42), (4.43) and (4.44) are schematically demonstrated in Fig. 4.18. For large values of θ_a compared to that of θ_S, (4.44) is nearly identical to (4.42).

The authors of all equations assume implicitly that θ_S is a simply measured, reliably quantified value. However, when θ_S is measured in the field, this assumption is rarely fulfilled and the resulting inaccuracy can easily destroy all the refinements of the above theoretical approaches.

In some instances, when experimental values of θ near saturation indicate the existence of a bi-modal porosity (Othmer et al., 1991), it may be more appropriate to consider that the SWRC is composed of two curves as shown in Fig. 2.4 and 5.12. Bi-modality is evident when a smooth curve drawn through experimental data exhibits three inflection points instead of one.

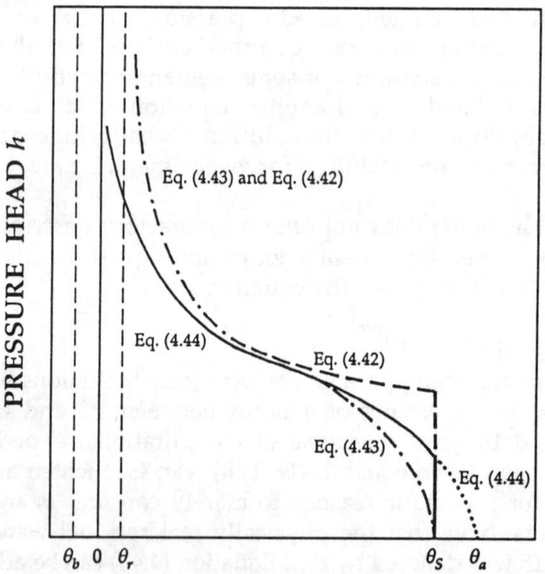

Figure 4.18. SWRC represented by either (4.42) of Brooks and Corey (1964) or by (4.43) of van Genuchten (1980) as well as by (4.44) which is a modification of that of van Genuchten proposed by (Šír et al., 1985). Note that the value of θ_b may indeed be less than zero.

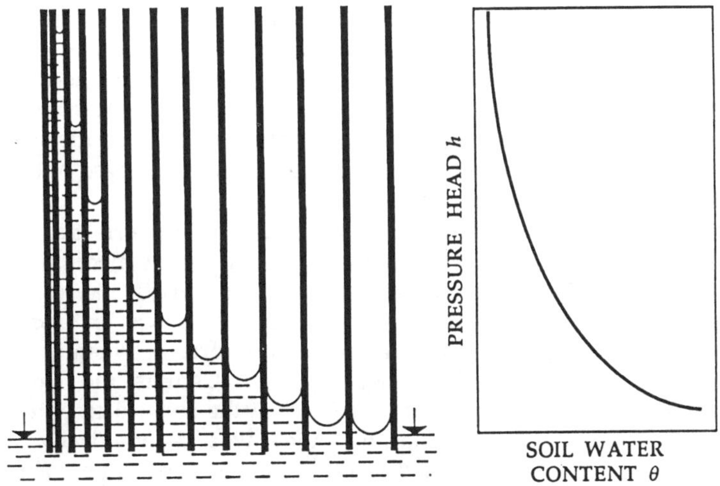

Figure 4.19. Model of SWRC consisting of parallel capillary tubes (left) and the resulting SWRC (right).

4.3.3 Models of Soil Water Retention Curves

The simplest model of the soil porous system is a bundle of parallel capillaries, see Fig. 4.19. A refinement of this model considers the capillaries as being rosary (ink-bottle) tubes. Still another refinement allows hysteresis owing to the behavior of contact angles.

When the scale of the soil water pressure head on the axis of the SWRC is replaced by that of the radius of the capillaries according to (4.11), we obtain the summation curve of the pore size distribution. The derivative of this summation curve is the frequency curve of the pore size distribution, see Fig. 2.4. The frequency curve is advantageously expressed by equations of probability density functions (Brutsaert, 1966). Among the more common functions are the incomplete gamma distribution, the log-normal distribution and the first asymptotic distribution. Criteria for optimal selection of the applied probability density function are the same as those used in ordinary statistics, e.g. the χ^2 test or Kolgomorov-Smirnov test. A frequency curve can be used to discuss the quality of the soil pore space and the influence of society's activity upon its alteration.

Parallel capillary tube models are distinctly different from real porous systems of soils (Dullien, 1979). Some improvement is gained when the pores formed between the contacts of soil grains are considered. Arya and Paris (1981) assumed that the similarity between the shapes of the particle-size distribution curve and the SWRC is closely related to the value of the representative pore diameter in each of the defined particle-size classes. Tyler and Wheatcraft (1989) defined an empirical coefficient α that relates the fractal dimension of the pore space and expresses a measure of the tortuosity of the pore space. The main disadvantage of models based upon particle-size is their negligence of the internal architecture of soil. The porous system of a soil is formed by a certain configuration of particles into aggregates and the aggregates further being arranged in a definite way. However, when the particle-size distribution of a soil is measured, the macro- and micro-aggregates are destroyed and individual particles separated.

Fatt (1956) was probably the first to realize that imperfections in parallel capillary models required a new approach to model porous media. He proposed an empirical, two-dimensional network of capillaries having randomly distributed radii. This lattice type network of pores was extended to a 3-dimensional network and for the solution of equilibria between the soil water potential and θ, percolation theory was applied (Chatzis and Dullien, 1977; Wardlaw et al., 1987; and Diaz et al., 1987). With the soil water potential being identified with the capillary potential, the model was hypothetically placed into the pressure apparatus. We shall first use the simplest type of percolation model which is a 2-dimensional square network where the pores are represented by mutually interconnected perpendicular cylindrical tubes of various radii. The radii are scattered randomly in the network with a prescribed distribution function $\rho(r)$. Inasmuch as the number of segments n approaches infinity in percolation theory, models of the theory usually contain segments $n > 1000$. Here for the sake of clarity in Fig. 4.20, we use only 72 segments.

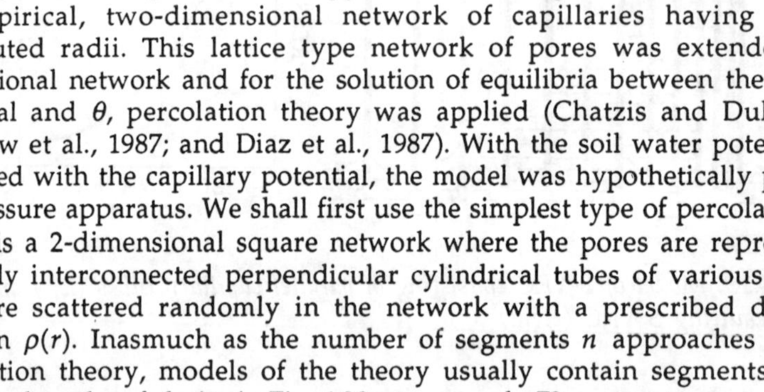

Figure 4.20. Pre-percolation (left) and percolation stage (right) in a 2-dimensional model consisting of capillary tubes of randomly distributed radii. Thick lines denote capillary tubes filled with air when the model is placed in a pressure plate apparatus.

This 2-dimensional model is conceptually first completely saturated with water and placed in a pressure plate apparatus. Similar to the method described in section 4.3.1, the air pressure is gradually and incrementally increased. When the pressure head h is below the percolation threshold $|h| < h_p$, water is replaced only in a very small portion of pores of radii r defined by (4.11) as

$$r \geq r_h = \frac{2\,\sigma \cos \gamma}{|h|\,\rho_W\, g}. \tag{4.45}$$

The great majority of pores having radii $r \geq r_h$ is surrounded by pores having radii of $r < r_h$. At this small initial value of pressure head h, the cluster of air-filled pores close to the top surface of the model is negligible and the model is not effectively drained. When the air pressure in the apparatus reaches the percolation threshold h_p, water is replaced by air in a continuous cluster of pores, all of radii $r \geq r_{hp}$, see Fig. 4.20. Isolated groups of water-filled connections of radii $r > r_{hp}$ remain undrained because they are surrounded by air-filled connections leaving them isolated and without a continuous path to the semipermeable membrane of the apparatus and to the free water pool. When the air pressure is further increased in our model, no additional drainage is realized and this is denoted as the post-percolation stage. A SWRC of such a model is step-like.

When the model still consisting of perpendicular capillary tubes is extended to a 3-dimensional net, drainage does not stop at one percolation stage but proceeds further when the external pressure is increased. The SWRC of the 3-dimensional net does not yield a unique step-like form similar to that of the 2-dimensional model. Owing to the three dimensionality of the net, clusters of undrained pores change and alter their configuration with a portion of undrained water remaining. Similarly, when water enters into a dry, air-filled model, clusters of pores remain filled with air in spite of the fact that they should be filled with water at the given pressure if water had free access to all pores. This procedure adequately explains the mechanistic part of hysteresis. With the occurrence of the clustering of entrapped water or air, the restrictive access of pore water or pore air to the outside pool of free water or free air is demonstrated.

Another type of 3-dimensional model consists of spherical pores interconnected by capillary tubes (throats). They are arranged in a cubical net, see Fig. 4.21. The throats may or may not be correlated to the spherical pores.

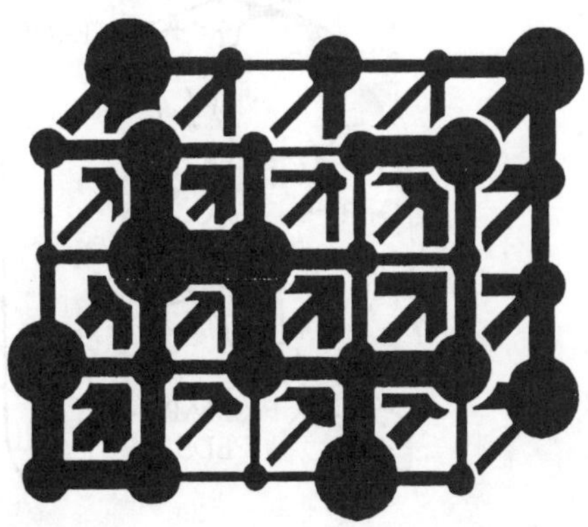

Figure 4.21: A 3-dimensional model consisting of spherical pores interconnected by capillary tubes (throats) whose radii are randomly distributed.

With radii of pores randomly distributed, the distribution function and skewness are described. An increase of the standard deviation increases the slope of the SWRC. Or, α and to a lesser extent n in (4.43) decrease their numerical values. The air entry value depends upon the skewness of this distribution curve (Rösslerová, 1993). The model can be expanded to a bi-modal porosity with hierarchical arrangement of sub models (peds) within the cubic net (Kutílek et al., 1992). This type of model corresponds nicely with the porous system found in the crumb structure of the A horizon of a soil. Numerical studies of this hierarchical arrangement confirm the existence of three inflection points on the SWRC. As it follows from studies of Rösslerová (1993), SWRC are more realistic to natural soils if bimodality is introduced. The slope of the SWRC is increased and the values of parameters α and n in (4.43) decrease owing to bimodality, especially for soils consisting of large peds. Moreover, the value of θ_r increases in such soils compared with that of monomodal soils. Bimodal soil systems manifest a broader hysteretic loop and generally, the amount of water or air residing in enclosed clusters is increased during the entire drainage or wetting process.

The configuration of pores between peds may be different from that within the matrix. In B horizons we frequently observe polyhedral or cubic structure which is modeled by cubes. The pores between the peds correspond approximately to a system of perpendicular slits having widths that are randomly distributed and defined by a prescribed distribution function.

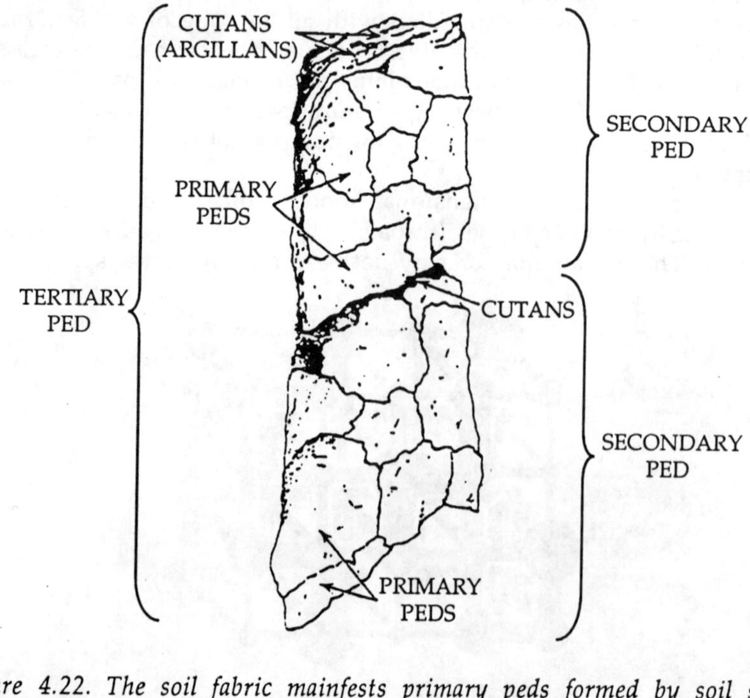

Figure 4.22. The soil fabric mainfests primary peds formed by soil solid constituents arranged into secondary and tertiary peds (Brewer, 1964).

The hierarchical arrangement of pores can proceed beyond the bimodal concept if we find primary, secondary and tertiary peds in a soil horizon, see Fig. 4.22. Generally, models of soil porous systems should properly mimic the morphological reality of the soil and the classification used in soil macro- and micro morphology (U. S. Soil Taxonomy, 1975; Brewer, 1976; and Jim, 1988).

Further progress in conceptually modeling of soil porous systems is achieved by the application of fractal geometry, especially when the soil is fragmented owing to aggregation. The self-similarity in porous systems of secondary and tertiary peds in Fig. 4.22 is distinct. Rieu and Sposito (1991) have discussed various types of models of fractal porous media with regard to real soils, see Fig. 4.23. The closest relation to micro- and macro-morphological descriptions of soils are the models denoted by (c) and (d) in Fig. 4.23. Model (c) is completely fragmented with self-similarity broadly developed. Model (d) is characterized by incomplete fragmentation, or by a clustering factor showing the extent to which peds stick together. Apparently, two avenues for model characteristics from soil analysis are available. Using aggregate size distribution data, Rieu and Sposito's (1991) simulated SWRC of an aggregated silty loam was close to measured data. Another avenue is the image analysis of micro- and macro-morphological studies usually presented as photos of either thin sections or of a topography of a plane. In both cases, the pores are distinctly identified. Macromorphological fractal analysis of staining patterns of macropores presented by Hatano and Boolting (1992) yielded fractal dimensions related to by-pass flow. Explanation of the term by-pass flow is presented in Chapter 5. Fractal interpretation of porous systems observed micromorphologically on thin sections has not yet been presented.

The above-mentioned refinement of models and procedures contribute to a more realistic conceptualization of the mechanical components of the total soil water potential in the wet portion of the SWRC. For the dry portion of the SWRC those improvements are practically irrelevant. Hence, we should concentrate on the force field at the molecular scale provided that we really need to interpret the dry part of the SWRC for a more accurate formulation of the processes.

Until now we dealt with rigid soils assuming that observations of the porous system were valid for both wetting and drying processes. For fine-textured soils we have already postulated macroscopic changes owing to swelling and shrinkage. We expect the volume of macropores as well as the distribution function of micropores to change when the opening and closure of cracks are observed. Schweikle (1982) provided evidence of such change for clay aggregates with summation curves of pores measured by mercury porometry at different values of soil water content. The distribution function of the pore radii in the aggregates obtained from the derivative of the summation curves (a) saturated with water and (b) drained to $h = -1.5 \cdot 10^{-4}$ cm (wilting point) are given in Fig. 4.24. We assume that full access of pores to outside pools of water or air exist. When the soil is saturated with water, the pore size distribution is approximately a mono-modal type with a slightly greater frequency of coarse pores and a smaller content of ultra fine pores. The mean equivalent radius or average pore size is of the order of 1 μm. When the soil was drained to h = -1.5

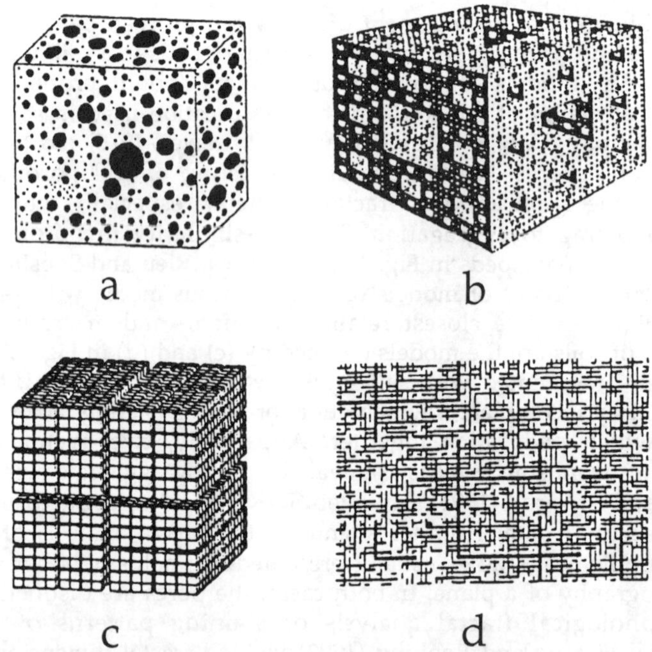

Figure 4.23. Fractal interpretation of models of porous media: a. a porous medium that is not necessarily fractal, b. the Menger sponge with fractal dimension D = 2.7268, c. a completely fragmented fractal porous medium with the number of self-similar elements N = 8 and D = 2.8 and d. a cross-section of a random, incompletely fragmented fractal porous medium with bulk fractal dimension D_r = 2.92 and D = 2.87 (Rieu and Sposito, 1991).

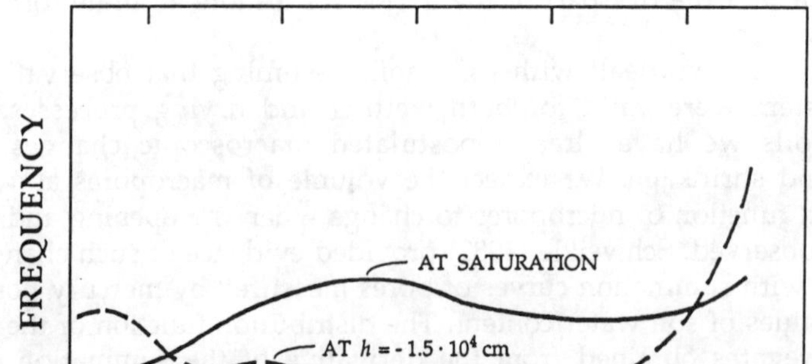

Figure 4.24. Frequency distributions of equivalent pore radii in aggregates of clay soil at two different soil water contents (h = 0 and h = -1.5 · 10^4 cm) using the data of Schweikle (1982) applied to a parallel capillary tube model.

MPa, the pore size distribution changed substantially. Shrinkage reduced the frequency of the average size pores and a large portion of these pores was transformed into coarse pores. A smaller portion of the average size pores was compressed and transformed to fine and ultra fine pores which created a bi-modal pore size distribution. However, from this scarce experimental evidence we cannot conclude if a transformation of the pore size distribution owing to soil water content change is significant in the majority of soils.

Inasmuch as the surface tension of water σ depends upon the temperature, we expect the SWRC to be effected by temperature. A temperature increase induces a decrease in the value of σ, and consequently the capillary height decreases. Accordingly, for a given soil water content, the value of the pressure head h increases (i.e. $|h|$ decreases). However, Novák (1975) and Hopmans and Dane (1986) report that the changes of the SWRC are greater than those expected just from $\sigma(T)$. Apparently, differences between theoretical and experimental values of $h(\theta, T)$ cannot be explained solely upon the basis of entrapped air as earlier assumed. A final conclusion based upon many experiments is that capillarity alone is not sufficient to explain water retention in soils even in the "wet" part of the SWRC. Therefore, all pore size distribution functions remain approximate if they are only based on capillary models.

PROBLEMS

1. A U-shaped capillary tube is partially filled with water. If the radii of the two branches are not equal ($r_1 > r_2$), show that the menisci in branch 1 and branch 2 can not be at the same elevation. Derive how the difference of elevation depends upon r_1 and r_2.
2. A core soil sample of 100 cm^3 is on a tension plate apparatus. At negative pressure heads h_i (cm) the equilibrium outflow V_i (cm^3) is read on the buret:

$-h_i$	0	10	20	30	40	50	60	80	100	120
V_i	31	31	30.3	29.2	27.5	25.5	23.3	21.5	19.1	17.8

Compute the soil water contents θ_i corresponding to h_i and plot the SWRC when $\theta_S = 0.48$.

3. Compute the total amount of water (m^3) drained from 1 ha of a homogeneous soil characterized by the SWRC from the previous problem when the ground water was lowered from the original depth of 40 cm to 120 cm below the soil surface.
4. A homogeneous loamy soil was covered by concrete so that neither infiltration nor evaporation could occur. What is the water content of the soil in contact with the concrete when the ground water table is 180 cm below the concrete? According to (4.42), $h_A = -10$ cm, $\lambda = 3.5$, $\theta_r = 0.05$ and $\theta_S = 0.45$.
5. The equivalent radii in Fig. 4.14 are for contact angle $\gamma = 0°$. How will the scale be changed if $\gamma = 60°$ owing to films of humic acids on the solid surfaces of the soil particles?

6. Assume a model of parallel capillary tubes will be replaced by a new model of parallel plates. What will be the new scale replacing the equivalent radii in Fig. 4.14? Instead of equivalent radii, plot the separation distance of the plates and consider $\gamma = 0°$.
7. What is the saturated pressure of water vapor above the meniscus in a capillary tube of radius 0.1 μm if the wetting angle $\gamma = 0°$?
8. Use Figs. 3.1 and 4.18 in order to derive the general shape of a clibration curve of soil water potential versus electrical resistance. What is the primary difference between this curve and that of the heat dissipation method?
9. Are we able to detect macropores from SWRC obtained from sand tank and tension plate methods? Explain.
10. If the soil water is in a state of equilibrium and a tensiometer reading of pressure head p_T = -350 cm at a depth of z = 20 cm, what will be the reading of a tensiometer at depth z = 50 cm?
11. What is the soil water pressure head measured by a tensiometer when the height of mercury in the tube of the liquid manometer is h_R = 20 cm above the free mercury level and h_2 = 30 cm and L = 80 cm? Symbols are defined in Fig. 4.11.
12. What is the value of h_R if the ground water level reaches the center of the porous cup of the tensiometer in the previous problem?
13. If a cylindrical capillary tube has an air-entry value of -15kPa, what is the height of capillary rise in this vertically oriented tube when one of its ends is inserted in free water?
14. Assuming a parallel tube model, what is the percentage of pores of equivalent radius in the range of 1 to 5 μm if the parameters of (4.44) are $\theta_S = 0.48$, $\theta_r = 0.065$, $\alpha = 0.012$ and $n = 1.37$?
15. In order to transform (4.43) to (4.42), what should be the values of parameters n, m and α in (4.43)?

5 HYDRODYNAMICS OF SOIL WATER

We introduced the concept of soil water potential in Chapter 4 where the system was studied at the equilibrium state. It was characterized by a driving force equal to zero and the value of the total potential being equal at all locations in the soil. Under such conditions, the water flux density is zero and both the soil water content θ and the soil water pressure head h are invariant with time. Here we derive equations describing the water flux when the soil system, not in equilibrium, manifests a total potential Φ not being constant in space.

5.1 BASIC CONCEPTS

The flow of water in soil can be described microscopically and macroscopically. On the microscopic scale, the flow in each individual pore is considered and for each defined continuous pore, the Navier-Stokes equations apply. For their solution we lack detailed knowledge of the geometrical characteristics of individual pores to obtain a solution for the REV. Even with this knowledge, a tremendous effort would be required necessitating voluminous calculations for even a relatively small soil domain. Nevertheless, this type of procedure is often applied in some theoretical investigations where the basic laws of fluid mechanics are invoked. In such studies the real porous system is usually defined by a model assuming great simplification of reality.

The macroscopic or phenomenological approach of water transport relates to the entire cross-section of the soil with the condition of an REV being satisfied. The rate of water transport through the cross section of the REV (the representative elementary area REA) is the flux. In order to emphasize the fact that water does not flow through the entire macroscopic areal cross section, the term flux density (or flux ratio, macroscopic flow rate et al.) is used to describe the flow realized through only that portion of the area not occupied by the solid phase and, by the air phase eventually when we deal later on with unsaturated soil. Moreover, we use the term flux density understanding that we actually mean the volumetric water flux density having the dimensions of velocity $[LT^{-1}]$.

Inasmuch as the principal equation derived for this macroscopic approach is Darcy's equation, the scale for which this approach is valid is often denoted as the Darcian scale. For soils, the area of this scale is usually in the range of cm^2 to m^2. Beyond this scale in either direction, larger or smaller, Darcian scale equations may not be realistic. Unless we state otherwise, equations will be derived and solved mainly for the Darcian scale related to a particular REV.

On the Darcian scale, water flow in soils is comparable to other transport processes such as heat flow, molecular diffusion etc. when the appropriate driving force is defined. For example, when the distant ends of a metal rod are kept at different temperatures, heat flow exists. Similarly, molecular diffusion depends upon a difference of concentration in two mutually interconnected pools. Soil water flow is conditioned by the existence of a driving force stemming from a difference of total potentials between two points in the soil. Laymen mistakenly suppose that the driving force of water flow in an unsaturated soil is related to differences in soil water content. This supposition, valid only for a few specified conditions, generally leads to erroneous conclusions.

Here, we first formulate basic flow equations for the simplest case of flow in a saturated, inert rigid soil. Afterwards, we deal with water flow in a soil not fully saturated with water. This latter type of flow is commonly called unsaturated flow while the former is called saturated flow. To be more precise, we should distinguish the former from the latter flows as those occurring at positive and negative soil water pressures, respectively. If the flow of both air and water in the soil system is simultaneously considered, we speak of two phase flow. Initially, we assume that the concentration of the soil solution does not affect the soil water flow. Subsequently, our discussion is extended to swelling and shrinking soils. Finally, we examine linked or coupled flows together with some specific phenomena of transport at temperatures below 0°C.

All equations that we derive are supposed to be applicable to not only analytical and approximate mathematical solutions of the components of the soil hydrological system but to all deterministic models of soil hydrology.

5.2 SATURATED FLOW

We assume that water is flowing in all pores of the soil under a positive pressure head h. In field situations the soil rarely reaches complete water saturation. Usually it is quasi-saturated with the soil water content $\theta_W = mP$ where m has values of 0.85 to 0.95 at $h \geq 0$, and P is the porosity. Entrapped air occupies the volume $P(1 - m)$. And for this discussion of saturated flow, the impact of entrapped air is not considered.

5.2.1 Darcy's Equation

For the derivation of Darcy's equation we shall discuss a simple experiment demonstrated in Fig. 5.1. The soil is placed in a horizontal cylinder connected on both sides with vessels containing water maintained at a constant level in each vessel by an overflow valve. If the water level on the left side is higher than that on the right side, water flows to the right. The rate of discharge $Q = V/t$ is simply measured by the volumetric overflow V in time t. The flux density q [LT^{-1}] (macroscopic flow rate) is

$$q = \frac{V}{At} \tag{5.1}$$

where A is the cross-sectional area of the soil column perpendicular to the direction of flow. Sometimes, the term q is also called the Darcian flow rate. The mean water flow rate (velocity) in the soil pores v_p is

$$v_p = q / P . \tag{5.2}$$

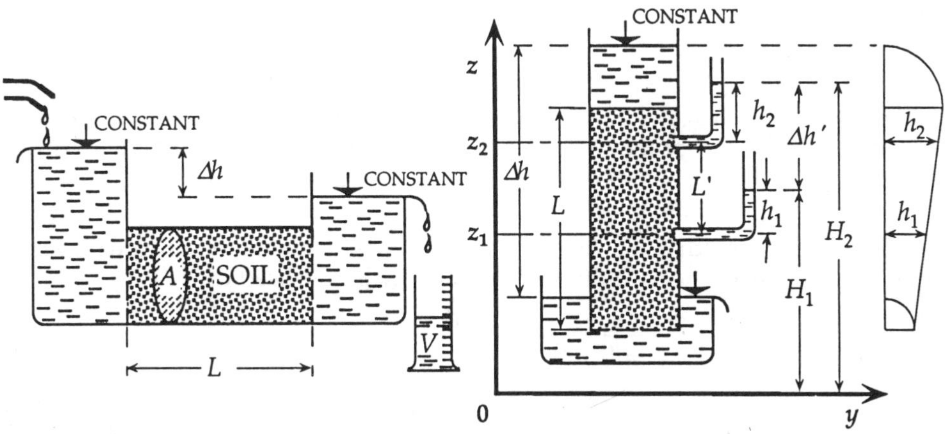

Figure 5.1. Simple steady flow experiments on saturated soil columns. On the right, the soil placed in a cylinder is provided with piezometers for measuring pressure heads h_1 and h_2 at depths z_1 and z_2, respectively. The total potential head $H = h + z$.

In 1856, Darcy experimentally demonstrated for columns of sand a linear relationship between the flux density q and the hydraulic gradient I_h. In our experiment shown on the right side of Fig. 5.1

$$q = K_S \frac{\Delta h}{L} = K_S \frac{\Delta h'}{L'} = K_S I_h \tag{5.3}$$

where $\Delta h/L$ or $\Delta h'/L'$ is the hydraulic gradient I_h, Δh the difference between water levels on both ends of the soil column of length L and $\Delta h'$ the difference between water levels in the piezometers separated by the distance L' in the direction of flow. Both Δh and $\Delta h'$ are considered the hydraulic head drop along the soil. Inasmuch as $\Delta h/L$ is dimensionless, K_S has the dimension of q [LT^{-1}]. When we read piezometer levels h_1 and h_2 at elevations z_1 and z_2, respectively, we have in terms of the total potential H

$$q = - K_S \left(\frac{H_2 - H_1}{z_2 - z_1} \right) \tag{5.4}$$

where the total potential head H (= $h + z$) is related to a unit weight of water. In a more general way (5.4) becomes

$$q = - K_S \, grad \, H \, . \tag{5.5}$$

Equation (5.5) states that the flux density is proportional to the driving force of the water flow which is the gradient of the potential. Inasmuch as K_S is a constant for a given soil, we write $\phi^* = K_S H$, and hence,

$$q = - grad \, \phi^* \tag{5.6}$$

where ϕ^* is $K_S H$. The negative sign in the above equations means that water flows in the direction of decreasing potential or against the positive direction of z in Fig. 5.1. The value of K_S depends upon the nature of the soil and is numerically equal to the flow rate when the hydraulic gradient is unity. Values of K_S commonly range from less than 0.1 cm·day^{-1} (10^{-8} m·s^{-1}) to more than 10^2 cm·day^{-1} (10^{-5} m·s^{-1}).

In layered soils we have to specify the direction of the flow relative to that of the layering. When the flow is parallel to the layers, the total flux is the sum of the fluxes for each of the individual layers, see Fig. 5.2. Hence, $Q = Q_1 + Q_2 + Q_3$ and

$$q(b_1 + b_2 + b_3)d = (q_1 b_1 + q_2 b_2 + q_3 b_3)d. \qquad (5.7)$$

For a column of width $d = 1$, length L and thickness b composed of three layers each of thickness b_i, the total flux density for a hydraulic head drop Δh is

$$q = \left(\frac{K_1 b_1 + K_2 b_2 + K_3 b_3}{b_1 + b_2 + b_3}\right)\frac{\Delta h}{L}$$

or

$$q = K_S' \frac{\Delta h}{L}. \qquad (5.8)$$

Here, the apparent hydraulic conductivity K_S' is the arithmetic mean of the individual values for each layer.

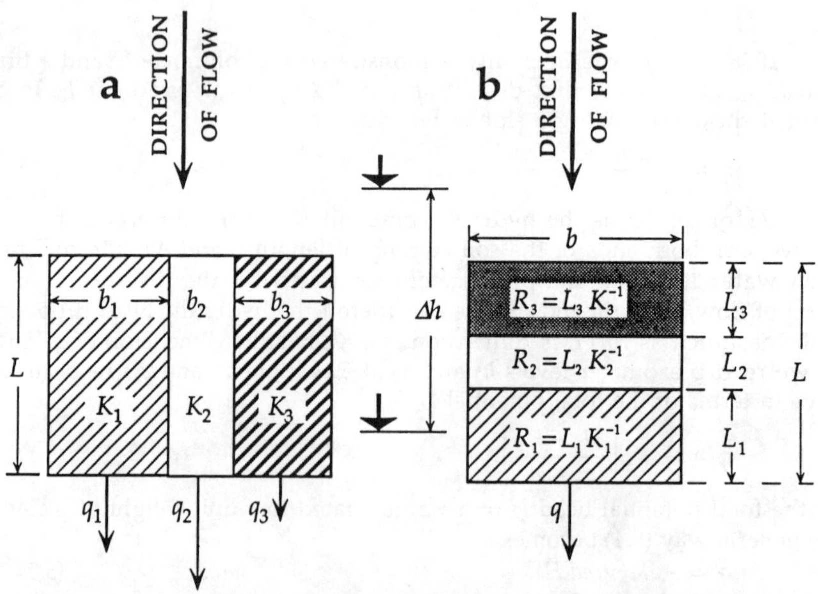

Figure 5.2. Flux density q in layered soils: a. Direction of flow is parallel to layering. b. Direction of flow is perpendicular to layering.

When water flows perpendicular to the layering, we introduce, analogous to an electrical resistance, the hydraulic resistance of each layer $R_i = L_i / K_{Si}$ having units of time. In Fig. 5.2 the flow combined from the three layers is

$$q = \frac{\Delta h}{R_1 + R_2 + R_3}. \tag{5.9}$$

With the total resistance of the system $R = \Sigma R_i$ we obtain the harmonic mean or the apparent hydraulic conductivity $K_S'' = L/R$.

When the flow is at an angle < 90° to the layers, the difference of K_S in each of the layers causes a change of the direction of streamlines (Zaslavsky and Sinai, 1981, and Miyazaki, 1990), see Fig. 5.3.

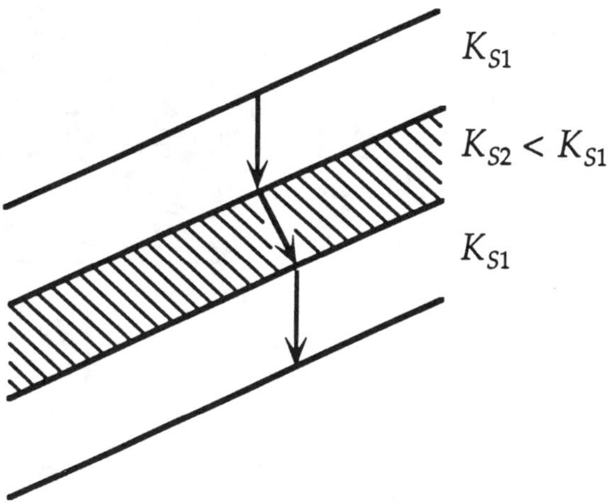

Figure 5.3. Change of the direction of flow in a soil with inclined direction of layers.

5.2.2 Saturated Hydraulic Conductivity

Inasmuch as the soil water potential H can be expressed in three modes, the dimension of the hydraulic conductivity is not necessarily [LT^{-1}]. From (5.5) we obtain for the three dimensions of H three different dimensions of K_S. Although expressing K_S in units of velocity is usually more convenient, any one of the following sets of units is occasionally preferred.

H		*grad H*		K_S	
$J \cdot kg^{-1}$	[L^2T^{-2}]	$J \cdot kg^{-1} \cdot m^{-1}$	[LT^{-2}]	s	[T]
Pa	[$ML^{-1}T^{-2}$]	$Pa \cdot m^{-1}$	[$ML^{-2}T^{-2}$]	$m \cdot s^{-1} \cdot \rho_W^{-1} \cdot g^{-1}$	[$M^{-1}L^3T$]
m	[L]	dimensionless		$m \cdot s^{-1}$	[LT^{-1}]

The empirical, intuitive derivation of Darcy's equation (5.5) can be theoretically justified from Navier-Stokes equations applied to an REV of a model of a porous medium and scaled with a characteristic length. In order to obtain (5.5), inertial effects were neglected and the density and viscosity of water were assumed invariant (Bear, 1972; Whitaker, 1986). Scheidegger (1957) showed that K_S should be considered a scalar quantity for isotropic soils, and a tensor of rank 2 for anisotropic soils with the value of K_S dependent upon the direction of flow. When the tensor K_S is assumed to be symmetric, its principal axes, defined by six values, are identical to those of an ellipsoid of conductivity. If the gradient of the potential is not in the direction of a principal axis, the direction of flow is different from that of the gradient.

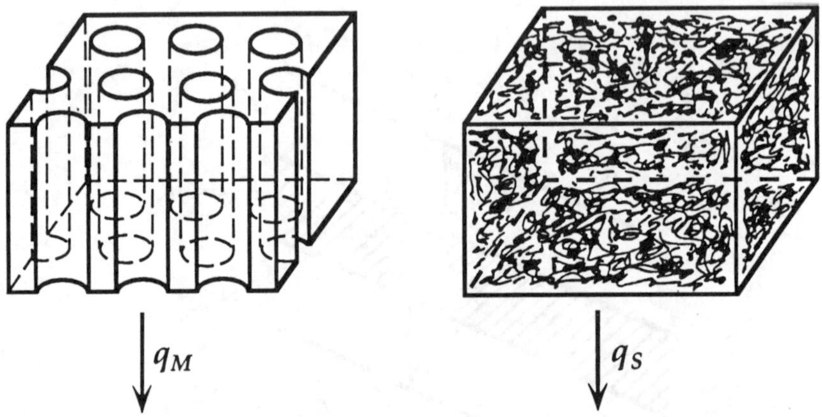

Figure 5.4. In Kozeny's model, the complicated soil porous system (right) is represented by a bundle of parallel capillary tubes of uniform radius (left). The flux density q, saturated hydraulic conductivity K_S, porosity P and surface area of pores A_m are the same in both the model and the soil.

From a theoretical treatment we can obtain a physical interpretation of the hydraulic conductivity. We develop here a modified and simplified model of Kozeny (cf. Scheidegger, 1957) consisting of a bundle of parallel capillary tubes of uniform radius. We assume that the soil and the model are identical with respect to porosity P, specific surface A_m [L^{-1}] and water flux density q [LT^{-1}], see Fig. 5.4. The mean flow rate v_p in a capillary of radius r is described by Hagen-Poiseuille's equation

$$v_p = \frac{\rho_W g r^2}{8\mu} I_h \tag{5.10}$$

where g is the acceleration of gravity [LT^{-2}], ρ_W the density of water [ML^{-3}], μ the dynamic viscosity [$ML^{-1}T^{-1}$] and I_h the hydraulic gradient [dimensionless]. With n being the number of capillaries of unit length x, the porosity of the model is

$$P = n\pi r^2 x / V_u \tag{5.11}$$

where V_u is the unit volume and the specific surface is

$$A_m = 2n\pi r x / V_u. \tag{5.12}$$

From (5.11) and (5.12) we obtain

$$r = \frac{2P}{A_m}. \tag{5.13}$$

And, from (5.2) and (5.10) we obtain

$$q = \frac{1}{2}\frac{\rho_W g}{\mu}\frac{P^3}{A_m^2} I_h . \tag{5.14}$$

Because soil pores are irregularly shaped and mutually interconnected, a shape factor c replaces 1/2 in (5.14). Letting

$$K_p = \frac{cP^3}{A_m^2} \tag{5.15}$$

we obtain

$$q = K_p \frac{\rho_W g}{\mu} I_h \tag{5.16}$$

which is identical to (5.3). Because the term K_p relates to the flow of any fluid through a soil, it is called the permeability [L^2]. The unusual dimension of K_p represents the cross-sectional area of an equivalent pore. Although now almost obsolete, the historical unit of 1 Darcy = 1 μm^2 was used for describing permeability.

Inasmuch as flow channels in the soil are curved compared with those of a capillary model, a tortuosity factor τ introduced in (5.15) yields the Kozeny equation

$$K_p = \frac{cP^3}{\tau A_m^2}. \tag{5.17}$$

The tortuosity τ is the ratio between the real flow path length L_e and the straight distance L between the two points of the soil. Because $L_e > L$, $\tau > 1$. In a monodispersed sand manifesting a value of $\tau \approx 2$, the flow path forms approximately a sinusoidal curve (Corey, 1977).

Equations identical or of similar type to (5.17) have been derived by many authors. If a model of parallel plates is used instead of capillary tubes and the slits are oriented in the direction of the laminar flow, we obtain the mean flow rate

$$v_p = \frac{\rho_W g d^2}{3\mu} I_h \tag{5.18}$$

where $2d$ is the distance between the plates. When B is the width of the plates, $P = 2ndBx/V_u$ and $A_m = 2nx(2d + B)/V_u$. Taking $x = 1$ and $B = 1$, we obtain $d = P(A_m - 2P)$ and hence,

$$K_p = \frac{c\,P^3}{\tau (A_m - 2P)^2}. \tag{5.19}$$

Looking at (5.17) and (5.19), we are reminded of the Kozeny-Carman equation

$$K_p = \frac{P^3}{5A_m^2 (1 - P)^2}. \tag{5.20}$$

Its derivation was shown in detail by Scheidegger (1957).

From (5.3) and (5.16) the relationship between K_S and any formulation of K_p is

$$K_S = K_p \frac{\rho_W g}{\mu}. \tag{5.21}$$

Kozeny's equation shows that K_S is sensitive to porosity. However, in his model the pore radii are considered uniform while those in real soils have broad distributions. For real soils, we subdivide the pores according to their radii into j categories each having an equivalent radius r_j. For $r_j > r_{j+1}$ the flux in each category $q_j(r_j^4, n_j)$ where n_j is the percentage of the j-th category in the whole soil, $q_j >> q_{j+1}$. Total flux $q = \Sigma\, q_j$ as shown for parallel layering. Thus, let us assume for $j = 1$, the percentage of the category of largest pores is eliminated by compaction. Although the porosity may be only marginally reduced, the value of K_S may be reduced by orders of magnitude. It is logical, therefore, that aggregation of a soil may increase K_S by orders of magnitude, yet the porosity may remain nearly the same. And, vice versa, soil dispersion or disaggregation substantially decreases K_S. For example, in a loess soil, the saturated hydraulic conductivity of its surface after a heavy rain decreases 3 to 4 orders of magnitude compared with its original value owing mainly to two processes – disaggregation and the blockage of pores by the released clay particles (McIntyre, 1958). Compaction of soil in the A-horizon and in the bottom of the plowed sub horizon causes a much greater decrease of K_S than that predicted from a decrease of porosity in the simple Kozeny equation because compaction reduces primarily the content of large soil pores associated with values of pressure head $h = 0$ to -100 cm.

Although the textural class of a soil may have a large influence on the value of K_S, any attempt to establish a correlation between the two attributes usually fails. Only for those soils and soil horizons of the same genetic development occurring in the same region and being similarly managed will a correlation between texture and K_S be manifested. On the other hand, a few generalities may exist. For example, the smallest values of K_S in each of the main textural classes can be approximated. In sandy soils, the minimum value of K_S is about 100 cm·day^{-1}, in silty loams about 10 cm·day^{-1} and in clays about 0.1 cm·day^{-1}. In peats, K_S decreases with an increasing degree of decomposition of the original organic substances. When the degree of decomposition of a peat is about 40 to 50%, the value of K_S diminishes to values of K_S typical of unconsolidated clays. Extreme drainage and concomitant drying of peat soils causing compaction and an increase of soil bulk density also reduce the magnitude of K_S. Moreover, because this drying increases hydrophobism, entrapment of air during wetting is enhanced and contributes even further to the decrease of K_S.

In loams and clays, the nature of the prevalent exchangeable cation plays an important role relative to the value of K_S, see Fig. 5.5. In vertisols, an increase of the percentage of exchangeable sodium (ESP) is accompanied by a decrease in K_S when the ESP reaches 15 to 20%, provided that the soluble salt content of the soil water is small. For example, with the electrical conductivity of the soil paste EC being 1 mS·cm^{-1} or less, the value of K_S can decrease two or three orders of magnitude. On the other hand, even for the same soil having a large ESP, if the concentration of soluble salts is increased substantially to an EC value of about 8 mS·cm^{-1} or more, the K_S value is not significantly affected. The value of ESP is closely related to the sodium adsorption ratio SAR of percolating water [SAR = Na/(Ca + Mg)$^{1/2}$]. If Ca is completely absent (i.e. only Mg appears

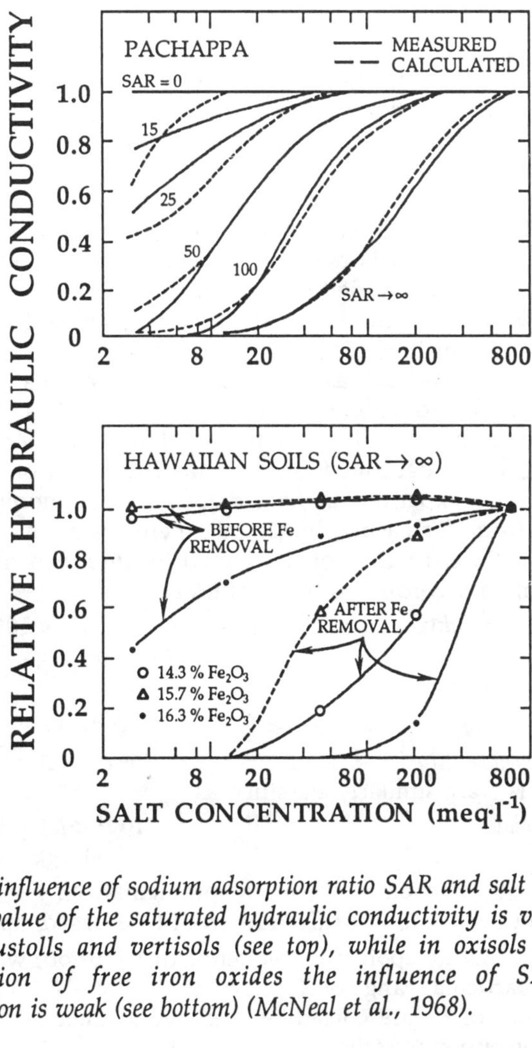

Figure 5.5. The influence of sodium adsorption ratio SAR and salt concentration upon the value of the saturated hydraulic conductivity is very strong in aridisols, ustolls and vertisols (see top), while in oxisols with a large accumulation of free iron oxides the influence of SAR and salt concentration is weak (see bottom) (McNeal et al., 1968).

in the denominator), the value of K_S is more greatly reduced than when Mg is absent (McNeal et al., 1968). When monovalent cations are considered in addition to Na, we find that potassium leads to a decrease of K_S but its influence is not as strong as that of exchangeable Na. It has been shown that large organic cations such as pyridinium cause the value of K_S to increase by several orders of magnitude in montmorillonitic clay while their impact on the value of K_S of kaolinitic clays is less significant (Kutílek and Salingerová, 1966). These variations are closely related to the degree of flocculation or peptization of the soil colloïdal particles that can be quantified with the value of the ζ-potential derived from double layer theory. Applying this theory, the decrease in K_S owing to the action of rain water (very small EC) is easily predicted for soils having large SAR values. These predictions are not necessarily successful for soils that differ pedologically. For example, a solution of high SAR value percolating through an Oxisol does not decrease the value of K_S even after reducing the solute concentration because the abundant free iron oxides prevent peptization and disaggregation of the soil particles. The value of K_S also depends upon the composition of the clay fraction. It decreases in the order kaolinite, illite and montmorillonite. And, soil organic matter has a profound impact upon the magnitude of K_S, owing to its cementing action that promotes aggregate stability. Bacteria and algae may reduce the value of hydraulic conductivity in long term laboratory tests and in some field conditions when sewage treatment effluents are either used for irrigation or disposed on special lands. These same effects can also exist in an irrigated soil well supplied with plant nutrients and sunshine. The process is not necessarily restricted to only anaerobic conditions inasmuch as some aerobic bacteria may cause a reduction in the value of K_S. The reduction is partly attributed to cells of bacteria and algae mechanically clogging the soil pores and to slimy, less pervious products of microbial activity being deposited on the walls and necks of the soil pores. For long term laboratory experiments, bactericides are commonly used to prevent the value of K_S from decreasing. In general, there are many factors influencing the value of K_S that are usually not considered in simplified models.

Soils classified according to their values of K_S are

very low permeability	$K_S < 10^{-7}$ m·s^{-1}
low permeability	$10^{-7} < K_S < 10^{-6}$ m·s^{-1}
medium permeability	$10^{-6} < K_S < 10^{-5}$ m·s^{-1}
high permeability	$10^{-5} < K_S < 10^{-4}$ m·s^{-1}
excessive permeability	$K_S > 10^{-4}$ m·s^{-1}

Geological materials are similarly classified as

compacted clays	$10^{-11} < K_S < 10^{-9}$ m·s^{-1}
gravel	$10^{-1} < K_S < 10^{1}$ m·s^{-1}

All such classification schemes above are problematical. For soils in a certain region, a more appropriate classification would be based upon the frequency distribution of K_S. Based upon that frequency distribution, we can identify sub regions where a particular range of K_S is expected.

When values of K_S are considered relative to their position within a soil profile, soils are grouped into these seven classes.

1. K_S does not change substantially in the profile.
2. K_S of the A-horizon is substantially greater than that of the remaining soil profile and no horizon of extremely low K_S exists.
3. K_S gradually decreases with soil depth without distinct minima or maxima.
4. K_S manifests a distinct minimum value in the illuvial horizon or in the compacted layer just below the plow layer.
5. Soil of high permeability with its development belonging to one of the first four classes covering the underlying soil of very low permeability.
6. Soil of very low permeability with its development belonging to one of the first four classes covering the underlying soil of very high permeability.

7. K_S changes erratically within the profile owing to extreme heterogeneity in the soil substrata.

The influence of the temperature upon the value of K_S can be examined with (5.21). Inasmuch as ρ_W is negligibly influenced by temperature, changes of $K_S(T)$ depend primarily upon the viscosity $\mu(T)$.

5.2.3 Darcian and Non-Darcian Flow

We have already mentioned that Darcy's equation is valid only for small rates when the inertial terms of the Navier-Stokes equations are negligible. For engineering purposes the upper limit of the validity of Darcy's equation given by (5.3) through (5.6) is indicated by the critical value of Reynolds' number for porous media

$$Re = \frac{q d \rho}{\mu} \tag{5.22}$$

where d denotes length. In sands, d is the effective diameter of the particles, or with some corrections, the effective pore diameter. Sometimes d is related to the permeability of the sand, e.g. $d = K_p^{1/2}$. However, in all soils other than sands, d is not at all definable and hence, (5.22) is not applicable. The difficulty in defining d is manifested by controversy in the literature regarding the assignment of critical values of Re. Most frequently, critical values of Re have been reported to range from 1 to 100. In this post linear region, the flow is often described by the Forchheimer equation (Bear, 1972)

$$\frac{dH}{dx} = aq + bq^2 \tag{5.23}$$

where a is the material constant analogous to K_S and b is functionally dependent upon the water flux density. This non linearity is caused primarily by inertia and by turbulence starting only at very large values of flux density, see Fig. 5.6. A more detailed theoretical discussion is given by Cvetkovič (1986).

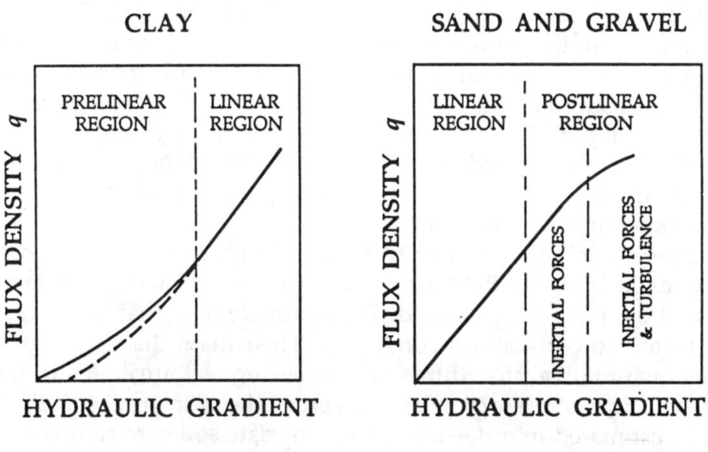

Figure 5.6. Deviations from the linearity of Darcy's equation.

Deviations from Darcy's equation have also been observed in laboratory experiments for very small flux density values. We define, therefore, the pre linear region of flow where q increases more than proportionally with I_h, see Fig. 5.6. This deviation from Darcy's equation, most often observed within pure clay having very large specific surfaces (e.g. 10^2 $m^2 \cdot g^{-1}$), has been explained by the action of three factors: a) Clay particles shift and the clay paste consolidates owing to the imposed hydraulic gradient and the flow of water, b) It is theoretically assumed that the viscosity of water close to the clay surfaces is different than that of bulk water or that in the center of the larger soil pores. According to Eyring's molecular model where the viscosity depends upon the activated Gibbs' free energy, the first two to five molecular layers have a distinct increased viscosity. Owing to the great value of the specific surface in clays, the contribution of the first molecular layers to the alteration of averaged viscosity may not be negligible. c) The coupling of the transfer of water, heat, solutes et al. may also contribute to the existence of the pre linear region (Swartzendruber, 1962; Kutílek, 1964 and 1972; Nerpin and Chudnovskij, 1967).

Deviations from Darcian flow are not frequently described or observed, and the post linear region is only rarely reached in sands and gravelly sands. There is not yet any field experimental evidence of the existence of a pre linear region. Darcy's equation is, therefore, either exact or at least a very good approximation entirely adequate for soil hydrology.

5.2.4 Measuring K_S

Saturated hydraulic conductivity is one of the principal soil characteristics and for its determination, only direct measurement is appropriate. Indirect methods, derived from soil textural characteristics which are sometimes combined with aggregate analyses, generally do not lead to reliable values. Considering soil texture as an example, soil water flow is totally independent from the laboratory procedure of dispersing, separating and measuring the percentage of "individual" soil particles which do not even exist "individually" in natural field soils. It has been shown in section 5.2.2 that the value of K_S is closely related physically to the porous system within a soil. Inasmuch as a quantitative description of this porous system is much more difficult than the measurement of K_S, direct measurement of K_S is preferred. When K_S is ascertained by water flux density and potential gradient measurements, we will speak about the determination of K_S. In order to avoid misunderstanding, when additional assumptions are used to evaluate these two quantities somewhat less directly, we will speak about the estimation of K_S.

Measuring is realized either in the laboratory on soil core samples previously taken from the field, or directly in the field without removing a soil sample. Field methods are preferred. They provide data that better represent the reality of water flow in natural conditions. Their main disadvantage is the lack of rigorous quantitative procedures for measuring soil attributes in the majority of field tests. For laboratory measurements, the size of the REV should be theoretically estimated in order that an appropriate soil core sampler be selected. In practice, because the REV is rarely determined, a standard core or cylinder size is used for most soils. As a result, larger numbers of samples are taken with appropriate statistical evaluation of the data in order to partly reduce the error associated with samples having smaller volumes than the REV. In soils without cracks and frequently occurring macropores a soil sample volume of 200 to 500 cm^3 is assumed satisfactory. In subsoils, a 100 cm^3 sample will sometimes suffice. Methods relying on "undisturbed" core samples are generally not applicable in stony soils, in forests and in soils that crack excessively upon drying.

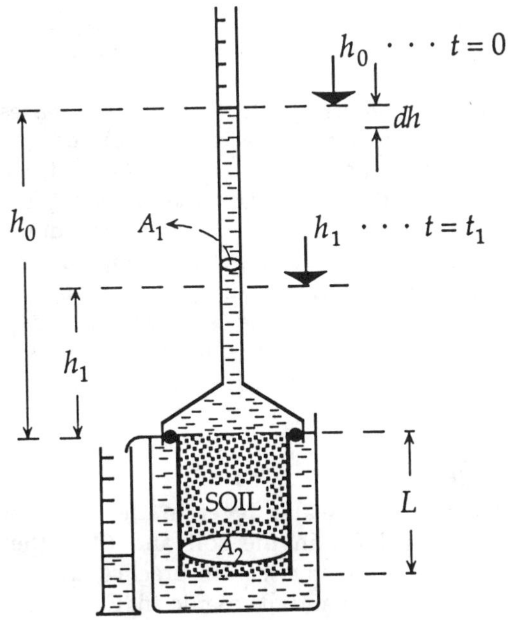

Figure 5.7. Falling head permeameter.

In the laboratory, the test is usually performed in a way similar to that demonstrated in Fig. 5.1. When both elevations of the water level are kept constant, i.e. in the vessel before water passes the soil ("upper level") and in the vessel after water has passed through the soil ("lower level"), (5.1) and (5.3) are applied. Apparatuses so constructed are usually called constant head permeameters. More frequently, the upper water level is allowed to fall, see Fig. 5.7. The fall of the water level dh in the measuring tube above the soil sample relates to the flux of water through the soil in time dt. Equating the two volumes, that moving in the measuring tube and that moving through the soil, we have

$$-A_1 dh = A_2 q\, dt \tag{5.24}$$

where A_1 is the cross-sectional area of the tube above the soil and A_2 is the cross-sectional area of the soil. Substituting q from (5.3) into (5.24) and recognizing that Δh in (5.3) is the total potential head difference h in Fig. 5.7, we have

$$\frac{A_1 L}{A_2 K_S}\int_{h_0}^{h_1}\frac{dh}{h} = -\int_0^{t_1} dt \,. \tag{5.25}$$

After integrating and rearranging,

$$K_S = \frac{A_1 L}{A_2 t_1}\ln\left(\frac{h_0}{h_1}\right). \tag{5.26}$$

The method is sometimes modified by having $A_1 = A_2$ and by keeping the bottom vessel without an overflow. Construction details of these apparatuses, usually referred to as falling head permeameters, are given in Klute and Dirksen (1986).

In the field, we generally deal with two cases relative to the determination of K_S. In the first case, a water saturated zone is formed by an unconfined aquifer close to the surface with the ground water level being not deeper than about 1.5 m. In the second case, the soil is not fully saturated.

There exists a number of field methods for determining K_S in the saturated zone. Among them, the auger-hole method and the piezometer method are the most commonly practiced.

In the auger-hole method, a hole is drilled to a depth well below the ground water level. After the water from the hole is rapidly pumped out, the rate of rise of the water level in the hole is registered. A general equation for the computation of K_S is simply

$$K_S = C\frac{dz}{dt} \tag{5.27}$$

where z is the depth of the water level in the hole measured from the hydrostatic ground water level in the soil and C is the shape coefficient dependent upon the geometry of the test. In practice, the derivative in (5.27) is replaced by the differences $\Delta z/\Delta t$ and the coefficient C is evaluated, e.g. according to the potential theory by Ernst (1950) quoted in Maasland and Haskew (1957). Except at the extreme bottom of the hole, it is assumed that the flow paths of water in the soil are perpendicular to the walls of the hole. In order that this assumption be fulfilled, pumping should not be repeated in short sequences of time. Estimation of C for less restrictive geometrical conditions has been described by Boast and Langbartel (1984).

The piezometric method is similar to the auger-hole method except that a tube driven into the augered soil serves as a lining for the hole. The rate of the rise of the water level in the lined hole is measured after pumping a portion of the water from the hole. The lining prevents inflow except through the bottom of the hole. For isotropic soils, K_S is computed from

$$K_S = \frac{\pi r^2}{C\,\Delta t}\ln\left(\frac{z_1}{z_2}\right) \tag{5.28}$$

where r is the radius of the tube inserted into the hole, z_1 and z_2 are the depths of water levels in the hole measured from the hydrostatic ground water level during the time interval Δt. The shape factor $C \approx 5.6r$ for a flat bottom and $C \approx 9$ for a hemispherical bottom. Note the similarity between (5.28) and (5.26). The falling head permeameter method modified here with a shape factor C applies to the piezometer method. When the lining does not reach the bottom of the hole, water penetrates additionally from the sides, and the value of the shape factor C increases in relation to L/r where L is the height of the unlined portion of the hole (Smiles and Youngs, 1965).

The piezometric method is ideally suited to measure K_S of anisotropic soils. Two measurements are required. For the first measurement, the lining reaches to the bottom of the hole and we compute with (5.28) the product of the vertical and horizontal hydraulic conductivities, K_V and K_H. In (5.28), K_S is replaced by $(K_V K_H)^{1/2}$ and we assume that the main axes of the hydraulic conductivity tensor are vertical and horizontal. For the second measurement the hole, deepened by the value L without pushing the lining tube deeper, allows inflow into the hole through both the unlined wall of height L and through the bottom. With a new shape factor C_2 in (5.28), the values of K_V and K_H, respectively, are simply calculated with data from the second experiment. A detailed description of methods of measurement of K_S in the field is given by Amoozegar and Warrick (1986). If the saturated soil or aquifer is of great thickness and the locality allows installation of several observation wells, pumping tests can be performed in order to obtain K_S. This method which deals with ground water hydraulics is frequently described in detail in the literature. Data obtained from these pumping tests are not generally compatible with data determined by the auger-hole method representing a weighted average K_S value in the domain of the cone of depression compared with the "point" data of the auger hole method.

The task of measuring K_S in an unsaturated zone is more difficult. Infiltration tests are most frequently used to measure K_S of the topsoil or other horizons near the soil surface. The value of K_S is estimated from readings of cumulative infiltration as a function of time. Alternatively, after a quasi-steady state infiltration rate is reached, its value together with that of the hydraulic gradient estimated from tensiometers installed at two or more depths are used in (5.5) to calculate the value of K_S. The details will be discussed in Chapter 6.

The auger-hole method is commonly modified to determine K_S below the soil surface of an unsaturated soil. Because the zone is not saturated, the modified procedure is opposite to that dealing with a saturated zone. Instead of water being removed from the saturated soil, water is poured into the hole from a marriotte flask. After a relatively short time of 15 to 30 minutes, quasi-steady flow is reached. The steady water flux density q is measured together with the constant height h of the water level above the bottom of the hole. Amoozegar and Warrick (1986) recommend that K_S be calculated according to Glover's solution

$$K_S = \frac{q\left[\sinh^{-1}(h/r) - \left(r^2/h^2 - 1\right)^{1/2} + r/h\right]}{2\pi h^2} \tag{5.29}$$

where r is the radius of the hole. This procedure is called the constant head well permeameter method. A similar device, the so-called Guelph permeameter (Reynolds and Elrick, 1985, and Elrick and Reynolds, 1992), is broadly used today owing to the efficiency of measurement.

5.3 UNSATURATED FLOW IN RIGID SOILS

By the term rigid soil we designate soils that do not change their bulk volume with a change of water content. We assume that unsaturated flow in soils is governed by the same laws that apply to saturated flow. For unsaturated flow we must consider the fact that a portion of the soil pores filled by air could indeed resaturate or drain. In our discussion of unsaturated flow, capillarity will be quoted as well as the term capillary rise frequently used in the literature. However, general mathematical formulations of physical phenomena should be independent of such simplifying ideas as soil capillaries and consequently, when we mention capillarity, it is just for the sake of modeling approximately some effects occurring in real soils.

5.3.1 Darcy-Buckingham Equation

A simple example of unsaturated flow demonstrated in Fig. 5.8 is analogous to the examples of experiments with saturated flow. The cylinder containing the soil has small openings within its walls leading to the atmosphere. Semipermeable membranes, permeable to water but not to air, separate the soil from free water on both sides of the cylinder. The pools of water are connected to the cylinder with flexible tubes. Full saturation of the soil is first achieved when both pools, lifted to the highest point of the soil, displace the soil air through the openings on the top side of the cylinder. At this moment, there is

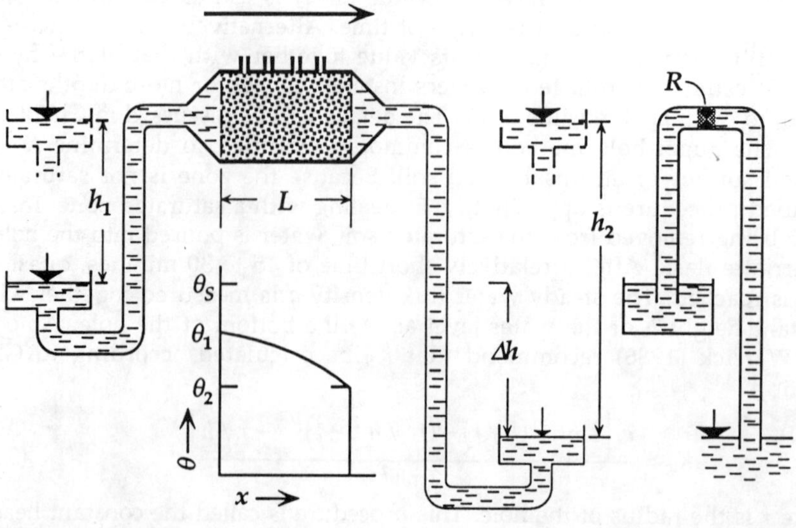

Figure 5.8. A simple steady flow experiment on an unsaturated soil column. At the right, an analogy of flow in a syphon with an installed resistance R.

no flow in the system and the soil is assumed water saturated. With the pool on the left side of the cylinder lowered to the position h_1 and the pool on the right side to position h_2, air enters into the soil through the openings as the soil starts to drain in a manner similar to a soil placed on a tension plate apparatus. The soil on the left side of the cylinder will be drained to a lesser extent than that on the right side with the soil water content distribution from left to right being nonlinear. Although water flows from the left pool to the right pool, the rate of flow is reduced significantly compared with that when the soil is water saturated. If the water level in each of the pools is kept at a constant elevation with time, steady flow will eventually be reached with the water content at each point within the soil remaining invariant. At this time, the flux density q will depend upon the hydraulic gradient and be governed by an equation similar to (5.3)

$$q = -K \frac{\Delta h}{L} \tag{5.30}$$

where K is the unsaturated hydraulic conductivity [LT^{-1}]. Inasmuch as the soil is not saturated and flow occurs primarily in those pores filled with water, the value of K will be smaller than that of K_S for the same soil. As for saturated flow we commonly take the potential related to the weight of water, i.e. in units of pressure head. For the majority of practical problems, all components of the total potential except those of gravity and soil water are neglected. Hence, (5.30) rewritten to allow the hydraulic conductivity to be a function of the soil water potential head h is

$$q = -K(h)\frac{dH}{dz} \tag{5.31}$$

and for two and three dimensional problems

$$q = -K(h)\, grad\, H. \tag{5.32}$$

Equation (5.33) is equivalent to Darcy's equation, and because Buckingham (1907) was the first to describe unsaturated flow dependent upon the potential gradient, equations such as (5.31) and (5.32) are called Darcy-Buckingham equations. The unsaturated hydraulic conductivity K is physically dependent upon the soil water content θ because water flow is realized primarily in pores filled with water. Because the relationship $\theta(h)$ is strongly influenced by hysteresis, $K(h)$ is strongly hysteretic. On the other hand, it follows from percolation theory that $K(\theta)$ is only mildly hysteretic.

Examples of $K(\theta)$ and $K(h)$ demonstrated in Fig. 5.9, show that the more permeable soil at saturation does not necessarily keep its greater permeability throughout the entire unsaturated region. It is also evident in Fig. 5.9 that the hysteretic behavior of $K(h)$ demands that, for a given value of h, the value of K is greater for drainage than for wetting.

The Darcy-Buckingham equation is adequate for describing unsaturated flow only if the soil water content is not changing in time. Unfortunately, this is seldom the case. When θ and q alter in time, we must combine (5.31) with the equation of continuity. The equation of continuity relates the time rate of change of θ to the spatial rate of change of q in a small elemental volume of soil. The resulting differential equation is strongly non-linear and its solution even for simple conditions is most difficult. Generally, (5.31) is in itself not satisfactory for the solution of such hydrologically important processes as evaporation, infiltration, drainage, subsurface flow etc. Exceptional situations or highly simplified flow conditions are usually the only problems described by the sole use of (5.31).

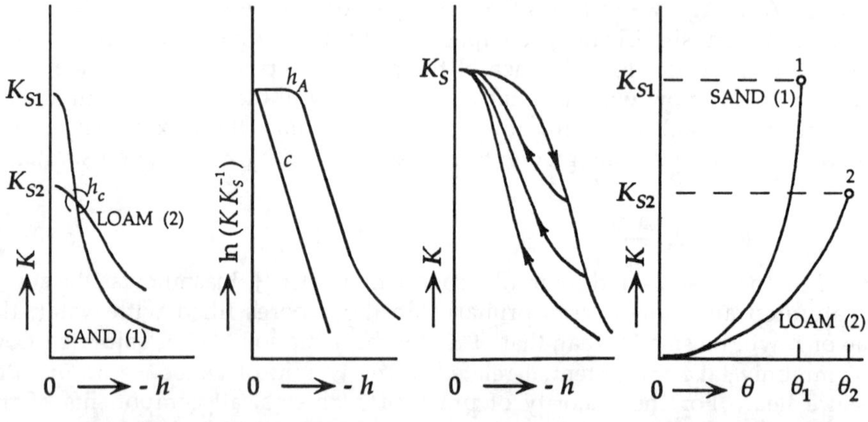

Figure 5.9. Dependence of unsaturated hydraulic conductivity K upon negative pressure head h (strongly hysteretic) and upon soil water content θ.

5.3.2 Unsaturated Hydraulic Conductivity

We distinguish two approaches for a physical interpretation of the measured hydraulic conductivity K. The first is based on the direct application of the Kozeny equation. The second uses the soil water retention curve to quantify the pore size distribution. With this quantification the Kozeny equation is used for sub-groups of pores. In addition to these two physical approaches, empirical formulations of $K(h)$ are used to merely express observed relationships.

Let us first apply the Kozeny equation. Inasmuch as only a portion of the pores is filled with water in an unsaturated soil, we replace the porosity P by the soil water content θ. We assume that the value of the tortuosity τ is described by Corey (1954, quoted by Corey, 1977) as

$$\frac{\tau_S}{\tau(\theta)} \approx \left(\frac{\theta - \theta_r}{P - \theta_r}\right)^2 \tag{5.33}$$

which is valid for sands where τ_S is the tortuosity in the saturated soil, $\tau(\theta)$ the tortuosity in the soil having water content θ and θ_r a residual water content. When it is assumed that the tortuosity owing to a change of soil water content is ignored in (5.17), Leibenzon (1947) derived the following expression

$$\frac{K}{K_S} = \left(\frac{\theta - \theta_r}{P - \theta_r}\right)^n \tag{5.34}$$

where the exponent n should have values ranging from 3.3 to 4. Averianov (1949) proposed that $n = 3.5$ is a good robust estimate. The value of exponent n is related to the pore size distribution and thus to the soil water retention curve SWRC, see Brooks and Corey (1964) who recommended $n = 2/\lambda + 3$ with λ read from (4.42) when P is replaced by θ_S in (5.34). Later on, Russo and Bresler (1980) found that values of 1 or 2 fit better than 3 in the exponent n. Childs and Collis-George (1950) obtained the equation

$$K = \alpha \frac{\theta^3}{A_m^2} \tag{5.35}$$

which is comparable to that of Deryaguin et al. (1956)

$$K = \alpha \frac{A_m d^3}{2\mu} \tag{5.36}$$

provided that we assume smooth walls. At small soil water contents and in soils having rough walls, the power function $K \alpha \theta^n$ remains but the exponent $n \neq 3$. A physical interpretation can be obtained using a fractal model of wall roughness (Toledo et al., 1990) The identity of both equations (5.35) and (5.36) is reached when the average thickness of the water film d is taken as functionally dependent upon θ for a given specific surface A_m. In both equations α is an empirical coefficient.

Inasmuch as $\theta(h)$ exists, the dependence of K upon h is also deducible with many empirical formulae quoted in the literature.

Gardner (1958) modified Wind's (1955) empirical proposal

$$K = a\,h^{-m} \tag{5.37}$$

to the relationship

$$K = \frac{a}{|h|^m + b} \tag{5.38}$$

applicable to $h = 0$ where a, b and m are empirical coefficients. Note that for $h = 0$, $a/b = K_S$.

Gardner's exponential relationship (1958)

$$K = K_S \exp(ch) \tag{5.39}$$

is frequently used in analytical solutions. If K/K_S is plotted against h on semi-log paper, a straight line is obtained. This relation usually fits the experimental data well in the range from $h = 0$ ($\theta = \theta_S$) to a certain h_{lim}, see Fig. 5.9. For soils manifesting a distinct air entry value h_A, Gardner and Mayhugh (1958) modified (5.39) to

$$K = K_S \exp\left[c\left(h - h_A\right)\right]. \tag{5.40}$$

The value of the empirical coefficient c with dimension [L^{-1}] is related to soil texture. and most frequently, $c = 0.1$ to $0.01\ cm^{-1}$. For δ-function soils in the Green and Ampt approximation of infiltration the value of c is numerically equivalent to the soil water pressure head $|h_f|$ at the wetting front, see Chapter 6. Both (5.39) and (5.40) have been broadly used in analytic and semi-analytic solutions , especially for steady flow problems as we show with some examples in Chapter 6 and as was fully reviewed by Pullan (1990). The reciprocal of c ($= \lambda_c^{-1}$) is sometimes used as one of the soil hydraulic characteristics. In such cases, λ_c is denoted as a microscopic capillary length (Bouwer, 1966; White, 1988).

Because (5.39) and (5.40) are valid in the wet range, (5.38) might be preferred in the dry range. For $h < h_{lim}$, we must use c_2 different from c to extend the applicability of the equation to the dry range. $K(h)$ is often defined as a composite function. For example, the range of $h_A > h > h_{lim}$, (5.39) applies and for $h < h_{lim}$, (5.37) or (5.38) applies in order to simulate the entire soil water regime in some instances.

From studies of capillarity in sands, Brooks and Corey (1964) obtained the frequently used relationship

$$\frac{K}{K_S} = \left(\frac{h_A}{h}\right)^m \tag{5.41}$$

where m depends upon the pore size distribution. Usually, m = 3 to 11.

Physical interpretation of $K(\theta)$ or $K(h)$ must include in addition to the total porosity, the distribution of the pore sizes. Recognizing from (5.10) that the flow rate in a cylindrical capillary is v_p (r^2), drainage of the largest pores drastically reduces the value of K in spite of the relatively small volume of those pores. Childs and Collis-George (1950) were the first to propose a method relating $K(\theta)$ to a pore size distribution function $f(r)$. Using the soil water retention curve to reflect $f(r)$, they obtained (5.35) as a simplified result. Their general approach attracted attention and was further developed and modified. We show those developments here.

In its simplest form the porous system is composed of j categories of pores with j = 1 for the category of smallest pores. In each category the pore radii are in ranges r_{j-1} to r_j. In each category the flux is $q_j(\bar{r}_j^4, n)$ where $\bar{r}_j$ is the mean radius and n is the percentage of the category and frequently $q_{j-1} < q_j$ even if $n_{j-1} > n_j$. Assuming $\nabla H = -1$, the unsaturated hydraulic conductivity $K = \sum q_j$. When the soil is only partially saturated with water, contributions of fluxes q_j, q_{j-1} et al. from the larger , empty pores of radii j, $(j\text{-}1)$ et al. do not exist, see Fig. 5.10. In a more exact derivation, we start with the mean flow rate v_p in pores of radius r according to the Hagen-Poiseuille equation

$$v_p(r) = ar^2 I_h \tag{5.42}$$

where $a = \rho_W g/8\mu$, see (5.10). The flux density in a porous system with a continuous distribution function of pores $f(r)$ and a tortuosity τ is

$$q = \frac{1}{\tau}\int_0^r v_p(r) f(r) dr. \tag{5.43}$$

Or, with (5.42) for I_h = 1, $q = K$ and

$$K = \frac{1}{\tau}\int_0^r ar^2 f(r) dr. \tag{5.44}$$

When $f(r)dr$ is approximated by $d\theta_E(h)$, i.e. by the derivative of the SWRC and for the relation between the pore radius and the pressure head ($r = c/h$), we obtain

$$K = \frac{ac}{\tau}\int_0^{\theta_E} \frac{1}{h^2(\theta_E)} d\theta_E(h). \tag{5.45}$$

For relative hydraulic conductivity K_r

$$K_r = K / K_S \tag{5.46}$$

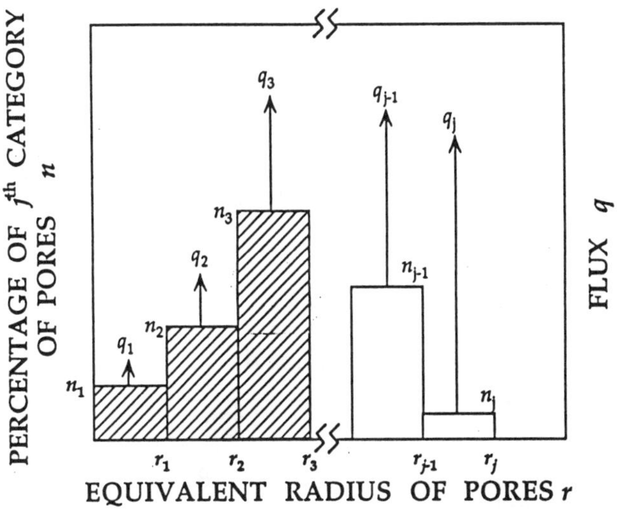

Figure 5.10. Contributions of soil pore categories with $r_1 < r_2 < \cdots r_j$ and their fluxes $q_1, q_2, \cdots q_j$ to the total flux $q = q_1 + q_2 + \cdots + q_j$. If grad $H = 1$, $q = K$ and $K_{r1 \cdots 3} \ll K_{rj}$.

and with the tortuosity from (5.33) modified to

$$\tau_S / \tau = \theta_E^{\ b}, \tag{5.47}$$

we have

$$K_r = \theta_E^b \int_0^{\theta_E} \frac{d\theta_E}{h^2(\theta_E)} \Big/ \int_0^1 \frac{d\theta_E}{h^2(\theta_E)}. \tag{5.48}$$

Various authors have not found a unique interpretation for exponent b in the above equation. Marshall (1958) and Millington and Quirk (1961) defined b as the probability of occurrence of continuous pores. For isotropic and homogeneous media $b = 2P_f$ with P_f denoting that portion of the porosity within which water is flowing. Marshall assumed $P_f = 1$ and hence, b was 2. Millington and Quirk used P_f as 2/3 and hence, b was 4/3. Burdine (1953) interpreting the tortuosity with (5.33) evaluated b as 2.

Inasmuch as the microscopic pore size distribution is used to characterize the macroscopic flux in a soil, Mualem (1976) classified such models as microscopic models. After evaluating about 50 soils on a macroscopic scale, Mualem decided that b was 0.5 and (5.48) is modified to

$$K_r = \theta_E^{0.5} \left[\int_0^{\theta_E} \frac{d\theta_E}{h(\theta_E)} \Big/ \int_0^1 \frac{d\theta_E}{h(\theta_E)} \right]^2. \tag{5.49}$$

If the van Genuchten soil water retention curve (4.43)

$$\theta_E = \frac{1}{\left[1 + (\alpha |h|)^n\right]^m} \tag{5.50}$$

and (5.48) are combined, we obtain

$$K_r(\theta_E) = \theta_E^b\left[1-\left(1-\theta_E^{1/m}\right)^m\right]^a \tag{5.51}$$

with $m = 1 - c/n$, and $n > 1$. These specific relations are suggested for simple evaluation of the integrals in (5.49). For the model of Burdine, $a = 1$, $b = 2$ and $c = 2$. For the model of Mualem, $a = 2$, $b = 0.5$ and $c = 1$. Let us note that Mualem's database consisted mainly of repacked laboratory soils and $b = 0.5$ does not hold for all field soils where the deviation may vary from less than -10 to more than 10 (van Genuchten et al., 1989).

Mualem's model of $K(h)$ is

$$\frac{K(h)}{K_S} = \frac{\left\{1-(\alpha|h|)^{n-1}\left[1+(\alpha|h|)^n\right]^{-m}\right\}^2}{\left[1+(\alpha|h|)^n\right]^{m/2}}. \tag{5.52}$$

Similarly, using the soil water retention curve $\theta_E = (h_A/h)^\lambda$ of Brooks and Corey (4.42) the relative hydraulic conductivity is

$$\frac{K(\theta_E)}{K_S} = \theta_E^{\,b+a/\lambda} \tag{5.53}$$

and

$$\frac{K(h)}{K_S} = \left(\frac{h_A}{h}\right)^{a+b/\lambda}. \tag{5.54}$$

For the model of Childs and Collis-George, $a = 2$ and $b = 2$. For that of Burdine, $a = 2$ and $b = 3$. For that of Mualem, $a = 2$ and $b = 2.5$.

The exponent $(a + b/\lambda)$ in (5.54) is identical to the exponent m in (5.41). Although the value of the exponent should theoretically be in a narrow range between 2.5 and 4.5, experimental data yield values that extend to about 11. This discrepancy can be explained by the over-simplification of the porous body in the model. In the derivation of the above equations, several approximations were made. First, the soil porous system was modeled by a bundle of cylindrical capillary tubes. Second, the pore size distribution function was approximated from the soil water retention curve. And third, the value of b was empirically evaluated. However, in spite of these approximations for the derivation of $K(\theta_E)$ and $K(h)$, the most problematic is the proper interpretation of the soil water retention curve close to θ_S.

A formal sensitivity analysis of (5.52) by Wösten and van Genuchten (1988) showed that differences in K_r increase with a decrease in h (i.e. with the soil drying) as the parameter α is altered, see Fig. 5.11. On the other hand, the influence of the exponent n also brings about a great potential error in the wet region. In Fig. 5.11, θ_E has been replaced by θ with the soil water retention curve (4.43) having the form

$$\theta = \theta_r + \frac{\theta_S-\theta_r}{\left[1+\alpha|h|^n\right]^m}. \tag{5.55}$$

Additional sensitivity analyses made by Šír et al. (1985) and Vogel and Císlerová (1988) show the role of an error $\delta h(\theta_E)$ in the experimental determination of

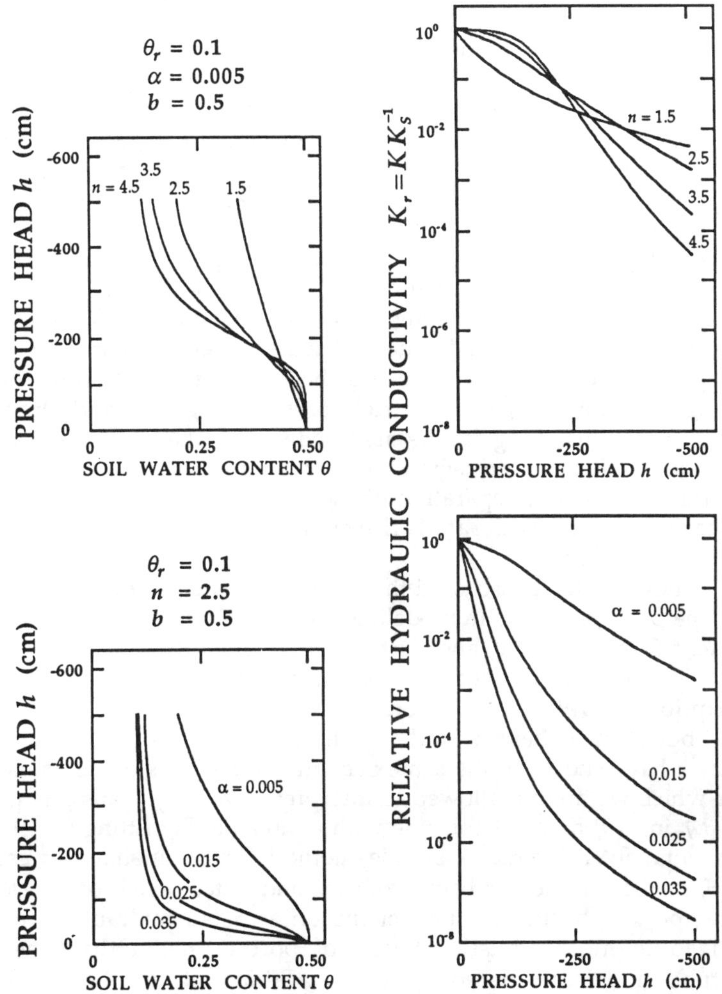

Figure 5.11. Sensitivity analysis of (5.50) and (5.52) by Wösten and van Genuchten (1988) shows the dependence of h(θ) and K(h) upon parameters α and n.

$h(\theta_E)$. If $\delta h(\theta_E)$ is a constant in the range $0 \leq \theta_E \leq 1$, the absolute error of $K(\theta)$ rises steeply with an increase of θ_E.

These derived equations are valid for the laboratory Darcian scale assuming microscopic homogeneity of the porous system. The assumption of microscopic homogeneity does not hold if the soil is aggregated, penetrated by plant roots and earthworms or dissected by fissures. In such cases the Darcian scale must be adapted to the pedon scale, i.e. to the reality of a field soil. On the pedon scale, microscopic homogeneity of the porous system may or may not exist.

The hydraulic conductivity function $K(h)$ of soil aggregates is about two orders of magnitude less than that of the bulk soil. The difference in $K(h)$ between aggregates of the soil and bulk soil decreases only slowly in the wet interval for $h > -800$ cm for well developed aggregates (Gunzelmann et al., 1987). In aggregated soils we deal with two different domains of velocity fields. One domain is related to interpedal pores and characterized by an accelerated flux. The second domain conducts water and solutes at relatively small flow rates and is found in intrapedal pores.

When the soil porous system is characterized by a bi-modal pore size distribution curve (Fig. 2.4), the relation $K(h)$ shows two distinct regions. For $0 > h > h_1$, only the by-pass pores belonging to the secondary peak are considered with Mualem's model applied to the soil water retention curve of the inter-pedal (by-pass) pores. For $h < h_1$ we use the remaining portion of the SWRC representing only the intrapedal (matric) pores, see Fig. 5.12 and Othmer et al. (1991). Hence, two matching factors are needed. For the region $0 > h > h_1$ the matching factor is K_S. For the region $h < h_1$ it is a measured value of $K(h < h_1)$. Although this mechanistic separation of the two porous systems uses the same basic equations, the accelerated fluxes through the by-pass pores are conveniently described.

Up to now we have discussed the problems related to $K(\theta)$ in a wet soil. In a dry soil, the probable errors in modeling $K(\theta)$ are related to θ_r. The residual soil water content θ_r in θ_E of (5.50) and further on in other $K(\theta)$ models leads to $K(\theta \leq \theta_r) = 0$. This zero value of hydraulic conductivity for $\theta > 0$ is in agreement with our description of SWRC in Section 4.3 where we used θ_{Wr} to denote the boundary between coherent and incoherent water phase distributions. However, we have shown in the same Section 4.3 that θ_r is obtained as a fitting parameter which we are not allowed to interpret physically. Thus, the physically observed θ_{Wr} in $K(\theta)$ may not coincide with θ_r obtained by fitting (4.42) or (4.43) to experimental SWRC data. A simple method for independently estimating θ_{Wr} for $K(\theta)$ models has not yet been proposed and tested on a broad scale.

Attempts to physically interpret the unsaturated hydraulic conductivity function lead to more realistic models of porous media than the parallel capillary tube model considered up to now. From soil morphology and from macroscopic measurements of the SWRC, the topological structures within a porous system can be deduced by either percolation theory or procedures of fractal geometry - or by a combination of both. The soil water flux is then related to the flow within individual pores and their fractal dimensions. Subsequently, the flow is formulated by the Hagen-Poiseuille equation with a procedure formally analogous to that developed by Childs and Collis-George (see e.g. Rieu and Sposito, 1991).

In addition to capillary pores, soil contains macropores where water is not influenced by meniscus forces. Such macropores originate owing to the growth and decay of plant roots, activities of soil edaphon and shrinkage in loam and clay soils.

Macropores play a special role in the flow of water especially during infiltration. When the soil water pressure is positive or when an unsaturated soil is ponded with water, water flows in the so-called "macropores". The

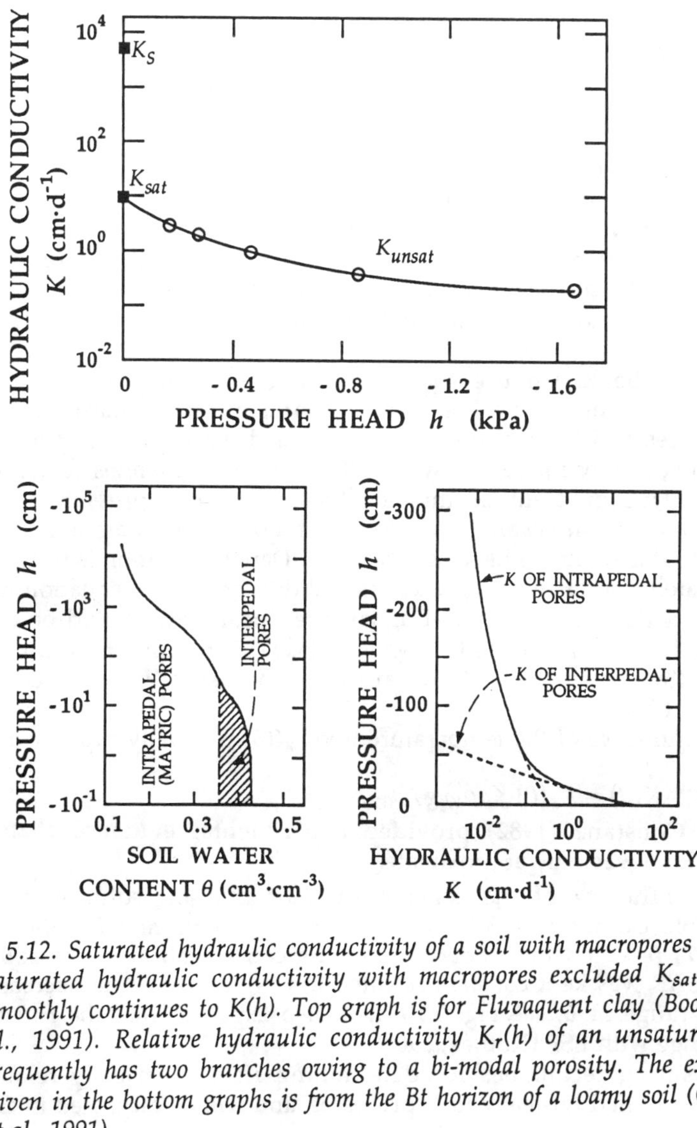

Figure 5.12. Saturated hydraulic conductivity of a soil with macropores K_S and saturated hydraulic conductivity with macropores excluded K_{sat} which smoothly continues to K(h). Top graph is for Fluvaquent clay (Boolting et al., 1991). Relative hydraulic conductivity $K_r(h)$ of an unsaturated soil frequently has two branches owing to a bi-modal porosity. The example given in the bottom graphs is from the Bt horizon of a loamy soil (Othmer et al., 1991).

mechanism of the flow in this case may be different from that of the capillary porous system. Water may flow either along the walls of the pores like a thick film, or through the entire cross-sectional area of the pore. When water conduction in cracks is combined with absorption, the kinetic wave approximation (German and Beven, 1985) can be used. The flux density is restricted just to macropores and it is generally reduced by absorption. The theory describes the transformation of both flux density and front velocity when water is transported in macropores as a pulse. It is applicable only under the provision that the macropores do not change during the transport of water. The theory cannot be used for modeling water flow in the fissures of shrinking-swelling soils.

When the soil is fully water-saturated and the flux exists at positive pressure, the saturated hydraulic conductivity comprises the flux in macropores together with flow in the soil matrix. If the flux in macropores is hindered, or if macropores are absent with the soil matrix not changed and remaining still water-saturated, the resulting value of the hydraulic conductivity may be decreased by two to four orders of magnitude (Boolting et al., 1991; Liu et al., 1994).

In this book, we use the term macropores only for pores without capillarity. In some of the literature a confusion exists inasmuch as coarse capillary pores and by-pass pores are also called macropores just to emphasize the large flux in those pores. However, if capillarity is manifested with the flow realized by the gradient of the negative soil water pressure, the Darcy-Buckingham equation is still appropriate with no need to replace it.

Up to this point we have assumed that Darcy's equation is fully applicable to unsaturated flow. However, when the validity of Darcy's equation is doubted for saturated flow in clays, non-Darcian pre linear flow should be even more pronounced for unsaturated flow in clays. Experiments indicating this possibility (Swarzendruber, 1963) have been theoretically explained (Bolt and Groenevelt, 1969).

The influence of the temperature upon $K(\theta)$ is usually expressed by $\mu_W(T)$ in

$$K(\theta) = K_r(\theta)\, K_p \rho_W g / \mu_W . \tag{5.56}$$

However, Constanz (1982) provided experimental evidence that in some instances (5.56) was only approximate.

The influence of the concentration of the soil solution and of the exchangeable cations is similar to that already mentioned for K_S. Dane and Klute (1977) reported that a decrease of the concentration in the soil solution when the SAR was kept constant resulted in roughly the same decrease of K in the whole range of θ, see Fig. 5.13. It is also expected that the function $K(\theta)$ would change with ESP (Kutílek, 1983).

Measuring techniques for determining $K(\theta)$ are usually related to the solution of specified unsteady flow processes, and will be discussed in Chapter 7.

5.3.3 Richards' Equation

Equation (5.32) is fully applicable to steady unsaturated flow when $\nabla \cdot q = 0$, $dq/dt = 0$ and $d\theta/dt = 0$. In practical situations, unsteady flow frequently exists with $d\theta/dt \neq 0$. In these situations, two equations are needed to describe the flux density and the rate of change of θ in time. The flux density is described by the Darcy-Buckingham equation and the rate of filling or emptying of the soil pores is described by the equation of continuity. Consider the prism element having edges of length Δx, Δy and Δz given in Fig. 5.14. The difference between the

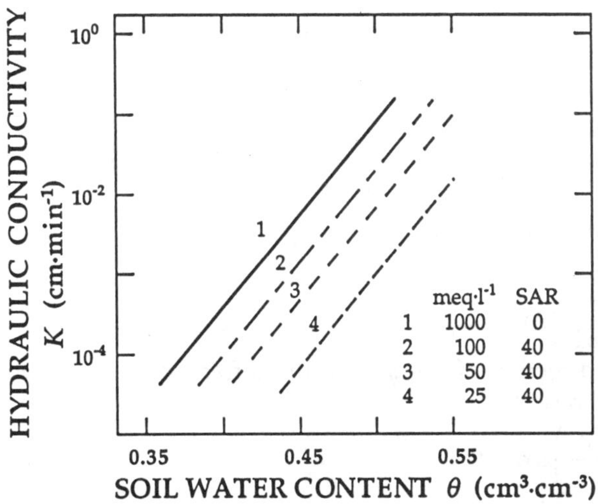

Figure 5.13. The influence of SAR and the concentration of the soil solution upon $K(\theta)$ according to Dane and Klute (1977).

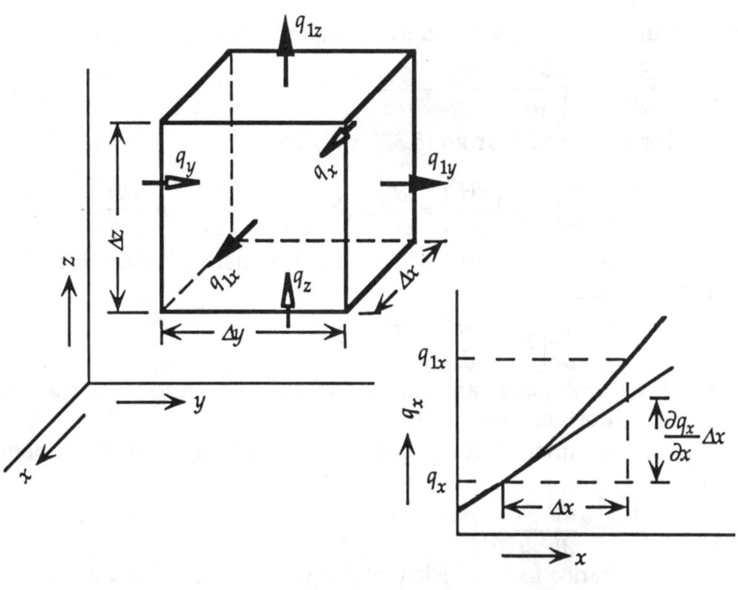

Figure 5.14. Derivation of the equation of continuity (5.62).

volume of water flowing into the element and that flowing out of the element is equal to the difference of water content in the element in time Δt. The rate of inflow (macroscopic) in the direction of the x axis is q_x. If we assume the change in q_x is continuous, the rate of outflow is $[q_x + (\partial q_x/\partial x)\Delta x]$. The inflow volume is $q_x \Delta y \Delta z \Delta t$ and the outflow volume is $[q_x + (\partial q_x/\partial x)\Delta x]\Delta y \Delta z \Delta t$. The difference between inflow and outflow volumes is

$$\left\{q_x \Delta y \Delta z \Delta t - \left[q_x + (\partial q_x / \partial x)\Delta x\right]\Delta x \Delta y \Delta z\right\} \tag{5.57}$$

or

$$-\left(\frac{\partial q_x}{\partial x}\right)\Delta x \Delta y \Delta z \Delta t. \tag{5.58}$$

Similarly in the direction of the y axis, the difference between the inflow and the outflow volumes is

$$-\left(\frac{\partial q_y}{\partial y}\right)\Delta x \Delta y \Delta z \Delta t \tag{5.59}$$

and that in the direction of the z axis

$$-\left(\frac{\partial q_z}{\partial z}\right)\Delta x \Delta y \Delta z \Delta t. \tag{5.60}$$

The sum of the above differences equals the change of the water content of the element. Provided that $\theta(t)$ has a continuous derivative for $t > 0$,

$$\frac{\Delta\theta}{\Delta t}\Delta x\, \Delta y\, \Delta z\, \Delta t = -\left(\frac{\partial q_x}{\partial x} + \frac{\partial q_y}{\partial y} + \frac{\partial q_z}{\partial z}\right)\Delta x\, \Delta y\, \Delta z\, \Delta t\,. \tag{5.61}$$

Taking the limit as $t \to 0$, we obtain the equation of continuity

$$\frac{\partial\theta}{\partial t} = -\left(\frac{\partial q_x}{\partial x} + \frac{\partial q_y}{\partial y} + \frac{\partial q_z}{\partial z}\right). \tag{5.62}$$

If we insert for q_x, q_y and q_z from (5.32), we have

$$\frac{\partial\theta}{\partial t} = \frac{\partial}{\partial x}\left[K(h)\frac{\partial H}{\partial x}\right] + \frac{\partial}{\partial y}\left[K(h)\frac{\partial H}{\partial y}\right] + \frac{\partial}{\partial z}\left[K(h)\frac{\partial H}{\partial z}\right] \tag{5.63}$$

provided that the soil is isotropic. In one-dimensional form for $H = h + z$ the above equation becomes

$$\frac{\partial\theta}{\partial t} = \frac{\partial}{\partial z}\left[K(h)\frac{\partial h}{\partial z}\right] + \frac{\partial K}{\partial z}. \tag{5.64}$$

Equations (5.63) and (5.64) are called Richards' equations in the name of the author who first derived them (1931).

If the soil is either wetting or drying, θ will be uniquely dependent upon only h and

$$\frac{\partial\theta}{\partial t} = \frac{d\theta}{dh}\frac{\partial h}{\partial t}. \tag{5.65}$$

Hence, the capacitance form of Richards' equation is obtained as

$$C_W(h)\frac{\partial h}{\partial t} = \frac{\partial}{\partial z}\left[K(h)\frac{\partial h}{\partial z}\right] + \frac{\partial K}{\partial z} \tag{5.66}$$

where soil water capacity $C_W = d\theta/dh$ $[L^{-1}]$ is illustrated in Fig. 5.15. An alternative development using

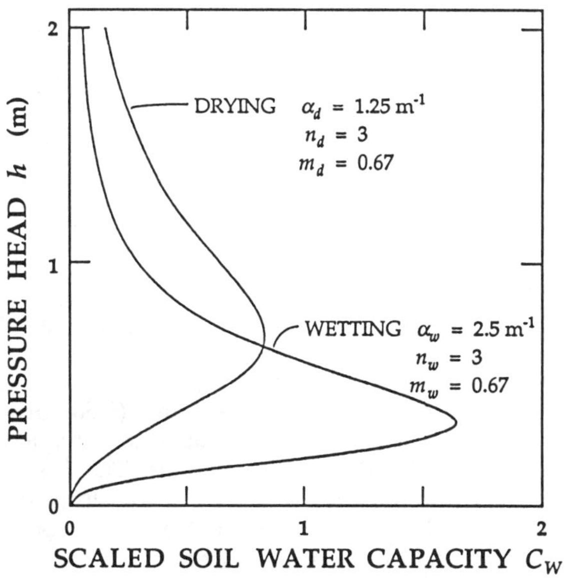

Figure 5.15. Soil water capacity C_W as a function of pressure head h for drying and wetting (Luckner et al., 1989).

$$\frac{\partial h}{\partial z} = \frac{dh}{d\theta}\frac{\partial \theta}{\partial z} \tag{5.67}$$

leads to the diffusivity form of Richards' equation

$$\frac{\partial \theta}{\partial t} = \frac{\partial}{\partial z}\left[D(\theta)\frac{\partial \theta}{\partial z}\right] + \frac{dK}{d\theta}\frac{\partial \theta}{\partial z} \tag{5.68}$$

where the Darcy-Buckingham equation has the diffusivity form

$$q = -D(\theta)\frac{\partial \theta}{\partial z} + K(\theta) \tag{5.69}$$

and the soil water diffusivity D is the term derived from

$$D(\theta) = K(\theta)\frac{dh}{d\theta}. \tag{5.69a}$$

The main reason for the derivation of either the capacitance equation (5.66) or the diffusivity equation (5.68) is the reduction of the number of variables from 4 to 3.

Both equations (5.66) and (5.68), strongly non-linear owing to functions $C_W(h)$, $K(h)$ and $D(\theta)$, are sometimes called Fokker-Planck equations. The name of (5.68) was derived from its resemblance (when its second term on the right hand side is omitted) to that for molecular diffusion. The units of D in (5.68) are identical to those of the diffusion coefficient. Many analytical and semi-analytical solutions for the diffusivity equation for various boundary conditions are known from the theories of diffusion (Crank, 1956) and heat flow (Carslaw and Jaeger, 1959). They have been profitably applied for the solution of many processes of unsaturated flow in soils. Whenever there is a region of positive pressure in the soil (5.68) is not applicable and (5.66) should be used.

Sometimes, Kirchhoff's transformation

$$U = \int_{h_0}^{h} K(h)\, dh \qquad (5.70)$$

is used with (5.64) to yield

$$\frac{C_W(h)}{K(h)}\frac{\partial U}{\partial t} = \frac{\partial^2 U}{\partial z^2} - \frac{1}{K(h)}\frac{dK}{dh}\frac{\partial U}{\partial z} \qquad (5.71)$$

or

$$\frac{\partial \theta}{\partial t} = \frac{\partial^2 U}{\partial z^2} - \frac{\partial K}{\partial z}. \qquad (5.72)$$

Because the last term of (5.64), (5.66) and (5.68) originated from the gravitational component z of the total potential H, it is frequently referred to as the gravitational term of the Richards' equations. The first term of the right hand side of each of those equations expresses the flow of water in the soil owing to the gradient of the soil water (matric) potential component h. In some instances when the gravitational term is neglected, the solution of the resulting non-linear diffusion equation with its non-constant diffusivity

$$\frac{\partial \theta}{\partial t} = \frac{\partial}{\partial z}\left[D(\theta)\frac{\partial \theta}{\partial z}\right] \qquad (5.73)$$

offers approximate results. If the flow is horizontal, solutions of (5.73) are exact.

5.3.4 Soil Water Diffusivity

The most common $D(\theta)$ relationship is demonstrated in Fig. 5.16. With the exception of the region of very small soil water contents less than θ_H ($h < -10^5$ cm), the curve steeply rises with θ. Soil water diffusivity $D(\theta)$ in the wet range above θ_H is typically less steep in its relation to θ as compared with $K(\theta)$. In this wetter range of θ, D changes about five orders of magnitude compared with seven orders of magnitude for K.

In the dry region of $0 \le \theta < \theta_H$ with a great portion of pores filled with air, water vapor flow is enhanced while liquid water flow is limited to that of very thin water films on the soil solid surfaces. The rate of liquid flow, strongly dependent upon the thickness of the film, has already been demonstrated by (5.36). Here, the vapor flux exceeds the liquid flux. A more detailed discussion on water vapor flux will be given in Section 5.3.5. Now, we shall study in detail the monotonically rising part of $D(\theta)$, i.e. for $\theta > \theta_H$.

Among the well known and frequently used empirical equations is the exponential form (Gardner and Mayhugh, 1958)

$$D = D_o \exp\left[\beta(\theta - \theta_o)\right] \qquad (5.74)$$

where D_o corresponds to θ_o and β ranges approximately between 1 and 30. Or,

$$D = \alpha \exp\left[\beta(\theta_E^{*})\right] \qquad (5.75)$$

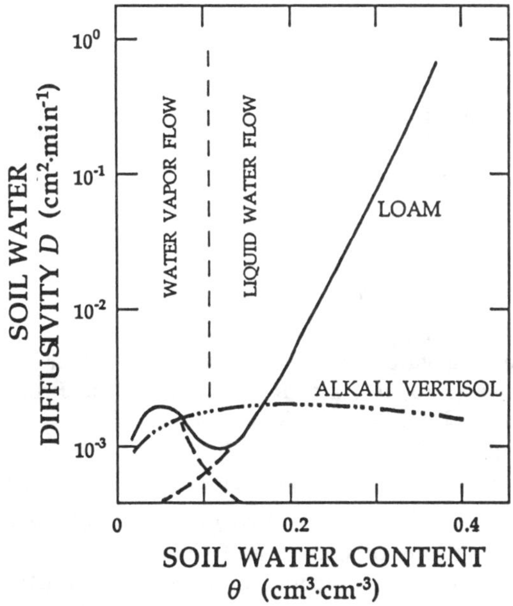

Figure 5.16. Dependence of soil water diffusivity D upon the soil water content θ. For liquid flow the function is nearly exponential for the majority of soils except for swelling clays. The secondary peak in the dry region near θ = 0 is owing to the dominance of water vapor flow.

where θ_r is replaced by θ_H in θ_E^{*} and at θ_H, $D = \alpha$. A physically more exact equation should be derived from the soil water retention curve and from $K(\theta)$. Using (4.43) and (5.51) in (5.69a), van Genuchten (1980) obtained

$$D(\theta_E) = \frac{K_S(1-m)\theta_E^{1/2-1/m}}{\alpha m(\theta_S-\theta_r)}\left[\left(1-\theta_E^{1/m}\right)^{-m}+\left(1-\theta_E^{1/m}\right)^{m}-2\right]. \quad (5.76)$$

If the simpler (4.42) is used instead of (4.43) we have

$$D(\theta_E) = \frac{K_S h_A \theta_E^{(b-1)+(a-1)/\lambda}}{\lambda(\theta_S-\theta_r)} \quad (5.77)$$

with the values of a and b being those given earlier for (5.53) and (5.54).

In some clays, mainly alkali Vertisols, the value of D decreases with an increase of θ, if the soil is confined and not allowed to swell, see Fig. 5.16 (Kutílek, 1983, 1984). For some undisturbed soils as well as for disturbed repacked soil columns in the laboratory (Clothier and White, 1981), D does not vary as strongly with θ as discussed above. If $(D_{max} - D_{min})$ is less than one-half an order of magnitude, the linearized form of (5.73)

$$\frac{\partial\theta}{\partial t} = \overline{D}\frac{\partial^2\theta}{\partial x^2} \tag{5.78}$$

serves as an excellent approximation where the mean-weighted diffusivity $\overline{D}$ for the wetting process (Crank, 1956) is

$$\overline{D} = \frac{5}{3(\theta_o - \theta_i)^{5/3}}\int_{\theta_i}^{\theta_o}(\theta-\theta_i)^{2/3}D(\theta)d\theta \tag{5.79}$$

and for the drainage process is

$$\overline{D} = \frac{1.85}{(\theta_o - \theta_i)^{1.85}}\int_{\theta_i}^{\theta_o}(\theta-\theta_i)^{0.85}D(\theta)d\theta \tag{5.80}$$

where θ_i is the initial soil water content and θ_o is θ at $x = 0$ for $t > 0$.

Soils manifesting values of D that are constant or nearly so are called "linear soils" because (5.78) is a linear equation. If the Brooks and Corey soil water retention curve (4.42) is used, $K_r(\theta_E)$ is described by (5.53) and $D(\theta_E)$ by (5.77). The condition of a "linear soil" is satisfied if in these equations $[\lambda = -(a-1)/(b-1)]$ or $[a = b = 1]$. If the first condition is applied to the Burdine equation, we obtain $h = h_A\,\theta_E^2$ and $K_r = \theta_E^{-1}$. Neither of these equations describe physical reality. Similarly, equations of Childs and Collis-George or those of Mualem lead to unacceptable results. The second condition leads to (Kutílek et al., 1985)

$$h = h_A\,\theta_E^{-1/\lambda} \tag{5.81}$$

$$K_r = \theta_E^{1/\lambda+1} \tag{5.82}$$

and

$$D = -\frac{h_A}{\lambda}\frac{K_S}{(\theta_S - \theta_r)} \tag{5.83}$$

This discussion shows the restrictions in the definition of a strictly linear soil when we require that D has a constant value and K_r is linearly dependent upon θ_E. If the second condition is not satisfied, we speak of "linear" soils. In general, there exists a family of "linear soils" described by the above equations. If $\lambda = 1$, the hydraulic conductivity function (5.82) is quadratic and meets the requirements of the solutions of Burgers' equation (Clothier et al., 1981).

Concluding, we should keep in mind that the soil water diffusivity is used in Richards' equation in order to reduce the number of variables. It has no direct physical meaning and is only defined mathematically, see (5.69). Moreover, inasmuch as $D(\theta)$ is dependent upon the derivative of the soil water retention curve, it has different values for wetting and drying processes. The temperature dependence of $D(\theta)$ is in accordance with changes of surface tension and viscosity with T. However, its prediction is only approximate owing to some not well understood phenomena that associates the temperature dependence of $h(\theta)$ and $K(\theta)$.

5.3.5 Diffusion of Water Vapor

In section 5.3.4 we have already shown that the relative maximum in the $D(\theta)$ relationship in the dry region is caused by water vapor flow. Indeed, the soil water diffusivity D contains two components: D_L the diffusivity of liquid water, and D_G the diffusivity of water vapor, i.e. the gaseous phase. Hence, $D = D_L + D_G$ (Philip, 1957a). Jackson (1964) derived D_G as analogous to the earlier introduced soil water diffusivity

$$D_G = D_p \frac{d\rho_{rG}}{d\theta} \tag{5.84}$$

where ρ_{rG} is the relative density (concentration) of water vapor and D_p the diffusion coefficient of water vapor in soil which is approximated by

$$D_p = D_a \alpha \left(P - \theta\right)^{\mu} \tag{5.85}$$

where D_a is the diffusion coefficient of water vapor in free air and α and μ are factors that account for the tortuosity and complexity of the soil porous system. Detailed information about (5.85) is provided by Currie (1960).The term $d\rho_{rG}/d\theta$ is actually the slope of the adsorption isotherm and its inflection point corresponds to the relative maximum of $D_G(\theta)$. The water vapor diffusivity rises to this maximum from $\theta = 0$ and reaches this value at a relative water vapor pressure $p/p_o = 0.3$ to 0.4 in the majority of soils. The maximum value of D_G, having a wide range between 10^{-4} to 10^{-3} $cm^2 \cdot s^{-1}$, depends upon soil texture, mineralogy of the clay fraction and organic matter content, see Fig. 5.17. At greater soil water contents D_G decreases as D_L increases. Values of D_L exceed those of D_G at $p/p_o \approx 0.5$ to 0.8. In terms of the average thickness of the adsorbed water films on the soil solid surface, D_G reaches a maximum value after the first molecular layer is completed and before or at least when the second molecular layer is formed. D_L exceeds D_G when about 4 to 6 molecular layers of adsorbed water exist.

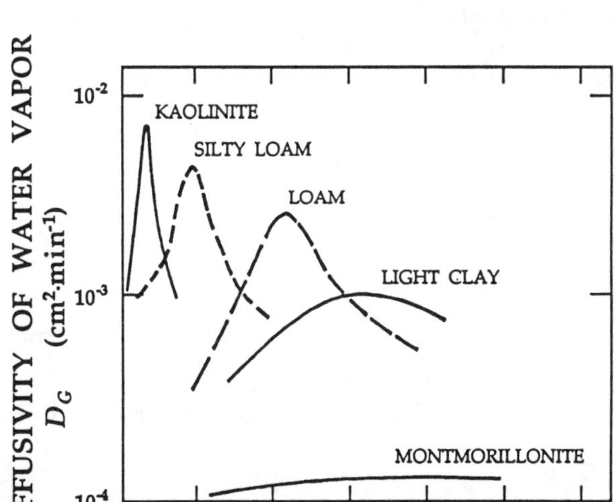

Figure 5.17. Diffusivity of soil water vapor versus soil water content close to θ = 0 for soils and clay minerals (Kutílek, 1978).

For structured soils, (5.84) and (5.85) are far from reality. For more realistic descriptions 2- or 3-modal porosity models should be considered (Currie and Rose, 1985). The effect of such consideration is similar to that of a bi-modal porosity model upon $K(h)$ in the wet region.

It is worthwhile to repeat here at the end of this chapter the principal theoretical gain of all discussions. Without a knowledge of the hydraulic functions of soils – $h(\theta)$, K_S, $K(\theta)$ and $D(\theta)$ – a quantitative description of water flow in soils is not feasible.

5.4 TWO PHASE FLOW

Up to now we have assumed that the presence of air in soil pores offers no resistance to water flow. Because the fluidity of the soil air is greater than that of water by an order of magnitude, its motion relative to that of water is implicitly neglected. The assumption of one phase flow – water only – is not appropriate when the free entrance or escape of air is blocked by small passages owing to either a layer of a greater water content or of lesser permeability. If the entrapped air is continuous, it forms a "barrier" against the flow of liquid water and the flux density of water is reduced. In such cases the flows of both air and water should be solved as mutually influencing each other (Elrick, 1961). For the sake of simpicity we are here neglecting water vapor flow. In wet soil where the flow of air can play an important role, the contribution of water vapor flux to the liquid water flux is negligibly small. In a dry soil at about $h < -10^5$ cm where the water vapor flux is important, the influence of air flow upon water flow is not expected. We are still dealing with isothermal conditions. We discuss therefore only the interactions of the flow of the liquid water phase and air as a complex gaseous phase.

In the theory of two phase flow, the phases (air and water) are assumed to behave like two immiscible liquids. The basic flow equations of Darcy-Buckingham rewritten to fit both phases (air and water) express the pressure gradient as the driving force are

$$q_A = -\lambda_A \left(\frac{\partial p_A}{\partial z} - \rho_A g \right) \qquad (5.86)$$

and

$$q_W = -\lambda_W \left(\frac{\partial p_W}{\partial z} - \rho_W g \right) \qquad (5.87)$$

where the indices A and W denote air and water, respectively, p is pressure $[ML^{-1}T^{-2}]$, ρ the density $[ML^{-3}]$ and g the acceleration of gravity $[LT^{-2}]$. The fluid mobility λ is

$$\lambda = \frac{K_p K_r}{\mu} \qquad (5.88)$$

where K_r is the relative conductivity of the particular fluid, K_p the permeability $[L^2]$ and μ the dynamic viscosity. The dependence of both K_{rA} and K_{rW} upon the water content of a sand determined experimentally by Touma and Vauclin (1986) is given in Fig. 5.18. The curve $K_{rA}(\theta)$ is not a mirror image of $K_{rW}(\theta)$. The

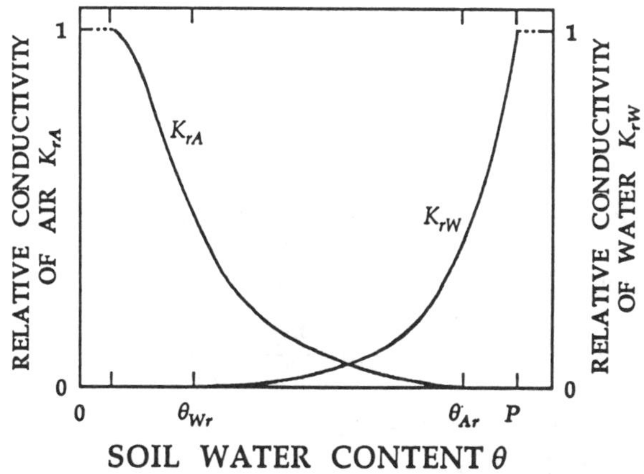

Figure 5.18. Relative conductivities of air K_{rA} and K_{rW} as related to the soil water content θ (Touma and Vauclin, 1986).

decrease of K_{rW} with a decrease in θ is more rapid than the decrease of K_{rA} with an increase in θ. At θ_{Wr}, the value of K_{rW} is zero. Similarly, at $\theta_{Ar} < P$, the value of K_{rA} is zero. The continuity equation for air is

$$\frac{\partial(\rho_A \theta_A)}{\partial t} + \frac{\partial(\rho_A q_A)}{\partial z} = 0 \tag{5.89}$$

with θ_A (= P - θ) the volumetric air content. The continuity equation for water is

$$\frac{\partial(\rho_W \theta)}{\partial t} + \frac{\partial(\rho_W q_W)}{\partial z} = 0 \ . \tag{5.90}$$

The difference between the pressure of the two fluids is the capillary pressure $p_c = p_A - p_W$, and for water taken as the incompressible fluid

$$\rho_W = \rho_{W_o},$$

we have

$$\frac{\rho_A}{\rho_{A_o}} = \frac{p_A}{p_{A_o}} \tag{5.91}$$

where the index o denotes the value at the reference atmospheric pressure. Inasmuch as the pressure head h was more convenient for one-phase flow, we use here

$$h_A = \frac{p_A - p_{A_o}}{\rho_W g} \tag{5.92}$$

and

$$h_W = \frac{p_W - p_{A_o}}{\rho_W g} \tag{5.93}$$

where $h_c = h_A - h_W$. Hence, (5.86) and (5.87) transcribe to the more familiar forms

$$q_A = -K_A(\theta)\left(\frac{\partial h_A}{\partial z} - \frac{\rho_A}{\rho_W}\right) \tag{5.94}$$

and

$$q_W = -K_W(\theta)\left(\frac{\partial h_W}{\partial z} - 1\right) \tag{5.95}$$

with $h_A - h_W = h_c \equiv h$. The air conductivity K_A ($= \rho_W g \lambda_A$) and K_W [identical to the earlier notation $K(\theta)$] each have the dimension [LT^{-1}]. Analogous to Richards' equations (Touma and Vauclin, 1986) we have for air

$$\frac{\partial(\rho_A\theta_A)}{\partial t} = \frac{\partial}{\partial z}\left[\rho_A K_A(\theta)\left(\frac{\partial h_A}{\partial z} - \frac{\rho_A}{\rho_W}\right)\right]. \tag{5.96}$$

When we substitute $\theta_A = P - \theta$ with

$$\rho_A = \rho_{A_o}\left(1 + h_A / h_o\right)$$

and $C_W = d\theta/dh_c$ we obtain the capacitance equation

$$\rho_A C_W \frac{\partial h_W}{\partial t} + \left[(P-\theta)\frac{\rho_{A_o}}{h_o} - \rho_A C_W\right]\frac{\partial h_A}{\partial t} = \\ = \frac{\partial}{\partial z}\left[\rho_A K_A(\theta)\left(\frac{\partial h_A}{\partial z} - \frac{\rho_A}{\rho_W}\right)\right]. \tag{5.97}$$

Similarly, for water

$$\frac{\partial \theta}{\partial t} = \frac{\partial}{\partial z}\left[K_W(\theta)\left(\frac{\partial h_W}{\partial z} - 1\right)\right] \tag{5.98}$$

and

$$C_W\left(\frac{\partial h_A}{\partial t} - \frac{\partial h_W}{\partial t}\right) = \frac{\partial}{\partial z}\left[K_W(\theta)\left(\frac{\partial h_W}{\partial z} - 1\right)\right]. \tag{5.99}$$

When the air is continuously connected to and has the same pressure as that of the external atmosphere (taken as the reference), $h_A = 0$, $h_c = h_W$ and (5.97) is identical to (5.66).

The above theory offers reliable results if the pressure (or potential) drop of the water across the less permeable barrier is small, or if a barrier limiting the air flow is not reached by a wetting front in the case of infiltration. If a substantially less permeable layer exists for water, a steep water pressure gradient develops across the less permeable layer. In this circumstance, (5.66) suffices.

The solution or dissolution of air in soil water should follow the behavior of the individual gases of the air mixture according to Henry's law

$$C_i = k_i p_i \tag{5.100}$$

where C_i is the concentration of the i-th gas, k_i a constant dependent upon the temperature and the nature of the gas and p_i the partial pressure of the i-th gas. Considering the numerical values of k_i and p_i for gases composing the soil air, the solubility of nitrogen and oxygen are important. When the pressure of the soil water varies by an order of magnitude, the concentrations of dissolved N_2 and O_2 in the soil water vary substantially. For example, with an abrupt decrease in soil water pressure as water passes through a less permeable layer, its originally dissolved gases are released. Consequently, small air bubbles usually accumulate on the bottom boundary of a less permeable soil layer. By this mechanism the hydraulic resistance of the less permeable layer increases and is time dependent. A detailed quantitative understanding based upon experimental data is yet to be reported in the literature.

5.5 FLOW IN NON-RIGID (SWELLING) SOILS

When a soil swells or shrinks owing to a water content change, the previously developed equations must be modified. A complete theory has not yet been developed for flow accompanied by three-dimensional volume changes resulting in an opening of cracks when a soil dries or their closing when a soil wets. The theory developed up to the present time (Smiles and Rosenthal, 1968; Philip, 1969b) deals only with one-dimensional deformation. It is not applicable to three-dimensional volumetric changes associated with non steady flow of water that lead to the formation and closing of cracks. The one-dimensional theory describes wetting of artificially repacked soil columns in the laboratory. For a draining or drying process even in laboratory conditions, it is not applicable because the formation of cracks and transport in the cracked medium are not described. Owing to restrictions and the artificial character of processes described by the theory, we are not presenting it here except for the introduction of the concept of Lagrangian coordinates.

When a soil swells during wetting, Darcy's equation should be modified according to Gersevanov (1937) to relate the rate of water flow to the solid phase. Instead of Euler's coordinate system, Lagrange's coordinates need to be considered. For a one-dimensional treatment of swelling, the original proposal of Gersevanov was redefined by Smiles and Rosenthal (1968) to yield

$$\frac{dm}{dx} = \frac{1}{1+e}. \tag{5.101}$$

Upon integration the above equation becomes

$$m = \int_{-\infty}^{x} \frac{dx}{1+e} \tag{5.102}$$

or

$$m = \int_{-\infty}^{x} (1-P)\,dx. \tag{5.103}$$

Equation (5.101) states that the ratio of the material coordinate m to the Eulerian coordinate equals the ratio of the volume of the solid phase to the total volume of the soil. Fig. 5.19 demonstrates the material coordinate associated with the experiment that follows. A dry soil originally packed into a tube to the height denoted by $x = 0$, has its surface flooded with water. As water infiltrates into the swelling soil, the soil surface rises. After a certain time of infiltration we denote the depth of the wetting front x_f from the original position $x = 0$. Simultaneously, we measure the soil water content and the soil bulk density. From those measurements the porosity P is plotted against negative values of x

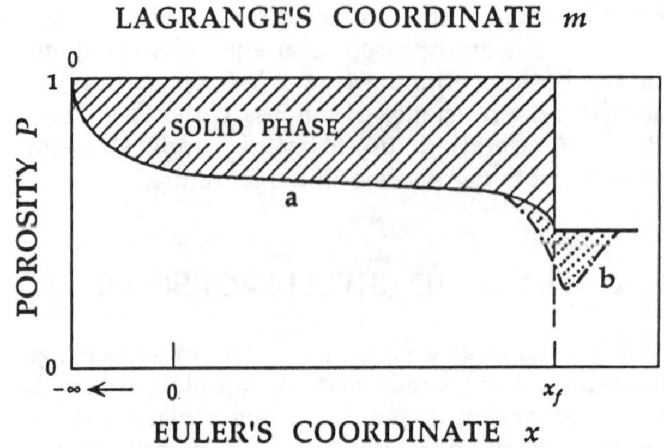

Figure 5.19. Material coordinate m in the solution for one-dimensional infiltration of water into: a. pre consolidated clay and b. the same clay without pre consolidation. The wetting front position is at x_f.

inasmuch as the soil surface rose compared to its original position at $x = 0$. The material coordinate m expresses the volume of the solid phase wetted by the water. Because we cannot predict the elevation to which the soil will swell, we define the lower limit of integration as $-\infty$. Note that for $P = 1$, the integral (5.103) yields a value of $m = 0$. If the soil is pre consolidated, we can assume that the value of $P = e(1 + e)$ is constant for $x > x_f$. If the soil is not pre consolidated, the soil is compressed ahead of the wetting front by the swelling pressure. This relative minimum P_{min} moves at the same rate as the wetting front with the value of P_{min} decreasing, i.e. the value of $(P_i - P_{min})$ increases with time and with surface load (Kutílek, 1984a). All of these effects are especially distinct in montmorillonitic clays of Vertisols and increase with an increase of ESP and a decrease of EC.

Darcy's equation has to be modified for Lagrangian coordinates and the hydraulic conductivity redefined, especially when we deal with non steady flow. Owing to a lack of an appropriate theory and the extremely complicated character of the system, we recommend simplified approaches related to individual cases of elementary hydrologic processes.

5.6 NON-ISOTHERMAL FLOW

Up to now we have described unsaturated flow for only isothermal conditions. Typically, such conditions are rare in soil hydrology. Nonetheless, most modeling of soil hydrology utilizes equations formulated for isothermal conditions as a good approximation for non-isothermal situations. However, for non-isothermal situations the equations must account for thermally driven processes.

Under non-isothermal conditions differences in temperature across a distance z produce a heat flux that alters water transport. More generally, each flow process influences every other flow process and hence, these coupled processes are studied with the theory of irreversible thermodynamics.

5.6.1 Coupled Processes

Coupled flow phenomena are described by phenomenological equations of Onsager (1931). The first flux J_1 is described by

$$J_1 = L_{11} X_1 + L_{12} X_2 + L_{13} X_3 + \cdots + L_{1n} X_n \qquad (5.104)$$

and similarly for the other flow processes,

$$J_i = \sum_{k=1}^{n} L_{ik} X_k \qquad (5.105)$$

where X_i is the conjugate driving force which produces the i-th flow and L_{ii} is the phenomenological "direct" coefficient which relates the flux to the driving force. The above equation shows that the flux J_1 may also be driven by forces X_2, $X_3, \cdots X_n$ if the coupling or cross coefficients $L_{12}, L_{13}, \cdots L_{1n}$ differ from zero. In general, L_{1k} shows the contribution of the k-th flux J_k to the first flux J_1 when a driving force X_k exists. On the other hand, if the i-th flow process is not related to any of the other n flow processes, $L_{ik} = 0$ for $k \neq i$ and we obtain

$$J_i = L_{ii} X_i \qquad (5.106)$$

which is the equation of heat flux, molecular diffusion or that of Darcy (5.5). The linearity holds only for slow processes not too distant from an equilibrium state. The coefficients $L_{ik} = J_i / X_k$ are flows per unit force (or conductances) and $L_{ik} = L_{ki}$. In this chapter when we deal with simultaneous fluxes of water and heat J_T, (5.105) becomes

$$J_W = L_{WW} X_W + L_{WT} X_T \qquad (5.107)$$

and

$$J_T = L_{TT} X_T + L_{TW} X_W. \qquad (5.108)$$

Note that the first term on the right hand side of (5.107) is the Darcy-Buckingham equation and the following term accounts for the additional contribution of water flow owing to the heat flow.

Equation (5.105) has validity for all coupled phenomena in soils when n flow processes occur simultaneously. The physical interpretation of L_{ik} and X_k in soils have been discussed in detail in the literature (e.g. Bolt and Groenevelt, 1972, and Groenevelt an Bolt, 1972).

5.6.2 Flow in Non-isothermal Conditions

Because the mechanisms of the influence of a temperature gradient upon the liquid and vapor water phases are not identical, separate terms are required to describe the flux of liquid and that of vapor. Here we follow a slightly modified procedure of Philip and de Vries (1957) to describe these coupled flows.

For the flux density of water vapor q_G we have

$$q_G = - D_{G\theta} \nabla\theta - D_{GT} \nabla T. \tag{5.109}$$

The dimension of D_{GT} is $[L^2T^{-1}grad^{-1}]$, usually in units $[cm^2 \cdot s^{-1} \cdot K^{-1}]$. For the flux density of liquid water q_L is

$$q_L = - D_{L\theta} \nabla\theta - D_{LT} \nabla T - K. \tag{5.110}$$

The dimension of D_{LT} is identical to that of D_{GT}.

The flux density of all of the pore water, both liquid and vapor, is

$$q = - D_{\theta} \nabla\theta - D_{GT} \nabla T - D_{LT} \nabla T - K \tag{5.111}$$

where $D_\theta = (D_{L\theta} + D_{G\theta})$ and $D_{G\theta}$ here is identical to D_G in (5.90). Such equivalent relations for the coupled coefficients D_{GT} and D_{LT} are not available. Combining (5.111) with the equation of continuity for a one-dimensional system with soil depth $z > 0$ measured from the soil surface, we obtain an equation similar to (5.68)

$$\frac{\partial\theta}{\partial t} = \frac{\partial}{\partial z}\left(D_\theta \frac{\partial\theta}{\partial z}\right) + \frac{\partial}{\partial z}\left(D_{GT} \frac{\partial T}{\partial z}\right) + \frac{\partial}{\partial z}\left(D_{LT} \frac{\partial T}{\partial z}\right) - \frac{\partial K}{\partial z}. \tag{5.112}$$

The temperature is described by

$$C_T \frac{\partial T}{\partial t} = \frac{\partial}{\partial z}\left(\lambda_T \frac{\partial T}{\partial z}\right) + \rho_G \chi \frac{\partial}{\partial z}\left(D_{GT} \frac{\partial T}{\partial z}\right) \tag{5.113}$$

where λ_T is the thermal conductivity $[W \cdot m^{-1} \cdot K^{-1}]$, χ the latent heat of vaporization $[J \cdot kg^{-1}]$ and C_T the volumetric heat capacity $[J \cdot K^{-1} \cdot m^{-3}]$.

Solutions of such practical problems of non-isothermal flow in layered soils require equations that are transcribed into forms analogous to the capacitance equation (5.66) with pressure head gradient replacing the soil water content gradient. Hence, (5.109) for one-dimension becomes

$$q_G = -D_{Gh} \frac{\partial h}{\partial z} - D_{GT} \frac{\partial T}{\partial z} \tag{5.114}$$

and (5.110) is

$$q_L = -\left(K_{Lh} \frac{\partial h}{\partial z} - 1\right) - D_{LT} \frac{\partial T}{\partial z}. \tag{5.115}$$

Note that $K_{Lh} = K(h)$ and D_{Gh} is not identical to $D_{G\theta}$. The equation of continuity is also modified and C_T and λ of (5.113) are specified once for the vapor phase and once for the liquid phase. For details of modifying the equations for numerical procedures, see e.g. Passerat de Silans et al. (1989).

5.6.3 Flow at Temperature T< 0°C

Although the transport of water at temperatures less than 0°C is generally described with theories outlined in the previous chapters, the hydraulic functions are greatly modified by the formation and presence of ice in soils.

In a frozen soil not far below 0°C we distinguish ice, liquid water and air within the mixture. The fundamental thermodynamic equation relates the relative Gibbs free energy to the freezing point as demonstrated in Fig. 5.20. Remembering that the definition of soil water potential is the equivalent of Gibbs free energy, the SWRC should be equivalent to the graph demonstrating

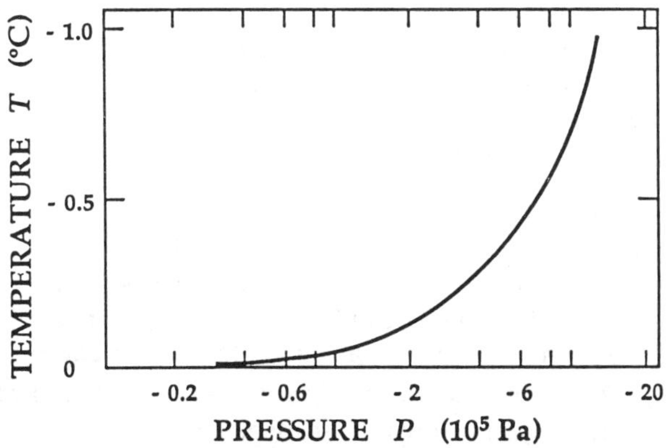

Figure 5.20. Theoretical relation between the temperature of water freezing and the free energy expressed as negative pressure.

the soil unfrozen water content $\theta_u(T)$. However, we must modify this simple consideration owing to the fact that frozen soils manifest varying degrees of rigidity. Soils rigid in their unfrozen state start to behave as non-rigid soils at $T < 0°C$. In the hydrostatics of frozen soils we must include a component in the total potential that accounts for the equilibrium ice pressure p_i. Moreover, the mechanism of the development of micro- and macro-lenses of ice is still not well known. Additionally, hysteretic concepts require modification owing to the deformation of the soil solid matrix and to the hysteretic behavior of liquid water and ice lenses. Hence, any final equilibrium state is substantially

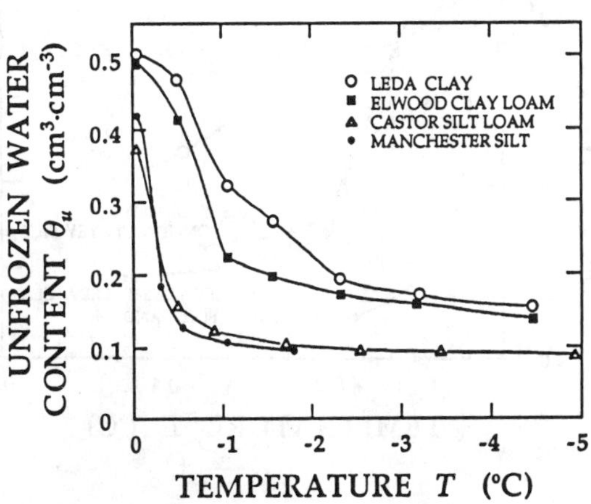

Figure 5.21. Unfrozen water content θ_u for various soils as a function of temperature (Patterson and Smith, 1981).

influenced by the path and direction of the temperature variation. Thus far, measurements of soil behavior show quantitative agreement with basic thermodynamic equations. For example, the content of unfrozen water θ_u decreases strongly with a drop of temperature below 0°C in the range of -0.5 to -2.0°C, see Fig. 5.21. With further decreases of temperature below this threshold range, the value of θ_u changes only slightly. Consequently, for the solution of problems of water flow below 0°C we need to directly measure $\theta_u(T)$. Additionally, the classical SWRC and $\theta_u(T)$ should be combined analogous to the procedure used for non-rigid soils at $T > 0$°C (Groenevelt and Kay, 1977). The temperature dependence of the hydraulic conductivity at $T < 0$°C is then similar to the water dependence of the unsaturated hydraulic conductivity function, roughly $K_u(\theta_u^3)$. Therefore, the steepest decrease of K_u occurs at temperatures only a few tenths of a °C just below 0°C, see Fig. 5.22.

Unsteady flow in partially frozen, unsaturated soil is accompanied by many side phenomena such as formation of ice lenses, frost heaving etc. Hence, an objective formulation of the basic transport equations analogous to Richards' equation remains unresolved except for some well-defined simple (or simplified) problems. Present-day equations of mass transport which consider heat flow and heat consumption and release are considered with a combination of Darcian and microscopic scales.

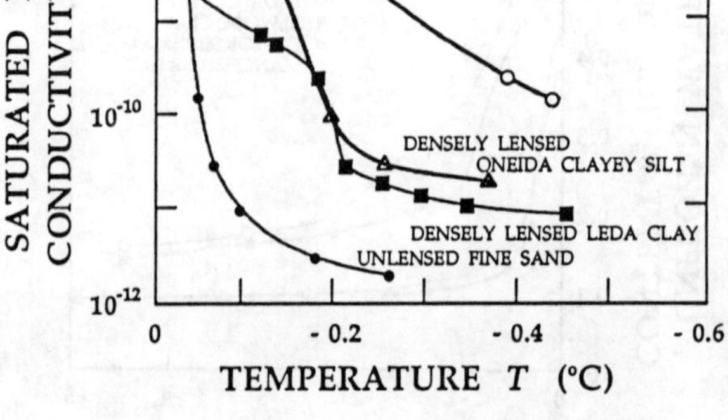

Figure 5.22. Hydraulic conductivity for various soils as a function of temperature. Before the temperature was decreased, the soil was saturated with water (Burt and Williams, 1976).

PROBLEMS

1. The permeability of a soil $K_p = 0.3\ \mu m^2$. What is the water flux density when the potential head gradient *grad* $H = 2$?
2. In the previous problem, what is the equivalent pore radius in the soil approximated by Kozeny's model?
3. Formulate the functional relationship $K(\theta)$ if the model of the soil leads to K_p described by (5.19)?
4. What is the direction of flow if the tensiometer readings of pressure head h_p are: h_p = -150 cm at soil depth z = 20 cm and h_p = -140 cm at z = 50 cm?
5. Compute the flux density in the previous problem if $K(h)$ is expressed by (5.39) with $K_S = 20\ cm{\cdot}day^{-1}$ and $c = 0.05\ cm^{-1}$.
6. If $K/K_S = \theta_E^m$ and $D = \alpha\theta_E^n$, what should be the relationship between m and n in order to obtain a physically realistic $\theta(h)$ retention curve? Interpret the physical meaning of the coefficient α.
7. If $q_z > q_{1z}$ in Fig. 5.14, insert <, > or = for:

$$\frac{\partial q}{\partial z}\ldots 0 \qquad \text{and} \qquad \frac{\partial \theta}{\partial t}\ldots 0$$

   Do the same if $q_z = q_{1z}$.
8. What are the dimensions of the terms in (5.62) and (5.71)?
9. For fluid mobilities λ_A and λ_W in (5.86) and (5.87), insert <,> or = in $\lambda_A \ldots \lambda_W$. Do the same for the conductivities at $\theta/\theta_S = 0.5$, $K_A \ldots K_W$.
10. Derive the dimensions in (5.86).
11. Derive the value of the Lagrangian coordinate m for the compressed part of the unconsolidated soil at the wetting front in a swelling soil (see Fig. 5.19). How is m related to x?
12. Consider that steady state water flow conditions exist in a 100-cm long horizontal column of homogeneous soil. A pressure head $h = 0$ is maintained at the left end of the column ($x = 0$) while h = -100 cm is maintained at the other end. Calculate the direction and magnitude of the flux density and graph $h(x)$ for each of the following $K(h)$ functions: a. $K = 2$, $K = (2 + 0.0199h)$, and $K = 2\exp(0.053h)\ cm{\cdot}hr^{-1}$.
13. Same as problem 12 except the column is in a vertical position with the pressure head $h = 0$ at the top (z = 0) and h = -100 cm at the bottom of the column.
14. For problems 12 and 13, show that the product of K and the derivative of h is everywhere along the column equal to a constant – the value of $K(h)$.
15. Consider that steady state water flow conditions exist in a vertical, homogenous soil column with a water table maintained at depth z = 300 cm. Assuming that $K(h)$ is described by (5.38) with values of a, b and m being 200 $cm^3{\cdot}d^{-1}$, 100 cm^2 and 2, respectively, calculate and graph $h(z)$ for evaporation rates of 0.0054, 0.005 0.002 and 0.001 $cm{\cdot}d^{-1}$ and for infiltration rates of 0.005, 0.001, 0.01, 0.1 and 1 $cm{\cdot}d^{-1}$. What is the maximum rate of evaporation when $h \rightarrow -\infty$?

6 ELEMENTARY SOIL HYDROLOGIC PROCESSES

In this chapter, for the sake of a lucid discussion of soil hydrology, we separate from the global hydrologic cycle those simple elementary processes which take place in the soil. The first group of elementary processes to be discussed are those to be described in the vertical direction and defined by simple boundary conditions. Here we discuss

-infiltration,

-redistribution of water within a soil following infiltration,

-drainage to a water table close to the soil surface,

-evaporation from a bare soil and

-evaporation and transpiration (evapotranspiration) from a vegetated soil surface.

Except for infiltration, all the above processes cause a water loss either from the entire soil profile or from at least a particular layer usually from the topsoil when 1-dimensional vertical flow is assumed. When two- and three-dimensional flow is considered owing to field circumstances, lateral subsurface and hypodermic flows may contribute to the water balance within the soil profile.

Meteorological situations actually control the extent of the elementary processes, and together with the water storage capacity of the soil profile, a particular stage of a hydrologic regime evolves over a long time period. If these stages are combined and averaged over still a longer span of time, we speak of a hydrologic regime of a soil. Analogously, as meteorological situations refer to weather during a period of weeks or months in a particular area, combined, long time averages are considered as the climate of the area.

6.1 PRINCIPLES OF SOLUTIONS

Our knowledge of elementary soil hydrologic processes stems partly from properly performed experiments and partly from mathematical solutions of equations describing physical processes. In each of the procedures we simulate the process either physically by an experiment or mathematically using either analytical or numerical methods. When we speak of properly performed experiments, we must experimentally impose exactly the initial and boundary conditions. Before imposing the boundary conditions, we first establish the initial conditions - the values of θ or h at all z of the 1-dimensional column at $t < 0$. When the initial condition demands a zero flux ($q = 0$), it is imperative that $dH/dz = 0$ along the entire column. When the initial value of the soil water content θ_i is assumed constant with depth ($d\theta_i/dz = 0$), a flux corresponding to a unit gradient of H exists, i.e. $q = -K(\theta_i)$ provided that a continuity of liquid water exists. If θ_i is very small, the downward flux may be negligibly small. Altering the value of the variable θ or h, respectively, at the boundary (i.e. at the topographical soil surface) induces a non-equilibrium condition that generates a soil water flux within the soil profile. Non-steady fluxes will then persist in the system until an equilibrium is reached with $q = 0$, or until the flux is invariant with time ($dq/dt = 0$) and a steady state flux density is reached. Another possibility for causing water flow in soil is by imposing a defined flux density $q(t)$ on the boundary $q_o(t)$.

One boundary of a 1-dimensional soil column is its topographical surface. The other boundary is that of a column having either a finite or infinite length. A finite column is used to manifest a field condition, e.g. a ground water table ($h = 0$) or a defined water content or water flux at its bottom end. If the column extends to infinity, we speak of a semi-infinite column with its lower boundary $z \rightarrow \infty$.

If we establish a new value of the variable θ or h on a boundary for $t \geq 0$, we obtain Dirichlet's (or a concentration) boundary condition (DBC). When a flux density is imposed for $t \geq 0$ on a boundary, we have Neuman's (or a flux) boundary condition (NBC). If θ or h is specified on one boundary and a flux is specified on the other boundary, we have a mixed boundary condition (MBC).

When we solve a steady flow problem which is characterized by $dq/dt = 0$ and $dq/dz = 0$, we do not define the initial condition because the flux and variables θ and h are independent of time. The solution is a particular value of the flux q for which a unique distribution of θ and h, respectively, along the z axis exists.

Formulations of non-steady flow problems yield distributions of θ or h, respectively in space and time. These distributions $\theta(z, t)$ and $h(z, t)$ for 1-dimensional problems are usually given as continuous functions $\theta(z)$ or $h(z)$ for specific times or time intervals. For DBC we are frequently searching additionally for the flux density at the boundary $q_0(t)$. In an experiment we must establish the initial θ_i or h_i at all depths for $t < 0$ and sustain θ_o or q_o at the boundary during the entire experiment. It is imperative that the experimental initial and boundary conditions are matched exactly with those described mathematically. Without such initial and boundary conditions, the experimental results cannot be properly evaluated and generalized.

In the mathematical treatment of problems, the initial and boundary conditions represent the limits of integration. Additionally, the properties of the soil need to be characterized. When we describe the process mathematically, we characterize the soil by its hydraulic characteristic functions K_S, θ_S, $h(\theta)$ and $K(\theta)$ or $K(h)$. $D(\theta)$ is obtained from $K(\theta)$ and $h(\theta)$ or is defined directly from primary measurements. Methods for obtaining the hydraulic functions will be discussed later in Chapter 7.

Mathematical solutions of the elementary hydrologic processes are either analytical or numerical. In *analytical procedures,* differential equations are usually integrated only after some kind of transformation. Many analytical procedures exist for somewhat trivial boundary conditions of unsteady flow in soils defined by simple hydraulic functions and for steady flow processes in homogeneous or distinctly layered soil columns. The analytical solutions typically involve infinite series or transcendental functions that are evaluated by numerical methods with the assistance of a computer. As a result, the calculated results are approximate in spite of having an exact analytical solution.

Close to analytical solutions are the *quasi-analytical solutions* for which a significant part of the procedure involves an analytical procedure. The overall equation, often reduced to one or more ordinary differential equations, is integrated using a convergent iterative scheme. Alternatively, we split the higher order partial differential equation (e.g. Richards' equation) into lower order differential equations that are solved separately. In such cases an auxiliary function is frequently assumed with its value being determined by consecutive iterations. Analytical and quasi-analytical solutions developed for some elementary soil hydrological processes fit soils with simple, special forms of hydraulic functions. Such solutions, even if they are not directly applicable to field situations, have a great advantage. They lead to a full understanding of the physical process and provide estimates of deviations, e.g. owing to an alteration of a boundary condition. Moreover, these solutions allow errors of estimation of approximate and numerical procedures to be quantified. *Approximate solutions* are frequently exact analytical solutions developed for a soil characterized by a simple or even oversimplified hydraulic function. Or, they are exact solutions for a very simple flow process that only approximates reality in the field.

Numerical methods used in the solution of soil hydrologic processes are procedures which enable us to replace a differential equation with a set of approximate algebraic equations solved with a computer. These approximate numerical procedures are (i) the method of finite differences and (ii) the method of finite elements. Although their theoretical derivations are based upon different mathematical approaches, there are many similarities between both methods.

In the method of finite differences the spatial domain within which we search for a solution is sectioned by a system of normals into small segments. In 1-dimensional problems we obtain line segments, in 2-dimensional problems rectangle segments and in 3-dimensional problems parallelepiped segments. For each node of these geometric segments, we determine the value of the differential function describing the problem.

We replace the derivatives at a point by differences of the variable over a small finite interval. This method for the function $h(z)$ is therefore the inverse of the definition of its derivative

$$\frac{dh}{dz} = \lim_{\Delta z \to 0} \frac{h(z+\Delta z) - h(z)}{\Delta z}. \tag{6.1}$$

Inasmuch as the approximation at one point depends to a certain degree upon approximations at neighboring points, local approximations are controlled by approximations applicable to the entire domain.

The continuous analytical equation is replaced by a set of algebraic equations with differences substituting for derivatives of functions obtained from Taylor's series expansions.

In finite element methods, the domain of the solution is subdivided into n smaller sub domains, i.e. finite elements. The simplest scheme is composed of triangles with a triangular pyramid erected over each node. The value of the base functions represents the approximate solution with time taken as the finite difference. Local approximation is the characteristic feature of the finite difference method. In contrast, the finite element method manifests a global view. Assuming that the solution is expressible by a set of basic functions, the most frequent procedures are those of Galerkin and Ritz which are both based upon minimizing a quadratic function.

For a full appreciation of numerical methods, the reader should study fundamental and applied concepts provided in the literature (e.g. Remson et al., 1971, Neuman, 1975, Pinder and Gray, 1977 and Vauclin et al., 1979).

Numerical methods offer *ad hoc* solutions which are successful for non-trivial boundary conditions and for soils having hydraulic functions not described by simple functions. Additionally, they can be advantageously used to study water flow in heterogeneous soils. Numerical methods offer a large number of solutions for studying soil parameters under a wide range of conditions. They may in some instances replace results of experimental studies which are difficult to carry out owing to (i) slowly changing conditions over excessively long time periods, (ii) excessive number of experiments needed for alternative initial and boundary conditions and (iii) some of the acting factors for some problems may not be experimentally separated and uniquely studied.

In spite of all their advantages, numerical studies cannot generally substitute for analytical solutions and experiments. Analytical methods are useful to test the validity of numerical procedures adopted as standard solutions, and they are particularly helpful to establish the behavior of solutions in the neighborhood of singularities. Experimental observations compared with numerical simulations demonstrate how well our simplified formulations capture the essence of complex, natural soil processes.

6.2 INFILTRATION

The term infiltration denotes the entry of water into the soil through its surface. The soil surface could be plane, concave or convex, and could be formed by the walls of a cavity of a defined shape such as a sphere, cylinder etc. The source of water can completely or only partially cover the entire surface. Equations describing infiltration are usually for 1-dimensional water flow in either the vertical or horizontal direction. A limited number of solutions exist for 2- and 3-dimensional infiltration processes. Here, we restrict our discussion to 1-dimensional, vertical infiltration. Implicitly, our solutions are valid for infiltration through a plane horizontal surface.

Hydrologically, the infiltration process separates rain into two parts. One part stored within the soil supplies water to the roots of vegetation and recharges ground water. The other part which does not penetrate the soil surface is responsible for surface runoff. Infiltration is therefore a pivotal point within the hydrologic cycle.

Being consistent with present-day terminology, we shall call the flux density of water across a topographical soil surface the infiltration rate. However, this infiltration rate is often confusingly described by terms such as infiltration velocity, infiltration capacity, infiltrability etc. in relation to a specifically imposed boundary condition and in accordance with only some kind of subjective criteria. For non-steady flows it is clear that the flux density is time dependent and moreover, when boundary conditions are changed, the flux density responds and is also time dependent. Thus, we could formulate a voluminous number of terms describing still the same phenomenon – the flux density across the top boundary of the soil.

We shall discuss separately steady and unsteady infiltration owing to the different hydraulic characteristics of both flows. Although steady infiltration is simpler to solve and to understand because only the Darcy-Buckingham equation is involved, unsteady infiltration is the dominant process in nature. We shall discuss unsteady infiltration in two sections according to the boundary conditions governing the type of infiltration. When the soil surface is instantaneously and excessively ponded as it is in an infiltration test performed with a ring infiltrometer, we have Dirichlet's boundary condition (DBC). When infiltration occurs under natural rainfall, we meet Neuman's boundary condition (NBC) for the full duration of the rain or for at least its initial occurrence. With these two types of flow being fundamentally different, we shall consider them separately.

6.2.1 Steady Infiltration

Steady infiltration is characterized by the condition that the flux density does not change with time nor with position in the unsaturated soil, i.e. $\partial q/\partial t = 0$ and $\partial q/\partial z = 0$. It follows from the equation of continuity (5.62) repeated here for 1-dimension

$$\frac{\partial \theta}{\partial t} = -\frac{\partial q}{\partial z} \tag{6.2}$$

that the soil water content does not change in time (i.e. $\partial \theta/\partial t = 0$ as well as $\partial h/\partial t = 0$). In order to satisfy the condition $\partial h/\partial t = 0$, we must define a non-variant hydraulic condition at the bottom of the soil column. The simplest practical provision is a constant ground water level at its bottom. Such conditions are simply demonstrated by the following process. A rain intensity q_R is constant in time ($\partial q_R/\partial t = 0$) and equals the infiltration rate as well as the flux density in the soil q provided that $q_R < K_S$. In this case, rainfall has been constant and infiltration has lasted long enough to allow the wetting front to reach the ground water level. We further assume that the ground water level is kept at a constant elevation by e.g. a drainage system. It is mathematically convenient to identify the origin of the z coordinate at the ground water level from which z increases positively upwards. As a result at $z = 0$, $h = 0$ and at the soil surface $z = Z$, $h = h_Z$ and $q = -q_R$. Some solutions derived for steady state conditions approximate non-steady infiltration after a long time has elapsed when $\partial q/\partial t \rightarrow 0$. For example, the development of a $h(z)$ or $\theta(z)$ profile in a crust-topped soil or in a soil with distinct horizons of different hydraulic functions and conductivities is practically identical for either steady state infiltration or the quasi-steady stage of non-steady infiltration after a long time.

6.2.1.1 Homogeneous Soil Profile

Here, inasmuch as the direction of flow is oriented downward while the positive direction of the z-axis is upward, the flux density $q < 0$. Equation (5.31) repeated here is

$$q = -K(h)\frac{dH}{dz} \tag{6.3}$$

where $H = (h + z)$. We obtain $q < 0$ when $dH/dz > 0$. If indeed, $dH/dz = 0$, we have a state of equilibrium with $q = 0$ and $h = -z$. The domain of $h(z)$ which satisfies (6.3) is therefore limited from the left side of the graph in Fig. 6.1 by $h = -z$ ($q = 0$) and from the right side of the domain by $q = -K_S$ with $dH/dz = 1$ and $dh/dz = 0$. For the determination of $h(z)$ we integrate (6.3) with an appropriate expression for $K(h)$. For an exponential expression of $K(h)$, see (5.39)

$$K = K_S \exp(ch)$$

we obtain for the limit $h = 0$ at $z = 0$

$$\int_0^z dz = -\int_0^h \frac{K_S\,dh}{q\exp(-ch) + K_S}. \tag{6.4}$$

After integration we have

$$z = \frac{1}{c}\ln\left[\frac{q + K_S}{q + K_S\exp(ch)}\right]. \tag{6.5}$$

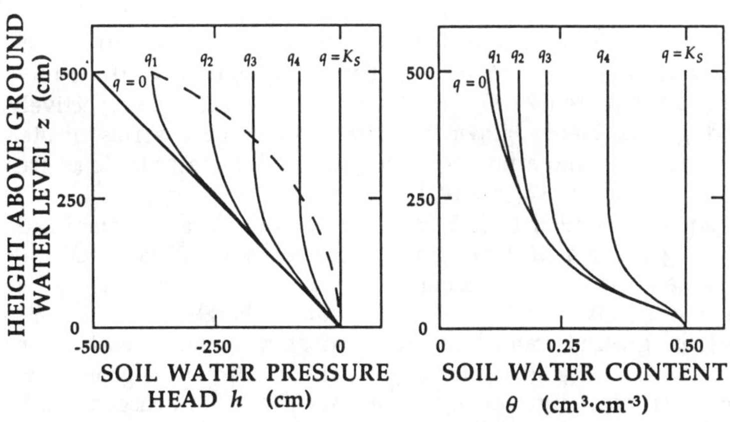

Figure 6.1. Steady flow infiltration. Pressure head profiles $h(z)$ and soil water content profiles $\theta(z)$ are plotted for flux densities $q_0 = 0$, $q_1 = -10^{-3}$, $q_2 = -10^{-2}$, $q_3 = -6 \cdot 10^{-2}$, $q_4 = -10^{-1}$ cm·h^{-1} and $q = -K_S$ in a soil characterized by $K_S = 2$ cm·h^{-1} and $c = 0.02$ cm^{-1} in (6.5) and $\theta_S = 0.5$, $\theta_r = 0.015$, $n = 1.8$, $m = 0.44$ $\alpha = 0.015$ in (4.43). The dotted line separates the domain where $dh/dz = 0$.

Solution (6.5) is represented graphically in Fig. 6.2. For measured values of K_S and c, we obtain values of z for a series of selected values of h using the appropriate values of K_S/q. We obtain $\theta(z)$ shown in Fig. 6.1 using the soil water

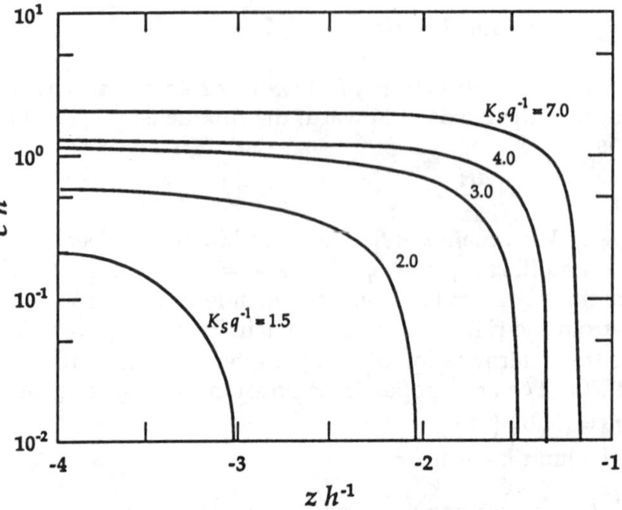

Figure 6.2. Steady flow infiltration. Graphical solution of (6.5) in dimensionless parameters (Bradáč and Kutílek, 1974).

retention curve SWRC. We see in the left graph that the depth of the zone having $dh/dz \approx 0$ increases with increasing absolute magnitudes of q. Also, within this zone $q \approx -K$ and $d\theta/dz \approx 0$. For example, for q = - 0.4 cm·h^{-1}, (h, θ) at z = 300 and 500 cm are (-79.98, 0.3396) and (-80.46, 0.3387), respectively, see the right hand graph. Hence, if we measure h or θ for a series of steady-state infiltration fluxes q_n into a homogeneous soil with a water table at great depth, we obtain $K_n(h) = |q_n|$ or $K_n(\theta) = |q_n|$, respectively.

An equation similar to (6.5) is easily derived for a soil manifesting an air entry value h_A when a $K(h)$ function described by (5.40) is used in (6.3). Soil water pressure head profiles calculated for h_A = -20 cm are shown in Fig. 6.3 for some values of q_n given in Fig. 6.1. Note that the height of the water-saturated zone above the ground water level is not constant but rises with an increase of $|q|$. And, if q = 0, the height of the saturated zone above the ground water level reaches only its minimum value, $z = - h_A$, see Fig. 6.3. Such observations are a graphical illustration that hydrostatic conditions differ markedly from those involving systems where water is flowing.

Many additional solutions of infiltration for other expressions of $K(h)$ have been reported by Kutílek (1984b).

6.2.1.2 Layered Soil Profiles

The simplest case is the crust-topped profile. Rainfall frequently destroys soil aggregates within a soil surface. Or, if infiltration lasts for a long time and the source of water is from a river or waste discharge carrying suspensions of clay or

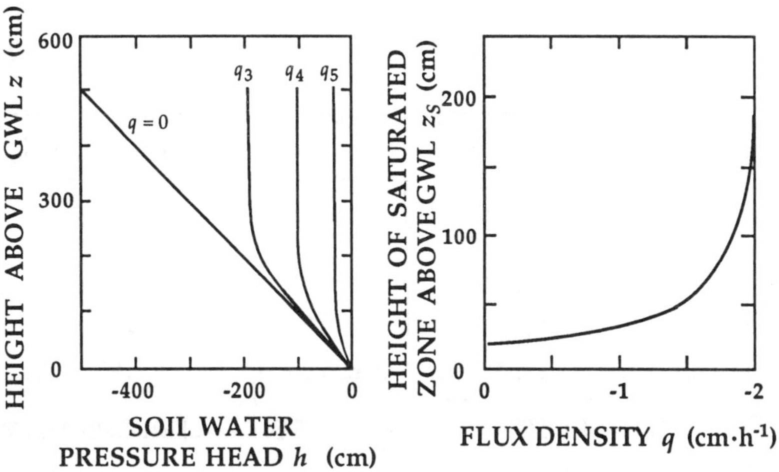

Figure 6.3. Steady flow infiltration. On the left, pressure head profiles h(z) if h_A = -20 cm, q_5 = -1.5 cm·h^{-1} and for other flux densities and soil characteristics identical to those in Fig. 6.1. On the right, the position of the saturated zone (elevation where h = -20 cm) depends upon the flux density.

fine particles, they are deposited on the soil surface or within the soil profile. Each of those processes denoted by sealing, crusting, collimation etc. results in the formation of a less permeable soil surface layer. Here for the sake of simplicity, we shall use the term crust for the result of all such processes. The characteristics of the crust will be denoted by the index 2, while those of the soil below the crust will be given the index 1, see Fig. 6.4. The origin of the z-axis is again identical with the position of the ground water level which is kept constant. The thickness of the soil between the ground water level and the crust is L_1, the thickness of the crust L_2 and the depth of water on the soil surface h_o. For steady-state flow, $q_1 = q_2$, and we have

$$-K_1\left(\frac{dH}{dz}\right)_1 = -K_2\left(\frac{dH}{dz}\right)_2 . \tag{6.6}$$

If $K_{S1} \gg K_{S2}$, and $K_1(h_I) \gg K_2(h_I)$ where h_I is the value of h at the interface, we have

$$\left(\frac{dH}{dz}\right)_1 >> \left(\frac{dH}{dz}\right)_2 \tag{6.7}$$

and because $H = (h + z)$,

$$\left(\frac{dh}{dz}\right)_1 >> \left(\frac{dh}{dz}\right)_2 . \tag{6.8}$$

This condition of a larger gradient of h occurring in the crust (layer 2) demands a sufficiently small value of h_I including $h_I < 0$. Because we assume that just below the interface in the subsoil (layer 1), $dH/dz \approx 1$, we can write

$$q = -K_1(h_I). \tag{6.9}$$

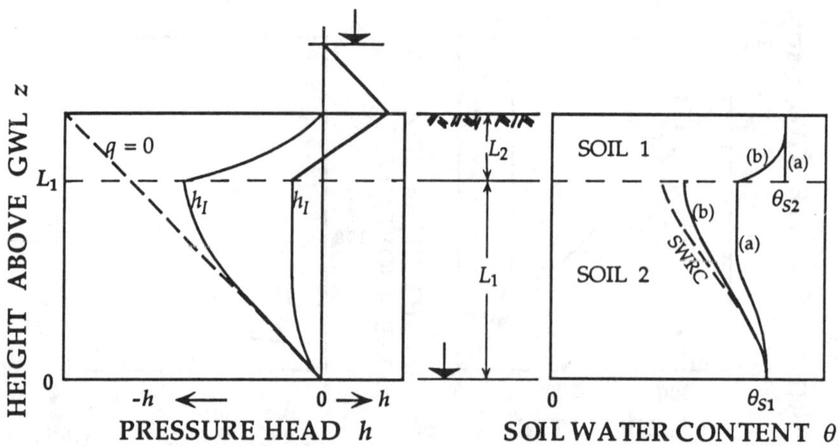

Figure 6.4. Steady flow infiltration into a crust-topped profile with h(z) on the left and θ(z) on the right for two cases. Case a., $h_I > h_{A2}$. Case b., $h_I < h_{A2}$. The pressure head at the boundary between crust and subsoil is h_I while h_{A2} is the air-entry value of the crust.

For the crust assuming it remains water-saturated,

$$q = -K_{S2}\left(\frac{h_o + h_I + L_2}{L_2}\right). \tag{6.10}$$

We also assume here that $h_{A2} = 0$ and $h_{A1} = 0$. The value of h_I is obtained by equating (6.9) and (6.10).

The criterion for $h_I < 0$ is derived from the total head loss between the free water level on the soil surface and that of the ground water - $(h_o + L_1 + L_2)$. Inserting this head loss into the modified Darcy equation (5.9) with the hydraulic resistance R_i for each of the two layers, we obtain

$$q = -\left(\frac{h_o + L_1 + L_2}{R_1 + R_2}\right). \tag{6.11}$$

The soil below the crust will be unsaturated if $|q| < K_{S1}$. From (6.11) it can be shown that the condition for unsaturation below the crust is

$$h_o < L_2\left(\frac{K_{S1}}{K_{S2}} - 1\right) \tag{6.12}$$

if $h_{A1} = 0$. For $h_{A2} > h_I$, the bottom part of the crust is also unsaturated, and $h(z)$ has a curved shape, see (b) in Fig. 6.4. For such cases, the above approach has to be modified, see e.g. Takagi (1960), Srinilta et al. (1966) and Bear et al. (1968). Kutílek (1984b) provides additional solutions when $dH/dz < 1$ below the interface.

Once $h(z)$ is known, we can determine $\theta(z)$ in the entire profile. Although h(z) has to be continuous, $\theta(z)$ is frequently discontinuous at the interface, see Fig. 6.4. When criterion (6.12) is fulfilled, the soil below the crust will be unsaturated if $h_{A1} > h_I$ provided the SWRC manifests h_A.

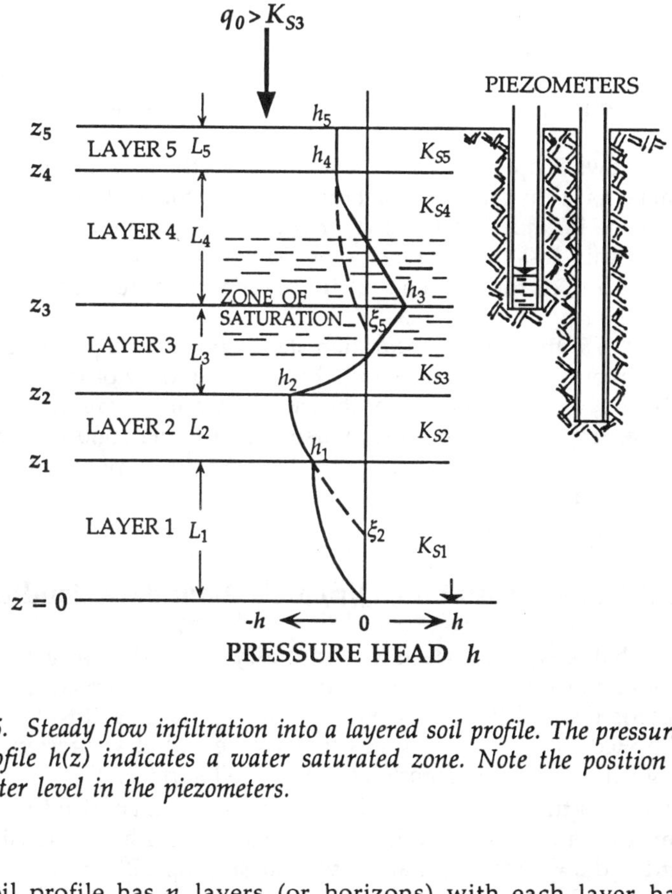

Figure 6.5. Steady flow infiltration into a layered soil profile. The pressure head profile h(z) indicates a water saturated zone. Note the position of the water level in the piezometers.

If a soil profile has n layers (or horizons) with each layer having its unique value of K_S and $K(h)$, we integrate in intervals identical with the height of the layers z_n. As a practical example, let us assume that the soil profile consists of 5 layers, see Fig. 6.5. The sequence of numbers again follows the positive direction of the z-axis with the layer in contact with the ground water level being 1, the next higher being 2 etc. Let us suppose that the soil has a strongly developed Bt horizon with a very small hydraulic conductivity, our layer 3. If $|q| > K_{S3}$ then for layer 3 we find that $(dH/dz)_3 > 1$, or $(dh/dz)_3 > 0$. For the other layers $|q| < K_{Si}, dH/dz \leq 1$ and $dh/dz \leq 0$, see Fig. 6.5. The distribution $h(z)$ can be found either analytically or with the graph in Fig. 6.2. We start with layer 1 as if it were a homogeneous profile to obtain h_1 on the boundary between layers 1 and 2. For layer 2 we find what would be the position of the ground water level to obtain h_1 for the given value of q. Let us call this position ξ_2 a substitute ground water level for soil layer 2, see Fig. 6.5. With ξ_2 we determine $h(z)$ between z_1 ad z_2, and at z_2, $h = h_2$. In layer 3 which has the very small hydraulic conductivity, we assume $h_2 \geq h_{A3}$ and $h(z)$ is linear. If indeed $h_2 < h_{A3}$, we proceed analytically to obtain the distribution $h(z)$, see Kutílek (1984b). Or, alternatively, we approximate reality by a linear relation as we did for $h_2 > h_{A3}$. Note that a linear increase of h for the sub layer with $h > h_A$ or for $h > 0$ is exact. Having obtained the value of h_3, we use in layer 4 Darcy's equation for the saturated flow which occurs in the domain $h \geq 0$. From the elevation z where $h = 0$, we follow the same procedure as that already described for a simple 2-layer soil profile.

It follows from the above analysis that the less permeable layer in a profile acts as a hydraulic resistance which causes the development of a saturated zone in and above this layer provided that the flux density is greater than K_S of this less permeable layer. The thickness of the saturated zone increases with q, or for a given q, it increases with a decrease of K_S in the less permeable layer.

In Fig. 6.5, the zone of saturation starts above the top boundary of the less permeable layer 3 and ends above the bottom boundary of layer 3 provided that $h_{A3} = 0$. For $h_{A3} \neq 0$ the thickness of the saturated zone is greater.

The example described above also explains the conditions for a pseudogley formation in layers 3 and 4 during long term steady rainfall, even without the presence of a permanent ground water table.

6.2.2 Unsteady Infiltration, Dirichlet's Boundary Condition (DBC)

Assuming that a soil surface is continuously flooded with a negligibly small depth of water at time $t \geq 0$, the surface soil will be water-saturated. Before the soil surface is flooded ($t \leq 0$), we assume that the initial soil water content $\theta = \theta_i$. Water supplied to the surface keeps the surface soil at saturation ($\theta = \theta_S$) but is never allowed to rise significantly above the soil surface.

Such a situation defines Dirichlet's boundary condition (DBC) for infiltration into a semi-infinite homogeneous soil. With z increasing positively downward and z = 0 identified at the soil surface, the DBC is

$$t \geq 0 \qquad z = 0 \qquad \theta = \theta_S, \tag{6.13}$$

or

$$t \geq 0 \qquad z = 0 \qquad h = h_o, \tag{6.14a}$$

$$t \geq 0 \qquad z = 0 \qquad h = 0. \tag{6.14b}$$

We use (6.13) when the diffusivity form of the Richards' equation (5.68) is solved and (6.14) for the capacitance form (5.66). Less frequently, the time dependent behavior of $\theta(t)$ or $h(t)$ are defined at the soil surface.

The initial condition for the simplest case is

$$t = 0 \qquad z > 0 \qquad \theta = \theta_i \tag{6.15}$$

Initial conditions were discussed in detail at the beginning of section 6.1. The initial condition is sometimes considered as a boundary condition with $t = 0$ taken as a boundary similarly to z = 0.

Boundary condition (6.14a) has the advantage that it specifies the depth of water flooding the surface, i.e. the pressure head on the surface. Flooded alluvium along a river, flooded infiltration in an irrigation basin or in a basin for tertiary sewage treatment are practical examples of that boundary condition. Or, late periods of some rainfall events are other examples of (6.13). If a basin is flooded at $t = 0$ without additional water being provided, the decreasing level of water in the basin equals the cumulative infiltration I. Inasmuch as I is a function of t, we have for the DBC at $z = 0$, $h(t) = [h_0 - I(t)]$ where $h = h_0$ at $t = 0$.

Infiltration caused by a DBC is frequently demonstrated with data from infiltration tests using double ring infiltrometers. The infiltration rate is measured by observing the decreasing water level within the inner ring, or even better, by measuring the inflow provided from a mariotte flask to the ring in order to keep a constant water level. The outer ring serves as a hydraulic buffer zone to minimize lateral flow below the inner ring. As a result, flow paths below the inner ring are nearly vertical, see Fig. 6.6. However, because a slight divergence of flow paths in the inner ring cannot be avoided, the measured data do not represent exactly 1-dimensional infiltration. Generally, the error is negligible compared with the inaccuracy of field experimentation and the spatial and temporal variability of the soil hydraulic functions provided that the soil is vertically homogeneous.

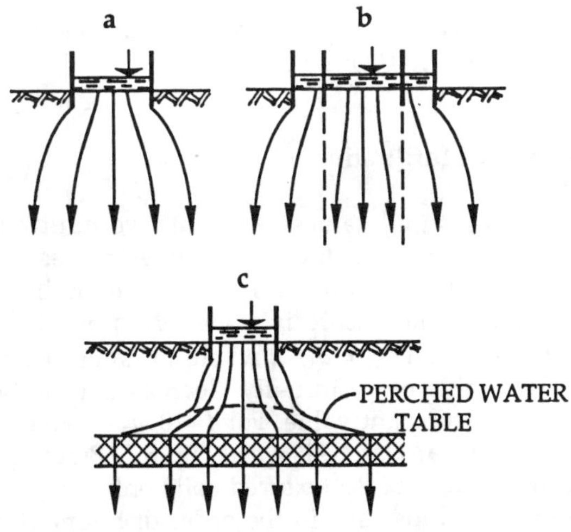

Figure 6.6. Flow paths when infiltration is measured a. by a single ring and b. by a double ring infiltrometer and c. is the influence of a less permeable layer (Bouwer, 1986).

When distinct soil layers exist in a profile, a strong divergence of flow paths occurs and the assumption regarding 1-dimensionality of the experiment is violated, see Fig. 6.6c. Hence, the measured data can be evaluated only for the period up to the time when the wetting front reaches the top boundary of the lesser permeable layer.

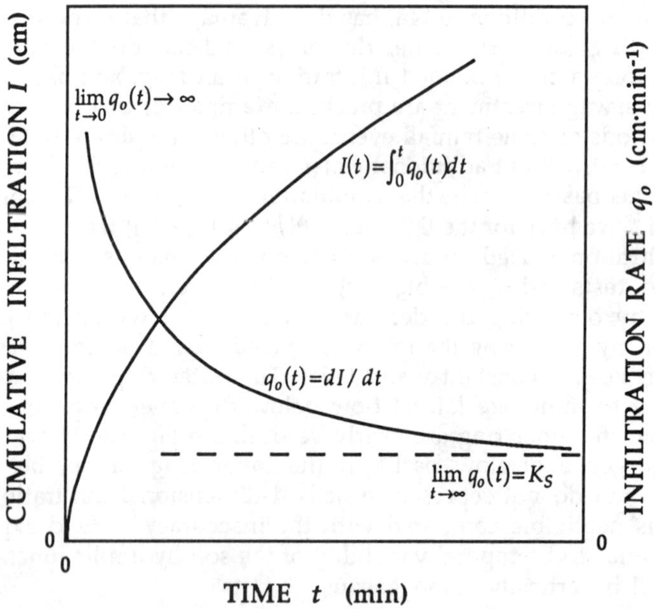

Figure 6.7. Time dependence of the cumulative infiltration I [L] and of the infiltration rate q_o [LT^{-1}] for DBC.

6.2.2.1 Characteristics of Infiltration

The primary data are measured values of cumulative infiltration I expressed as [L], usually in cm as a function of time. The values represent the total amount of water infiltrated into the soil surface from the beginning of the infiltration test at $t = 0$. A typical $I(t)$ relationship is a smooth, monotonically rising curve, see Fig. 6.7. The infiltration rate $q_o = dI/dt$ where the subscript o refers to the soil surface at $z = 0$. The value of q_o initially decreases rapidly with time and eventually approaches a constant value. For $t = 0$, $q_o \to \infty$, and for $t \to \infty$, q_o = constant. Theoretically, $q_o \to K_S$ as $t \to \infty$, see Fig. 6.7. Practically, the infiltration rate starts to be constant for coarse textured soils only after decades of minutes while that for fine textured loams is in the order of hours, depending upon the hydraulic functions of the soil and θ_i. Infiltration sometimes denoted as quasi-steady after this time limit will be discussed more fully in section 6.2.2.2. Steady infiltration into a crust-topped profile or into a layered profile can be successfully analyzed when 1-dimensional flow is guaranteed, e.g. by ponding water on a large area at $t = 0$ when the ground water level is at great depth or absent. The shape of $q_o(t)$ is empirically approximated by either a hyperbolic or exponential curve.

For a solution of the infiltration problem we first search for $h(z, t)$ and from it we obtain $\theta(z, t)$ from the SWRC. Some solutions provide $\theta(z, t)$ directly. Two examples of $\theta(z, t)$, one for sand and one for light clay are given in Fig. 6.8.

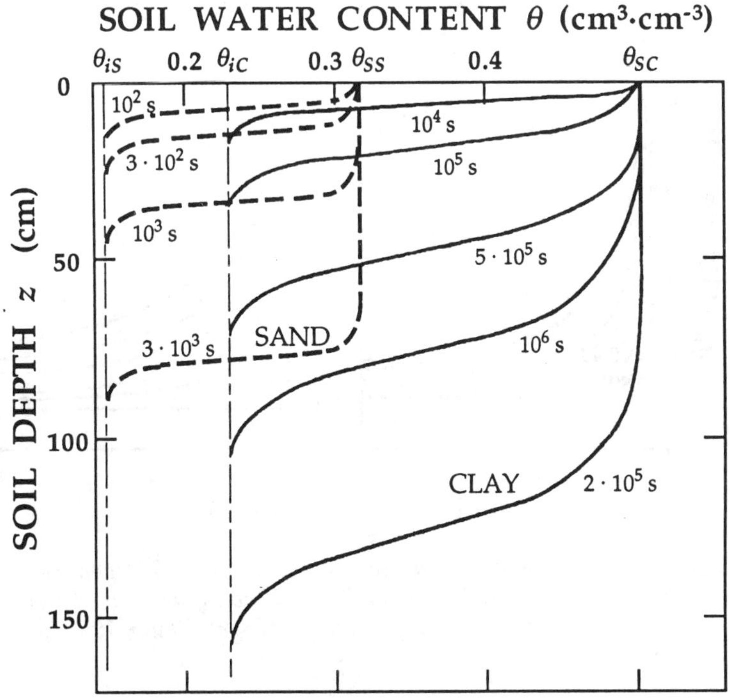

Figure 6.8. Soil water content profiles θ(z, t) in a sand and in a light clay for infiltration with DBC. θ_{is} is the initial water content of the sand, θ_{ic} the initial water content of the clay, θ_{Ss} the saturated water content of the sand and θ_{Sc} the saturated water content of the clay after Haverkamp et al. (1977).

The profiles are "piston-like", particularly for the sand. Where θ decreases steeply with z is called the wetting front. The rate of progress of the wetting front into the sand profile is more than two orders of magnitude greater than that into the clay profile. As the depth of wetting increases the shape of the wetting front becomes more gradual, especially for the clay. As infiltration proceeds, the shapes of $\theta(z)$ profiles for a given soil become nearly identical. Theoretically, the shapes are identical as $t \to \infty$.

Integration of the soil moisture profile at time t defines the cumulative infiltration at time

$$I = \int_{\theta_i}^{\theta_s} z\, d\theta \qquad (6.16)$$

which according to (6.15) will decrease as θ_i increases. The influence of the initial value of water content θ_i is demonstrated in Fig. 6.9 for the light clay. If $\theta_i / \theta_S \geq 0.95$, the infiltration rate q_o can be approximated by $q_o \approx K_S$. The influence of the depth of ponding on the soil surface upon $q_o(t)$ is illustrated in Fig. 6.9. For $0 \leq h_o \leq 2$ cm, the influence is negligibly small. For $h_o = 10$ cm, the value of

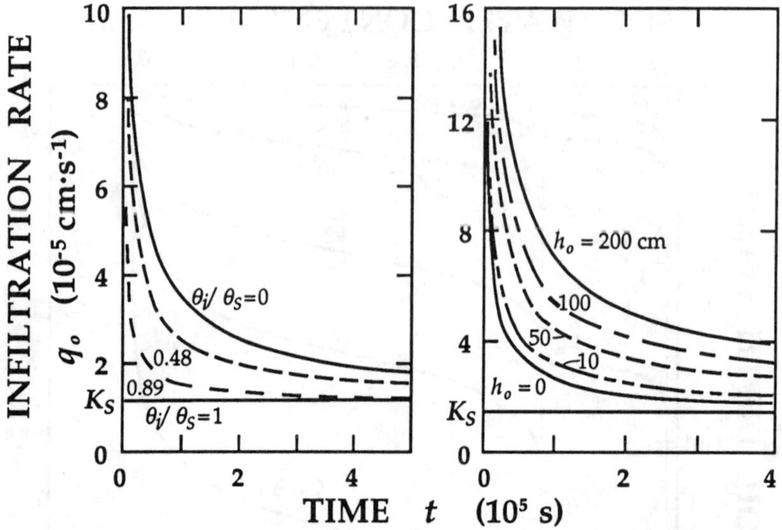

Figure 6.9. The influence of initial soil water content θ_i upon the DBC-infiltration rate q_o (left, Philip, 1957c) and the influence of the depth of ponding water h_o on the soil surface upon q_o (right, Philip, 1958) for Yolo light clay.

q_o is increased by 20% for large times and by more than 50% for short times in light clay. These relationships demonstrate how important it is to keep the value of h_o constant and as small as possible in experiments when the DBC (6.13) is applied.

Solutions to this type of infiltration can be divided into the three classes – (i) analytical and semi-analytical procedures, (ii) approximate solutions and (iii) empirical equations.

6.2.2.2 Analytical and Semi-Analytical Procedures

Richards' equation in its diffusivity form (5.68) is repeated here for the vertical coordinate oriented positively downward from the soil surface located at $z = 0$

$$\frac{\partial \theta}{\partial t} = \frac{\partial}{\partial z}\left[D(\theta)\frac{\partial \theta}{\partial z}\right] - \frac{dK}{d\theta}\frac{\partial \theta}{\partial z}. \tag{6.17}$$

This equation, sometimes denoted as the non-linear Fokker-Planck equation (Philip, 1969a), is non-linear owing to the strong dependence of D and K upon θ. The first term on the right-hand-side of (6.17) describes the transport of water owing to the initial degree of unsaturation of the soil profile. Therefore, as θ_i increases, the importance of this term decreases. The second term on the right-hand-side of (6.17) originates because the gravitational potential. Hence, it is called the gravitational term and describes the flow of water owing to the force of gravity.

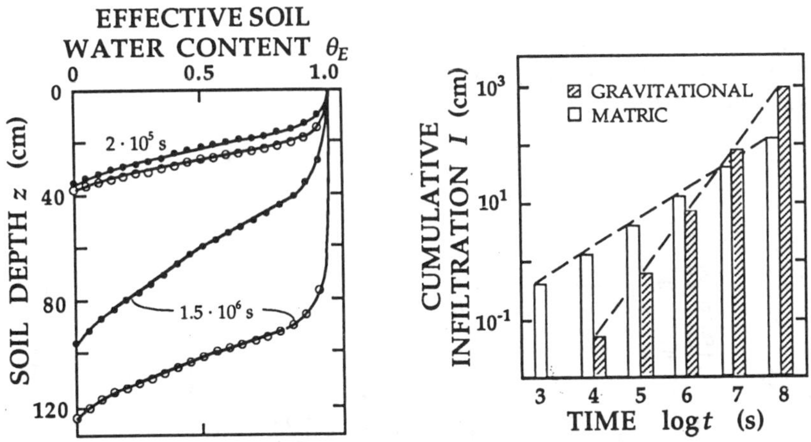

Figure 6.10. The influence of gravity upon the vertical DBC-infiltration into Yolo light clay. On the left, soil water content profiles for either vertical or horizontal infiltration. On the right, the partitioning of cumulative infiltration into matric and gravitational quantities (Kunze and Nielsen, 1982).

Philip's (1957b) solution of (6.17) is based upon the idea of separating the infiltration into its two components – those caused by the matric potential force and by the gravitational potential force. The idea is illustrated in Fig. 6.10. In the first step, he neglected the gravitational force and obtained a solution for horizontal infiltration in the form $x(\theta, t)$. Here, the dependent variable was changed to that of the horizontal axis x. Next, he assumed that the real $z(\theta, t)$ for vertical infiltration was the horizontal component $x(\theta, t)$ plus a correction , see Fig. 6.10. The correction owing to the gravitational force is time dependent. The influence of gravity upon infiltration is shown in Fig. 6.10 when θ_i is small. For short infiltration times its influence is very small, but with time it increases and for very large times, the force of gravity dominates the process. Hence, we first study horizontal infiltration.

Our horizontal soil column, initially at an unsaturated water content θ_i, has its end at $x = 0$ maintained at water saturation θ_S. Hence, for

$$t \geq 0 \qquad x = 0 \qquad \theta = \theta_S \tag{6.18}$$

$$t = 0 \qquad x > 0 \qquad \theta = \theta_i \tag{6.19}$$

we solve (6.17) without the gravitational term

$$\frac{\partial \theta}{\partial t} = \frac{\partial}{\partial x}\left[D(\theta)\frac{\partial \theta}{\partial x}\right]. \tag{6.20}$$

It is only here for a homogeneous soil (i.e. not layered) that the gradient of θ represents the driving force of the process. When D is a constant in (6.20), the solution is according to Carslaw and Jaeger (1959)

$$\frac{\theta - \theta_i}{\theta_S - \theta_i} = erfc\left(\frac{x}{2\sqrt{Dt}}\right). \tag{6.21}$$

When D is a function of θ, we transform (6.20) into an ordinary differential equation using the Boltzmann transformation. The transformed equation has a new variable η instead of the two original variables x and t. The new variable η defined by the Boltzmann transformation

$$\eta(\theta) = xt^{-1/2} \tag{6.22}$$

leads to

$$\frac{\partial \eta}{\partial t} = -\frac{1}{2} xt^{-3/2} = -\frac{\eta}{2t} \tag{6.23}$$

$$\frac{\partial \theta}{\partial t} = \frac{d\theta}{d\eta}\frac{\partial \eta}{\partial t} = -\frac{\eta}{2t}\frac{d\theta}{d\eta} \tag{6.24}$$

$$\frac{\partial \eta}{\partial x} = t^{-1/2} \tag{6.25}$$

and

$$\frac{\partial}{\partial x}\left[D(\theta)\frac{\partial \theta}{\partial x}\right] = \frac{\partial}{\partial \eta}\left[D(\theta)\frac{d\theta}{d\eta}\frac{\partial \eta}{\partial x}\right]\frac{\partial \eta}{\partial x} = \frac{\partial}{\partial \eta}\left[D(\theta)\frac{d\theta}{d\eta}t^{-1/2}\right]t^{-1/2}. \tag{6.26}$$

From the above (6.20) transforms to

$$-\frac{\eta}{2}\frac{d\theta}{d\eta} = \frac{\partial}{\partial \eta}\left[D(\theta)\frac{d\theta}{d\eta}\right]. \tag{6.27}$$

The transformed boundary conditions are

$$\eta = 0, \qquad \theta = \theta_S \tag{6.28}$$

and

$$\eta = \infty \qquad \theta = \theta_i. \tag{6.29}$$

The solution for which we search is simply $\theta(\eta)$, see Fig. 6.11. Measured soil water profiles $\theta[x(t_1)]$, $\theta[x(t_2)]$, $\theta[x(t_3)]$ etc. are thus transformed into the unique $\theta(\eta)$ relationship by merely dividing x by $t_1^{1/2}$ for the first profile, $t_2^{1/2}$, for the second profile etc. Note that for $t = 1$, $x \equiv \eta$. Hence, the physical reality of $\theta(\eta)$ is the soil water profile $\theta(x)$ when the infiltration time is unity.

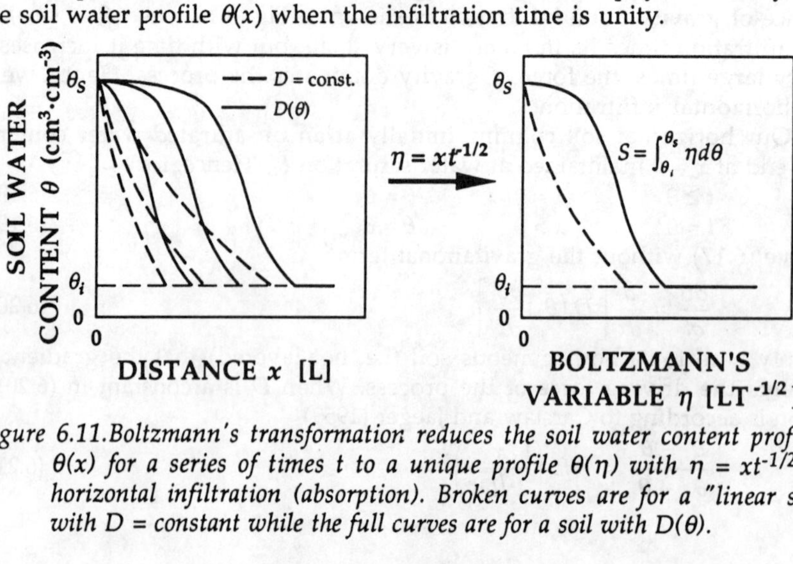

Figure 6.11. Boltzmann's transformation reduces the soil water content profiles $\theta(x)$ for a series of times t to a unique profile $\theta(\eta)$ with $\eta = xt^{-1/2}$ for horizontal infiltration (absorption). Broken curves are for a "linear soil" with D = constant while the full curves are for a soil with $D(\theta)$.

Philip (1960) and Kutílek (1984) have shown for which analytical expressions of $D(\theta)$ analytical solutions of (6.27) subject to (6.28) and (6.29) exist. Because it is exceptional that any of those analytical expressions accurately describe $D(\theta)$ of a real soil, an iterative procedure proposed by Philip (1955) is commonly used to calculate $\theta(\eta)$ from measured distributions of D versus θ.

With the content of infiltrated water being denoted as cumulative infiltration I,

$$I = \int_{\theta_i}^{\theta_s} x \, d\theta \tag{6.30}$$

or with (6.28),

$$I = \int_{\theta_i}^{\theta_s} \eta(\theta) \, t^{1/2} \, d\theta . \tag{6.31}$$

Inasmuch as $\eta(\theta)$ is unique for each soil, Philip (1957b) introduced the term sorptivity S [$LT^{-1/2}$]

$$S = \int_{\theta_i}^{\theta_s} \eta(\theta) \, d\theta \tag{6.32}$$

and

$$I = S \, t^{1/2}. \tag{6.33}$$

Because the infiltration rate

$$q_o = dI / dt, \tag{6.34}$$

we have

$$q_o = \frac{1}{2} S \, t^{-1/2}. \tag{6.35}$$

Here, we note that the sorptivity is physically the cumulative amount of water infiltrated at $t = 1$, and at that time, the infiltration rate has diminished to one-half the value of S. Sorptivity depends not only upon the $D(\theta)$ function but upon θ_i. The value of S decreases with increasing θ_i and as $\theta_i \rightarrow \theta_S$, $S \rightarrow 0$, see Fig. 6.12. When S is measured for a particular θ_{i1}, we can linearly interpolate between θ_{i1} and θ_S in order to obtain a first approximation of S for $\theta_{i2} > \theta_{i1}$. A more laborious, exact procedure is described by White and Broadbridge (1988). If $\theta_o < \theta_S$ is used in (6.18) instead of θ_S, we proceed in the same manner to derive S. However, the resulting value of $S(\theta_i, \theta_o)$ may indeed drop substantially from that of $S(\theta_i, \theta_S)$.

Sorptivity is an integral part of most investigations describing vertical infiltration. As a first approximation of the solution of (6.17) subject to (6.18) and (6.19), Philip used (6.22), the solution of (6.20) for horizontal infiltration, i.e. $z_1(\theta, t) = x(\theta, t)$. He corrected this approximation with the term y, i.e. $z = z_1 + y$. However, because an exact value of y cannot be obtained, its approximation y_1 defines another correction u, i.e. $y = y_1 + u$. Again, instead of an exact u we can only find still another estimate u_1 etc. Hence, Philip obtained the infinite series solution

$$z(\theta,t) = \eta_1(\theta)t^{1/2} + \eta_2(\theta)t + \eta_3(\theta)t^{3/2} + \cdots + \eta_n(\theta)t^{n/2} \tag{6.36}$$

where the functions $\eta_1, \eta_2, \eta_3, \cdots \eta_n$ are defined with $D(\theta)$, $K(\theta)$ and η_{n-1}. The procedure for computing terms η_n is described in detail by Kirkham and Powers (1972).

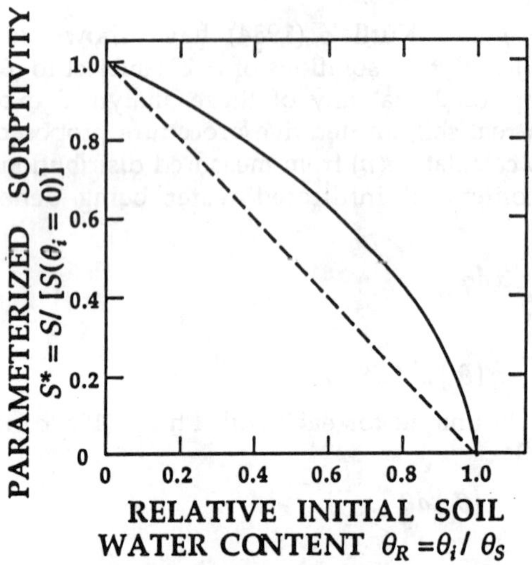

Figure 6.12. The dependence of the parameter S^* [$LT^{-1/2}$] upon the relative soil water content θ_i/θ_S. Broken curve is for a "linear soil" with D = constant while the full curve is for a soil with $D(\theta)$.

Inasmuch as the cumulative infiltration I according to (6.16) is

$$I = \int_{\theta_i}^{\theta_s} z\,d\theta, \tag{6.37}$$

Philip formulated analogously to the sorptivity equation (6.32) for horizontal infiltration, the following equation from (6.36)

$$I = S\,t^{\,1/2} + \left(A_2 + K_i\right)t + A_3 t^{\,3/2} + \cdots + A_n t^{\,n/2} \tag{6.38}$$

where

$$A_n = \int_{\theta_i}^{\theta_s} \eta_n(\theta)\,d\theta \qquad n = 2, 3, \cdots$$

and K_i is $K(\theta_i)$. Note that $K_i t$ expresses the cumulative water flow with dH/dz = -1 at $\theta = \theta_i$. Thus, we understand physically that boundary condition (6.19) can be kept only if we impose a steady flux density $q_o = K(\theta_i)$ for $z \geq 0$ within the semi-infinite column.

The series (6.38) converges for short and intermediate times of infiltration and the infiltration rate $q_o(t)$ obtained by differentiation is

$$q_o = \frac{1}{2}S\,t^{\,-1/2} + \left(A_2 + K_i\right) + \frac{3}{2}A_3 t^{\,1/2} + \cdots + \frac{n}{2}A_n t^{\,n/2-1}. \tag{6.39}$$

For large times, (6.38) does not converge. Inasmuch as the shape of the wetting front remains invariant at large times, the wetting front moves downward at a rate

$$v = \left(\frac{K_S - K_i}{\theta_S - \theta_i}\right) \tag{6.40}$$

while the infiltration rate for $t \to \infty$ is

$$q_o = K_S\,. \tag{6.41}$$

Equations (6.40) and (6.41), commonly called the infinite time solutions, are theoretically traveling wave solutions (Philip, 1969a).

The times for which (6.38) or (6.39) continue to converge was found to range broadly from 0.67 h for sand to 250 h for light clay (Haverkamp et al., 1988). Similarly, the times for which the infinite time solution is applicable varies widely from approximately 100 min for a silt loam (Nielsen et al., 1961) to about 10^5 min for light clay (Kunze and Nielsen, 1982). Piece wise solutions for 1-dimensional infiltration have been discussed by Philip (1987).

In order to obtain an intermediate time solution, Swartzendruber (1987) adjusted Philip's time series solution of q to apply between the limits $t \to 0$ and $t \to \infty$. Inasmuch as the solution for infiltration into linear soils as well as some approximate solutions lead to exponential forms of $I(t)$, Swartzendruber proposed intuitively the form

$$I = \frac{S}{A_o}\left[1 - \exp\left(-A_o t^{1/2} - B_o t - C_o t^{3/2} - \cdots\right)\right] + K_S t \tag{6.42}$$

where A_o, B_o, C_o, $\cdots$ are constants depending upon the soil hydraulic functions as well as θ_i and θ_S. The time derivative of (6.42) gives the infiltration rate

$$q_o = \frac{S}{A_o}\left[1 - \exp\left(-A_o t^{1/2} - B_o t - C_o t^{3/2} - \cdots\right)\right] \cdot$$

$$\cdot\left[\frac{A_o}{2} t^{-1/2} + B_o + \frac{3}{2} C_o t^{1/2} + \cdots)\right] + K_S. \tag{6.43}$$

Parlange (1971), realizing that Richards' equation originated from a combination of the Darcy-Buckingham equation and the equation of continuity, also obtained a solution. His original procedure consisting of iterative processes was gradually corrected by Císler (1974) and further modified (Parlange et al., 1982, and Parlange et al., 1985) to its present form (Haverkamp et al., 1990). The procedure, based upon an integral moment balance (Raats, 1988), uses a double integration of the equation of continuity (6.2). The starting point is

$$\int_0^\infty \left(q - q_\infty\right) dz = \frac{\partial}{\partial t} \int_0^\infty \left(\theta - \theta_\infty\right) dz. \tag{6.44}$$

When the diffusivity form of the Darcy-Buckingham equation was combined with (6.44), Haverkamp et al. (1990) using two approximation steps obtained two well behaved equations. Owing to this contribution of Parlange and Haverkamp, we call them the P-H equations. In their dimensionless form they are

$$I^* = \frac{\gamma}{q_o^* - 1} + (1 - \gamma)\ln\left[1 + \frac{1}{q_o^* - 1}\right] \tag{6.45}$$

and

$$t^* = (1 - 2\gamma)\ln\left[1 + \frac{1}{q_o^* - 1}\right] + \frac{\gamma}{q_o^* - 1} - \frac{1-\gamma}{q_o^*} \tag{6.46}$$

where

$$q_o^* = \frac{dI^*}{dt^*},$$

$$I^{*} = \left[I - K_i t\right]\frac{2(K_S - K_i)}{S^2 + 2K_i h_o\,(\theta_S - \theta_i)},$$

$$t^{*} = \frac{2(K_S - K_i)^2 t}{S^2 + 2K_i h_o\,(\theta_S - \theta_i)}$$

and

$$\gamma = \frac{2\,K_S(h_o - h_W)(\theta_S - \theta_i)}{S^2 + 2K_i h_o\,(\theta_S - \theta_i)}.$$

The value of h_W is the water entry value on the wetting branch of the SWRC. After I^* and t^* are computed for chosen values of q_o^*, all three terms are transformed back to the dimensional forms I, q_o and t.

The integral method can also involve a mass balance formulation where (Raats, 1988)

$$\int_0^t \left[q_o(t) - q_\infty\right] dt = \int_{\theta_i}^{\theta_S} z\,(\theta,t)\,d\theta \qquad (6.47)$$

which is also obtained by integrating the equation of continuity. Philip (1973) and Philip and Knight (1974) managed to reduce the number of independent variables of the infiltration problem by using a guessed shape for the ratio of flux densities q/q_o. With the main idea of this "flux-concentration relation" method being explained in more detail in section 6.2.3.2., only the principle is given here.

The parameterized flux density F related to the parameterized water content θ_R for *horizontal infiltration* is

$$F\,(\theta_R,t) = \frac{q}{q_o} \qquad (6.48)$$

or with (6.32)

$$F\,(\theta_R) = \frac{\int_{\theta_i}^{\theta} \eta\,(\theta)\,d\theta}{\int_{\theta_i}^{\theta_S} \eta\,(\theta)\,d\theta} \qquad (6.49)$$

where

$$\theta_R = \frac{\theta - \theta_i}{\theta_S - \theta_i}.$$

Relation (6.49) is the ratio of the partial and total sorptivities calculated from (6.32) and illustrated in Fig. 6.13. $F(\theta_R, t)$ is the guessed shape which is inserted into the diffusivity form of the Darcy-Buckingham equation to obtain

$$F\,(\theta_R,t) = -\frac{D\,(\theta)}{q_o}\frac{\partial\theta}{\partial x}. \qquad (6.50)$$

With

$$q_o = \frac{1}{2} S t^{-1/2}$$

the sorptivity is (Philip and Knight, 1974)

$$S = \left[\,2\int_{\theta_i}^{\theta_S} \frac{(\theta - \theta_i)\,D\,(\theta)}{F\,(\theta_R)}\,d\theta\right]^{1/2}. \qquad (6.51)$$

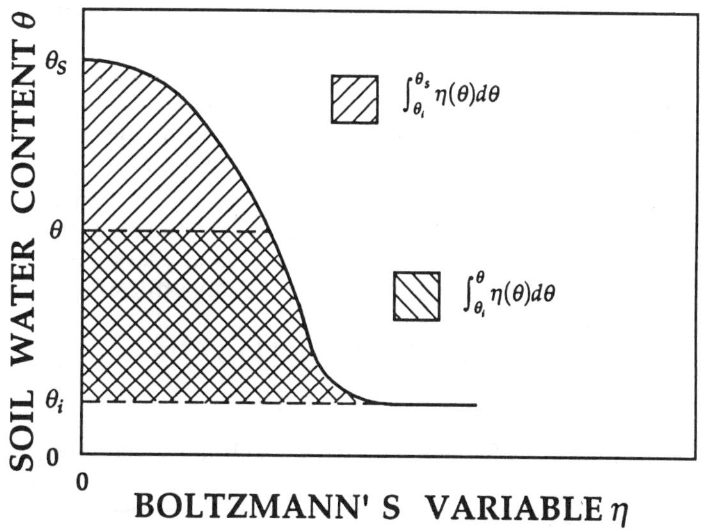

Figure 6.13. Flux concentration relation F(θ) is expressed as the ratio of these two integrals.

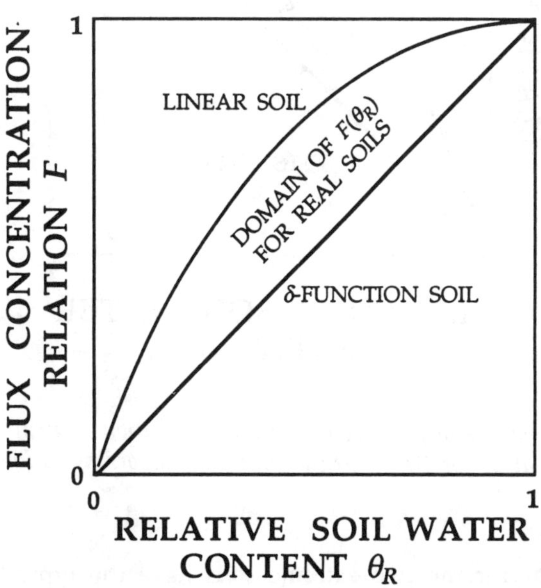

Figure 6.14. Dependence of the flux concentration relation F upon relative soil water content θ_R for DBC-horizontal infiltration according to Philip (1973).

If $F(\theta_R)$ is known, the above solution is available. Inasmuch as the time dependence of $F(\theta_R)$ is neglected, the solution is approximate. Philip (1973) has demonstrated that $F(\theta_R)$ can only exist within a relatively narrow domain. In Fig. 6.14 it can be seen that the domain is limited on one side by the curve which represents a linear soil having a constant D and from the other side by the straight line ($F = \theta_R$) which is descriptive of a soil having a $D(\theta)$ equal to a Dirac δ-function. For $\theta_R = 0$, $F = 0$, and for $\theta_R = 1$, $F = 1$. These two soils should represent the extremes of existence of $D(\theta)$ for real soils. Although the difference between the $F(\theta_R)$ relationships for these two soils appears slight in Fig. 6.14, their soil water content distributions for horizontal infiltration are strikingly different, see Fig. 6.16.

For vertical infiltration (6.48) is modified to

$$F(\theta_R, t) = \frac{q - K_i}{q_o - K_i} \tag{6.52}$$

and the $F(\theta_R)$ relationships for the two extreme soils are given in Fig. 6.15.

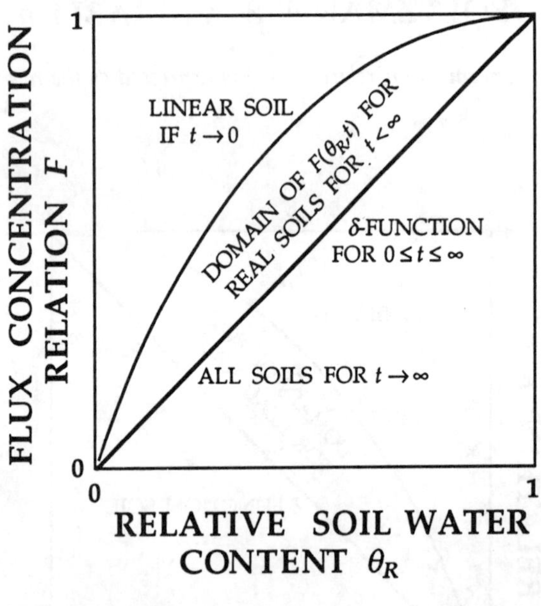

Figure 6.15. Dependence of the flux concentration relation F upon relative soil water content θ_R for DBC-vertical infiltration according to Philip (1973).

The solutions for the two extreme soils have the typical features of a theoretical treatment with no direct applicability to reality. Linear soil, characterized by a constant D and a K proportional to θ_R leads to h being proportional to $\ln\theta_R$ – a relationship for a SWRC which is not realistic for soils or other porous media. It would appear that for δ-function soils, infiltration

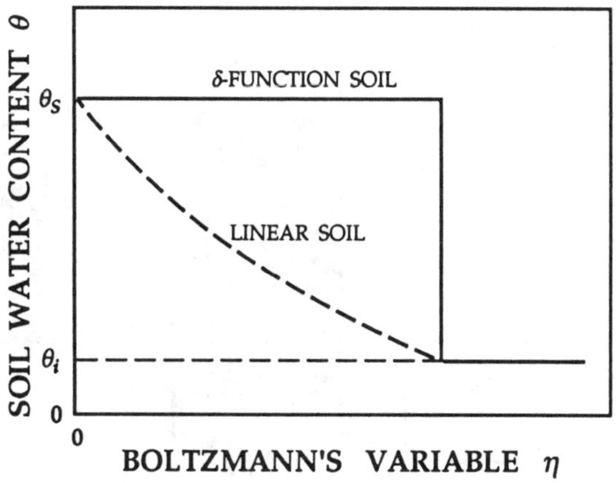

Figure 6.16. Soil water content profile $\theta(\eta)$ for DBC-horizontal infiltration for the "linear soil" (broken curve) and for a δ-function soil (full curve).

should be easily approximated owing to the steep rise of D with θ_R and that $F(\theta_R) = \theta_R$ looks like a good approximation for the ratio of flux densities. However, when the sorptivities computed for δ-function soils were compared to those obtained analytically for soils having a very steep $D(\theta)$, relative errors exceeded 20% (Kutílek and Valentová, 1986). Therefore, for most soils the iterative procedures proposed by Philip and Knight (1974) should be adopted.

In order to obtain solutions for models characteristic of real soils, proper functional relationships $D(\theta)$, $K(\theta)$ and $h(\theta)$ are required. To obtain them Richards' equation is linearized with transformations, e.g. those of Storm or Hopf and Cole. A review of such solutions identifying the authors of each transformation is provided by Raats (1990).

6.2.2.3 Approximate Solutions

The solution of the infiltration process, approximated physically or mathematically, is usually not kept wholly within either category but relies more heavily upon one or the other. A physical approximation is dominant in the procedure of Green and Ampt (1911) while mathematical approximations prevail in remaining, more recent procedures.

Green and Ampt (1911) simplified a real soil water profile of infiltration to a step-like profile, see Fig. 6.17. In this model, water penetrates into the soil like a piston which proceeds with time to greater depths. Below the abrupt, horizontal wetting front, the soil remains dry at its initial value of $\theta = \theta_i$. In the saturated upper part of the soil, flow is now simply described by Darcy's equation. If at time t the position of the wetting front is $z = L_f$ (the thickness of the soil saturated with water is also L_f), the infiltration rate is

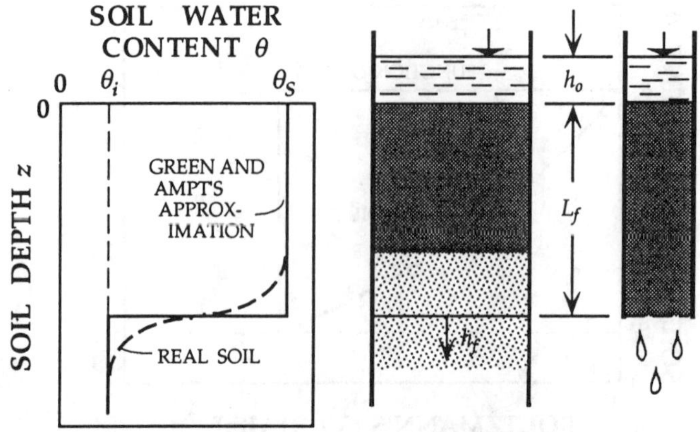

Figure 6.17. Solution of DBC-vertical infiltration according to Green and Ampt's approximation where the real soil water content profile θ(z, t) given by the broken curve is replaced by the step-like profile given by the full line.

$$q_o(t) = -K_S \left\{ \frac{h_f - [h_o - L_f(t)]}{L_f(t) - 0} \right\} = K_S \left\{ \frac{h_o + L_f(t) - h_f}{L_f(t)} \right\} \quad (6.53)$$

where h_o is the pressure head at the soil surface (i.e. the depth of water on the surface). Note that L_f is time dependent. The term h_f is the soil water pressure head at the wetting front owing to the unsaturated condition of the soil below z with $h_f < 0$. If there were no soil below $z = L_f$ and water was freely falling out of the saturated soil column ($h = 0$ at $z = L_f$), the water flux throughout the column of thickness L_f would be

$$q_o = q = K_S \frac{h_o + L_f(t)}{L_f},$$

see Fig. 6.17. Because there is dry soil below $z = L_f$, its unsaturated condition causes the flux to increase. Green and Ampt added the term h_f to the driving force to account for the extra force acting at the wetting front. Neuman (1976) has shown that

$$h_f = \frac{1}{K_S} \int_0^{h_i} K(h)\, dh \quad (6.54)$$

and

$$h_f = \frac{1}{K_S} \int_{\theta_S}^{\theta_i} D(\theta)\, d\theta. \quad (6.55)$$

Or, using the Parlange (1975) solution

$$h_f = \frac{1}{2} \int_0^{h_i} \left(\frac{\theta_S - \theta(h) - 2\theta_i}{\theta_S - \theta_i} \right) \left[\frac{K(h)}{K_S} \right] dh. \quad (6.56)$$

Obviously, without knowing the original publication of Green and Ampt, Budagovskij (1955) based his monographic study upon the same principle.

Theoretically, the procedure is based upon the expected shape and similarity of the $\theta(z, t)$ profiles. Philip (1957b, 1973) showed that the following Green and Ampt approximation is an exact solution only if $D(\theta)$ is expressed as a Dirac δ-function. Considering (6.53), we know $q_o = dI/dt$ and $I = L_f \Delta\theta$ where $\Delta\theta = \theta_S - \theta_i$, and hence,

$$q_o = \frac{dL_f}{dt} \Delta\theta. \tag{6.57}$$

The rate of progress of the wetting front is linearly dependent upon the infiltration rate for all t > 0. When the force of gravity is neglected, i.e. for horizontal infiltration, substituting (6.57) into (6.53), we have

$$\frac{dL_f}{dt} \Delta\theta = K_S \left(\frac{h_o - h_f}{L_f(t)} \right), \tag{6.58}$$

and after separating variables

$$\int_0^{L_f} L_f \, dL_f = \int_0^t K_S \left(\frac{h_0 - h_f}{\Delta\theta} \right) dt. \tag{6.59}$$

After integrating,

$$L_f = \left[2K_S \left(\frac{h_0 - h_f}{\Delta\theta} \right) \right]^{1/2} t^{1/2} \tag{6.60}$$

or with $L_f = I/\Delta\theta$,

$$I = \left[2K_S \left(h_0 - h_f \right) \Delta\theta \right]^{1/2} t^{1/2}. \tag{6.61}$$

Comparing (6.61) with (6.33), we obtain the sorptivity S

$$S = \left[2K_S \left(h_o - h_f \right) \Delta\theta \right]^{1/2}. \tag{6.62}$$

Equation (6.61) can also be used to estimate I during a brief initial period of vertical infiltration. Equation (6.62) defines approximately how S depends upon θ_i.

When gravity is not neglected, (6.53) becomes

$$\Delta\theta \frac{dL_f}{dt} = K_S \left\{ \frac{L_f(t) + \left(h_o - h_f \right)}{L_f(t)} \right\}. \tag{6.63}$$

After separating variables and integrating between the limits $(0, t)$ and $(0, L_f)$, we obtain

$$t = \frac{\Delta\theta}{K_S} \left\{ L_f - \left(h_o - h_f \right) \ln \left[1 + \frac{L_f}{\left(h_o - h_f \right)} \right] \right\}. \tag{6.64}$$

Notice that this solution does not allow $I(t)$ to be described explicitly. Such an implicit transcendental function is typical for all solutions embracing the Green and Ampt approach even when it is not apparent in some more sophisticated developments. When (6.64) is transformed with dimensionless terms

$$t^* = \frac{K_S t}{\Delta\theta \left(h_o - h_f \right)} \tag{6.65a}$$

and

$$I^* = \frac{I}{\Delta\theta\left(h_o - h_f\right)} \tag{6.65b}$$

we have

$$t^* = I^* - \ln\left(1 + I^*\right) \tag{6.66}$$

which can be evaluated graphically (see Fig. 6.24) or by computing t for a series of values of L_f with $I = L_f\Delta\theta$.

Because the procedure is simple, the Green and Ampt approximation has been widely used in research as well as in the solution of many practical engineering problems. It has also been applied to the description of infiltration into layered profiles and those having a crust surface. However, we have to keep in mind that real soils do not manifest a δ-function $D(\theta)$, and hence, the method offers results of disappointingly poor accuracy. For example, the error involved in predicting $I(t)$ or $q_o(t)$ can approach 30% in homogeneous soil profiles. Its use should be limited to those wanting only a convenient, rough estimate of infiltration.

Within the second category of approaches, Philip's (1957b) algebraic infiltration equation is the most common. This approximate equation is merely the first two terms of the series solution (6.38) with the cumulative infiltration $I(t)$ being

$$I = St^{1/2} + At \tag{6.67}$$

and the infiltration rate q_o (= dI/dt) being

$$q_o = \frac{1}{2}St^{-1/2} + A. \tag{6.68}$$

These equations like their parent time series solution (6.38), are applicable to relatively short times. The magnitude of A is ($A_2 + K_i + \varepsilon$) where ε is the truncation error for having used only the first two terms of (6.38). It was expected that A be related to K_S by a simple, sufficiently accurate relation $A = mK_S$. Although the most frequently used value of m is 2/3, its value ranges between 0.2 and 0.67 (Philip, 1987). However, detailed studies show that m depends upon both θ_i and time and sometimes exceeds a theoretical upper limit of 2/3. The error of estimate of K_S derived from A could theoretically reach about 30% in a relatively dry homogeneous soil (Kutílek et al., 1988). Therefore, (6.67) cannot be reliably used for estimating the value of K_S from infiltration tests.

Sorptivity S in (6.67) and (6.68) is an estimate of the theoretical value of sorptivity for a soil having initial water content θ_i. The truncation error influences the estimated value of the sorptivity to a lesser degree than that of A. Thus, S evaluated from the early stage of infiltration is considered a reliable value (Kutílek et al., 1988).

In order to reduce the truncation error, Kutílek and Krejča (1987) proposed to use three terms of the time series solution (6.38)

$$I = C_1 t^{1/2} + C_2 t + C_3 t^{3/2} \tag{6.69}$$

where C_1 is the estimate of sorptivity S, C_2 the estimate of ($A_2 + K_i$) and C_3 the value of ($A_3 + \varepsilon_1$) where ε_1 is the truncation error for having used three terms of (6.38). If we approximate the limiting time for which the truncated series (6.69) converges as the value of t when $dq_o/dt \to 0$, we have

$$t_{\lim} = \frac{C_1}{3C_3}. \tag{6.70}$$

And, if we make an additional approximation that $q_o(t_{lim}) \approx K_S$, we obtain the estimate

$$K_S \approx (3C_1C_3)^{1/2} + C_2. \tag{6.71}$$

Simplifying (6.42) in a manner similar to (6.67), Swartzendruber (1987) suggested using

$$I = \frac{S}{A_o}\left[1 - \exp\left(-A_o t^{1/2}\right)\right] + K_S t. \tag{6.72}$$

Substituting $4K_S/3S$ for A_o into the above equation, we obtain the two-parameter infiltration equation proposed by Stroosnijder (1976). On the other hand, if we consider only the first four terms of a series expressing the exponential term $\exp(-A_o t^{1/2})$, we obtain an equation identical to (6.69).

Equations (6.69) and (6.72) have similar disadvantages. Parameters C_2, C_3 and A_o are not simply calculated or predicted from known hydraulic functions K_S, $K(h)$ and $\theta(h)$. When those equations are used for estimating S and K_S from measured values of $I(t)$, the estimates are reliable only for a strictly homogeneous soil column with an initial condition $d\theta_i/dz = 0$. When the equations are applied to field data, significant, intolerable errors are sometimes apparent. For example, when K_S is being evaluated, physically unreal values of $K_S<0$ are sometimes obtained (Kutílek and Krejča, 1987, and personal communication from Krejča, 1989).

Brutsaert (1977) also began with the horizontal solution of Philip (1957a) and subsequently sought a correction for the gravitational force. He obtained

$$I = K_S t + \frac{S^2}{BK_S}\left\{1 - \frac{1}{\left[1 + \left(BK_S t^{1/2}\right)/S\right]^2}\right\} \tag{6.73}$$

and

$$q_o = K_S + \frac{1}{2} S t^{1/2}\left\{\frac{1}{\left[1 + \left(BK_S t^{1/2}\right)/S\right]^2}\right\}. \tag{6.74}$$

He considered values of $B = 1/3, 2/3$ or 1 each descriptive of physical reality, but for most practical purposes, recommended $B = 1$. Values of $I(t)$ computed with (6.73) are nearly identical to those from (6.38) when the hydraulic functions of the soil are known. And, inversely, estimates of S and K_S from the experimental data using (6.73) appear more reliable than those using other approximate equations based on comparative theoretical errors.

From this and the previous section, we conclude that both $I(t)$ and $q_o(t)$ can be quickly and reliably computed for trivial initial and boundary conditions with (6.45) and (6.46) or (6.73) and (6.74), respectively. The value of S is advantageously obtained using the approximate expression of Parlange (1975)

$$S^2 \approx \int_{\theta_i}^{\theta_s} (\theta_S + \theta - 2\theta_i) D(\theta) d\theta \tag{6.75}$$

or the iterative procedure described by Philip and Knight (1974) or by White (1989) for the solution of (6.51).

From the authors' experience, increasing the number of parameters brings a theoretical improvement especially when soil hydraulic characteristics are evaluated from infiltration tests. For example, if we deal with approximate equations based upon the infinite series solution, the truncation error is reduced with an increased number of terms. However, equations with three or more parameters are more vulnerable to the field soil not being homogeneous and the boundary conditions deviating from the trivial ones assumed in the theoretical development. This vulnerability is realized when a number of physically non-realistic parameters are obtained (e.g. a negative value of K_S in the inverse solution).

6.2.2.4 Empirical Equations

Historically, empirical equations have been used to describe a decreasing infiltration rate q_o as a function of time t. The shape of a smooth curve drawn through measured values of $q_o(t)$ was simply compared with that of an analytic function. Inasmuch as both equations and experiments were empirical, it is useless to try to physically interpret the coefficients of the equations. The coefficients have the character of fitting parameters only with no scientific merit (Haverkamp et al., 1988, and Kutílek et al., 1988). On the other hand, because of their popularity in the literature and their usage persists, we briefly present them here.

Kostiakov's (1932) equation of $q_o(t)$ is the hyperbola

$$q_o = c_1 t^{-\alpha} \tag{6.76}$$

with

$$I = \frac{c_1}{1-\alpha} t^{(1-\alpha)} \tag{6.77}$$

where c_1 and α are empirical coefficients. The value of c_1 should equal q_{o1}, the infiltration rate at one unit of time (usually 1 min), and $0 < \alpha < 1$. The equation does not describe infiltration at large times inasmuch as $q_o \rightarrow 0$ when $t \rightarrow \infty$. Mezencev (1948) overcame this inconvenience by shifting the q_o-axis

$$q_o = c_2 + c_3 t^{-\beta} \tag{6.78}$$

with

$$I = c_2 t + \frac{1}{1-\beta} c_3 t^{(1-\beta)} \tag{6.79}$$

where c_2, c_3 and β are empirical coefficients. With the shift, as $t \rightarrow \infty$, $c_2 \rightarrow q_{oc}$, the constant infiltration rate when quasi-steady infiltration is reached, and hence, $q_{oc} \approx K_S$. The infiltration rate after the first time unit $q_{o1} = (c_2 + c_3)$.

Horton's equation (1940) represents an exponential decay of $q_o(t)$

$$q_o = c_4 + c_5 \exp(-\gamma t) \tag{6.80}$$

with

$$I = c_4 t + \frac{c_5}{\gamma}\left[1 - \exp(-\gamma t)\right] \tag{6.81}$$

where c_4, c_5 and γ are empirical coefficients. In contradiction to the theory of infiltration for a DBC, q_o has a finite value at $t = 0$. As $t \rightarrow \infty$, $c_4 \rightarrow q_{oc}$ which yields a value of $c_4 \approx K_S$. With this approximation for K_S, the value of $c_5 = [q_o(0) - K_S]$ where $q_o(0)$ is q_o at $t = 0$. Inasmuch as Horton derived his equation for infiltration of a high intensity rainfall, the physical objection against a finite value of $q_o(0)$ is largely eliminated as we shall see in the next section.

Holtan's equation (1961) for a decay of q_o with I is

$$q_o = c_6 (W - I)^{\varepsilon} + c_7 \tag{6.82}$$

where c_6, c_7 and ε are empirical coefficients, $c_7 = q_{oc}$, W is soil water storage above an impeding layer and ε, not an integer, is most frequently greater than unity. Equation (6.82), incorporated into the USDA Hydrograph Laboratory model USDAHL, has empirical coefficients related to the soil mapping units in USA (Holtan and Lopez, 1971).

6.2.3 Unsteady Infiltration, Neuman's Boundary Condition (NBC)

When we describe rainfall infiltration, we consider that the REV is defined at the Darcian scale. Therefore, we do not describe individual raindrop events, but consider the rain as a continuous flux with the intensity of the rain q_r being the flux density passing either totally or at least partially through the surface of the soil. The boundary condition at $z = 0$ and $t \geq 0$ is formulated by the Darcy-Buckingham equation (5.31)

$$q_o = -K \frac{\partial H}{\partial z} \tag{6.83a}$$

or in diffusivity form (5.69)

$$q_o = K(\theta) - D(\theta) \frac{\partial \theta}{\partial z}. \tag{6.83b}$$

Condition (6.83), called Neuman's boundary condition, describes not only rainfall infiltration, but infiltration caused by sprinkler irrigation or by a special flux controlled technique, e.g. by a peristaltic pump providing a constant flux through a membrane placed upon the soil surface. For drip irrigation, we have a 2-dimensional problem with boundary conditions appropriately modified. Field measurements of infiltration with boundary conditions (6.83) are usually performed with rain simulators, see a review of Amerman (1983). Nozzles or hypodermic needles are used to produce drops similar to raindrops at a certain height above the soil surface. Regardless of how boundary condition (6.83) is achieved, the initial condition is kept the same as that for the DBC.

6.2.3.1 Description of the Process

We denote this description into the three categories (i) constant rain intensity $q_r > K_S$, (ii) constant rain intensity $q_r < K_S$ and (iii) rain intensity $q_r(t)$. In all three categories soil water profiles $\theta(z)$ at intermediate times do not resemble $\theta(z)$ during early stages of infiltration. The distinguishing feature is that the soil water content $\theta_o(t)$ increases at the surface with time.

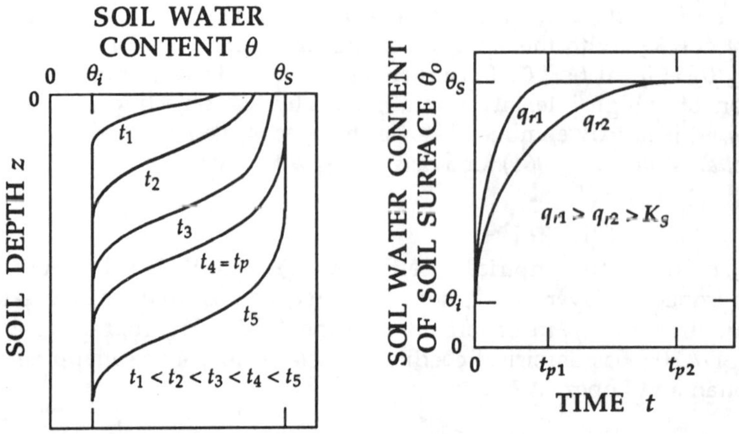

Figure 6.18. Constant rainfall infiltration with rain intensity $q_r > K_S$. Soil water content profiles $\theta(z, t)$ on the left, and the soil water content of the soil surface $\theta_o(t)$ increasing with time on the right. Note that ponding time t_p depends upon q_r.

Constant rain intensity $q_r > K_S$. The value of the soil water content of the surface θ_o increases steeply with time until it reaches θ_S, see Fig. 6.18. The greater is q_r, the steeper is $\theta_o(t)$. If rain continues, water ponds on the surface and the start of ponding is called ponding time t_p. If surface runoff is prevented, the depth of water on the surface $h_o(t)$ increases with time and $h_o > 0$ for $t > t_p$. With the increase of h_o being time dependent, $dh_o/dt < q_r$. The shape of the soil water profile at $t < t_p$ depends upon both q_r and the hydraulic functions of the soil, see Fig. 6.22. For $t = t_p$ the thickness of the saturated zone L_f extending below the soil surface is (Rubin and Steinhardt, 1964)

$$L_f = a \left(\frac{h_A K_S}{q_r - K_S} \right) \tag{6.84}$$

where a is an empirical parameter. For $t \geq t_p$ the soil water profile $\theta(z, t)$ resembles the profile with a DBC, i.e. with water ponded on the soil surface. Hence, we specify the boundary conditions as follows

$$q_r = K(\theta) - D(\theta)\frac{\partial \theta}{\partial z} \qquad z = 0 \qquad 0 < t < t_p \tag{6.85}$$

$$\theta = \theta_S \qquad z = 0 \qquad t \geq t_p \tag{6.86}$$

or, more exactly

$$h = 0 \qquad z = 0 \qquad t = t_p \tag{6.87}$$

and either

$$h = 0 \qquad z = 0 \qquad t > t_p \tag{6.88}$$

or

$$h = h_o(t) \qquad z = 0 \qquad t > t_p \tag{6.89}$$

Ponding time t_p separates the infiltration event into two different periods. The first is governed by the NBC (6.85) while the second is governed by the DBC (6.86) or (6.87) to (6.89).

Constant rain intensity $q_r < K_S$. For the entire duration of infiltration, the flow is governed by NBC (6.85) where the term t_p does not exist. On the soil surface the value of q_o is alwys less than that of q_S and hence, t_p cannot appear.

For *non-constant rain intensity* $\theta_r(t)$, we formulate the functional dependence $\theta_r(t)$ in the NBC (6.85).

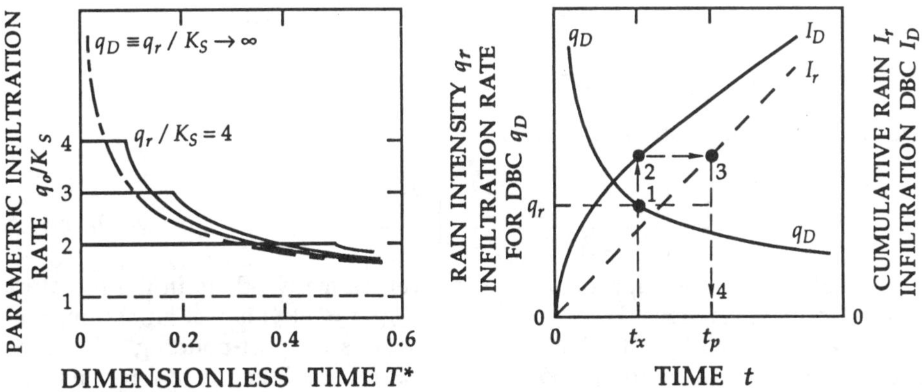

Figure 6.19. Infiltration rate versus time for several constant rainfall intensities $q_r > K_S$ (left). The estimation of ponding time t_p for a given q_r when infiltration rate q_D and cumulative infiltration I_D for DBC are known (right).

6.2.3.2 Approximate Solutions

Constant rain intensity $q_r > K_S$. For an approximate intuitive derivation of t_p we compare infiltration for a DBC with that for a NBC. The parameters will be indexed by D for DBC and by N for NBC. Rubin (1966) has shown that the ponding time t_p decreases with an increase of q_r and that $t_p > t_x$ where t_x is the intersection of q_r and q_D, see Fig. 6.19. In order to satisfy boundary conditions (6.85) and (6.86) as the NBC transforms to DBC, we assume the soil water profiles $\theta(z, t_p)_N$ and $\theta(z, t_x)_D$ are identical. Hence, we are searching t_x and t_p which satisfy two conditions. The first one is $I_N(t_p) = I_D(t_x)$, or

$$\int_0^{t_p} q_r(t)\,dt = \int_0^{t_x} q_D(t)\,dt. \qquad (6.90)$$

The second condition is

$$q_r(t_p) = q_D(t_x) \qquad (6.91)$$

and for a constant value of q_r we obtain

$$t_p = \frac{1}{q_r}\int_0^{t_x} q_D(t)\,dt \qquad (6.92)$$

where t_x is the time at which q_r and q_D intersect. Graphical interpretation of (6.90) and (6.91) is given in Fig. 6.19. With $q_D(t)$ expressed by Philip's approximate algebraic equation (6.68), we obtain

$$t_p = \left(\frac{S}{A}\right)^2 \frac{2Q_r^* - 1}{4Q_r^*\left(Q_r^* - 1\right)^2}. \tag{6.93}$$

where $Q_r^* = q_r/A$. By approximating $F(\theta_R) = \theta_R$, Kutílek (1980) also analytically derived this equation, see section 6.2.3.3. Similarly, $q_D(t)$ from other approximate solutions can be used to calculate t_p.

At $t \leq t_p$ the infiltration rate $q_o = q_r$. At $t > t_p$ the infiltration rate can be approximated by shifting $q_D(t)$ by $(t_p - t_x)$. With (6.92) and with

$$t_x = \left(\frac{S}{A}\right)^2 \frac{1}{4\left(Q_r^* - 1\right)^2} \tag{6.94}$$

we obtain the infiltration rate $q_o(t)$ for $t > t_p$

$$q_o = \frac{1}{2} S \left[t - \frac{S^2}{4A^2 Q_r^*\left(Q_r^* - 1\right)}\right]^{-1/2} + A. \tag{6.95}$$

All of the above approximations as well as many others in the literature have two disadvantages. First, the equality $I_N(t_p) = I_D(t_x)$ is just an assumption teoretically derived by Mls (1980) and the same is for post-ponding infiltration rates when they are computed as simple translations of the rate q_D. Second, the simple explicit formulation of $q_D(t)$ is, in reality, only an approximation. Therefore, the purpose of our above discussions was to illustrate as simply as possible the nature of infiltration under two different boundary conditions. The equations can be used for rough engineering estimations provided that the soil surface quality is not altered during the process, see section 6.2.4.1.

Constant rain intensity $q_r < K_S$. Here, the value of the soil water content on the surface $\theta_o(t)$ increases similarly to the first case but its limiting value is $\theta_f < \theta_S$. Inasmuch as $dH/dz \to -1$ at the soil surface as $t \to \infty$, we obtain a quasi-steady infiltration rate $q_o \equiv q_r = K(\theta_f)$. And, the value of $\theta_o(t)$ approaches θ_f asymptotically in time, see Fig. 6.20. The shapes of the soil water profile vary with both time and infiltration rate. In Fig. 6.22 we see the importance of the hydraulic characteristics of the soil for a constant infiltration rate at different times. This example from Broadbridge and White (1988) is for two different forms of the SWRC which are derived from the solution having an empirical coefficient C. The influence of the value of C upon the shape of the SWRC is shown in Fig. 6.22. The details of the solution are described by the authors.

Non-constant rain intensity $q_r(t)$. When $q_r(t)$ is strongly time dependent as it is in a great majority of heavy rainfalls, we obtain an estimate of t_p using the procedure described for a constant value of $q_r > K_S$. Note that t_x is no longer the intersection of $q_r(t)$ with $q_D(t)$ as it was derived for constant rain intensity infiltration. Equations (6.90) and (6.91) can be solved iteratively. Or, if $q_r(t)$ is capable of being expressed as a probability distribution function, we can solve them analytically. Regardless of the procedure, we recall that the solution remains only an approximation owing to the use of $q_D(t)$. Figure 6.21 shows the example of the graphical solution of (6.90) and (6.91) for ponding time t_p. By simply shifting q_D by $(t_p - t_x)$ we obtain estimates of the infiltration rate for $t > t_p$ and the hydrologically effective rainfall. The same priciple was applied for a

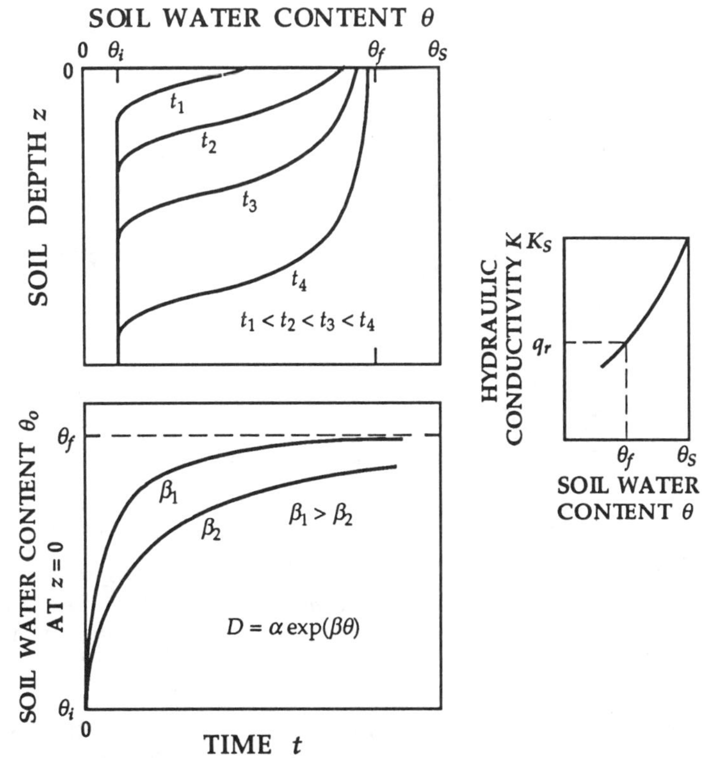

Figure 6.20. Constant rainfall infiltration with rain intensity $q_r < K_S$. Soil water content profiles $\theta(z, t)$ on the top. Soil water content of the soil surface $\theta_o(t)$ increasing with time on the bottom. The asymptotic value of θ_f at $t \rightarrow \infty$ follows from the condition $K(\theta_f) = q_r$.

histogram of rainfall intensity. Details regarding the construction or computation are described by Peschke and Kutílek (1982).

White et al. (1989) proposed the approximate analytic solution

$$t_p = \frac{1}{\bar{q}_r} M \frac{S^2}{K_S} \ln\left[\frac{q_r(t_p)}{q_r(t_p) - K_S}\right] \tag{6.96}$$

where $\bar{q}_r$ is the mean q_r during the time interval $(0, t_p)$ and $0.5 \le M \le 0.66$. More than a decade earlier Parlange and Smith (1976) had derived a very similar expression. Both expressions have features like those of the Green and Ampt solution. In order to avoid redistribution, these expressions require $q_r(t)$ not to decrease, see section 6.3.

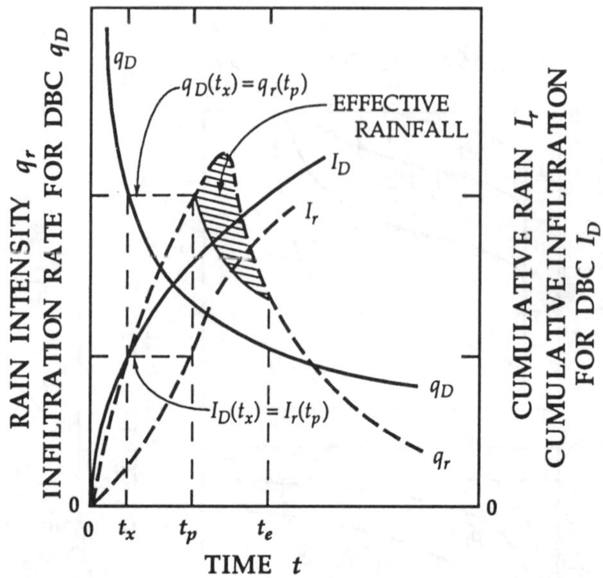

Figure 6.21. Time-dependent rainfall infiltration. Equations (6.90) and (6.91) were applied for estimating the ponding time t_p. With the exception of the time interval (t_p - t_e), the actual infiltration rate equals the rain intensity $q_r(t)$ if the excessive rain runs off. For $t_p < t < t_e$ the infiltration rate is obtained by shifting $q_D(t)$, i.e. the DBC infiltration rate.

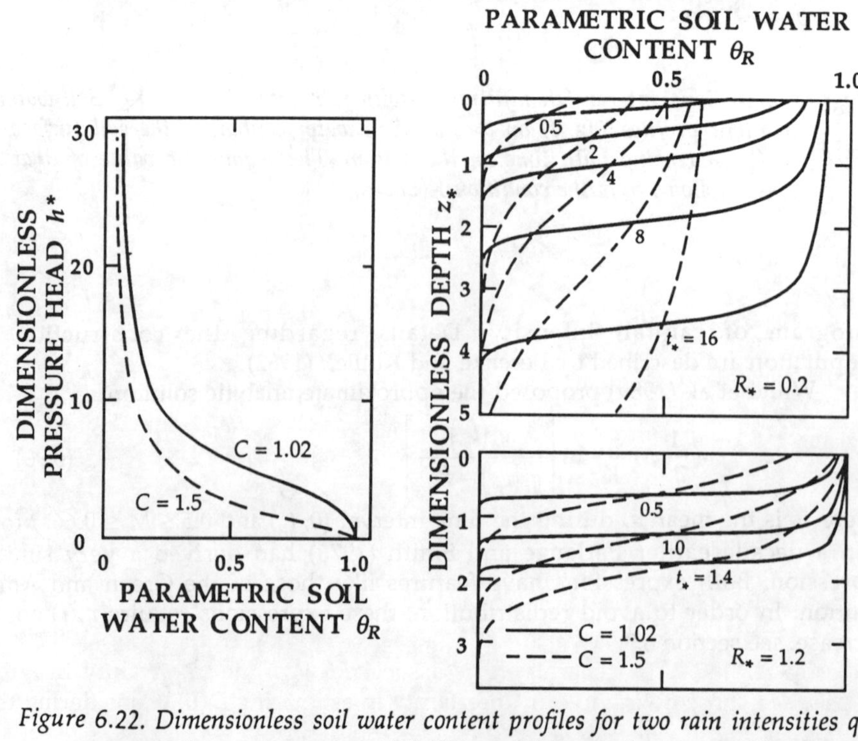

Figure 6.22. Dimensionless soil water content profiles for two rain intensities q_r expressed as $R_* = (q_r - K_i)/(K_S - K_i)$ and for two different soils characterized by the parameter C in the SWRC of Broadbridge and White (1988).

6.2.3.3 Analytical Solutions

If we neglect early approximate solutions based upon the Green and Ampt approach (e.g. Mein and Larson, 1973), four scientists eventually achieved an analytical solution for t_p. Parlange (1972) provided the initial effort which was subsequently modified by Philip (1973) and by Philip and Knight (1974). Philip and Knight utilized the concept of "flux concentration relation" [for convenience, (6.52) is repeated here]

$$F(\theta_R, t) = \frac{q - K_i}{q_o - K_i}$$

and neglected its time dependence which is even weaker than that for the DBC. When the above expression for $F(\theta_R, t)$ is inserted into the diffusivity form of the Darcy-Buckingham equation (5.69)

$$q = -D(\theta)\frac{\partial\theta}{\partial z} + K(\theta) \tag{6.97}$$

we obtain with $q_o = q_r$

$$F(\theta_R)(q_r - K_i) - [K(\theta) - K_i] = -D(\theta)\frac{\partial\theta}{\partial z}. \tag{6.98}$$

Integrating from z = 0 yields

$$z = \int_{\theta_i}^{\theta_o(t)} \frac{D(\theta)\,d\theta}{F(\theta_R)(q_r - K_i) - [K(\theta) - K_i]} \tag{6.99}$$

where $\theta_o(t)$ remains unknown. Integrating the equation of continuity (5.62) with q_r constant between the limits (0, t) and (0, z) gives

$$-(q_r - K_i)t = \int_{\theta_o(t)}^{\theta_i} z\,d\theta. \tag{6.100}$$

Combining (6.99) and (6.100) and integrating leads to

$$(q_r - K_i)\,t = \int_{\theta_i}^{\theta_o(t)} \frac{(\theta - \theta_i)D(\theta)\,d\theta}{F(\theta_R)(q_r - K_i) - [K(\theta) - K_i]}. \tag{6.101}$$

From (6.101) the evolution of the water content of the soil surface $\theta_o(t)$ is ascertained. When we know θ_o for a particular time t, we compute the soil water profile θ(z) from (6.99). If $q_r > K_S$ we compute the ponding time t_p from (6.101)

$$t_p = \frac{1}{q_r - K_i}\int_{\theta_i}^{\theta_s} \frac{(\theta - \theta_i)D(\theta)\,d\theta}{F(\theta_R)(q_r - K_i) - [K(\theta) - K_i]}. \tag{6.102}$$

A detailed step-by-step development of the procedure shows that the authors treat component equations of the Richards' equation and integrated the Darcy-Buckingham equation by using a guessed shape of the flux concentration relation $F(\theta_R)$.

Proper judgment for an appropriate value of $F(\theta_R)$ is critical. Philip's (1973) calculations of $F(\theta_R)$ for horizontal infiltration into linear and δ-function soils lead to approximations for early stages of infiltration limited to values of $q_r \gg K_S$. Another approximation for a constant flux infiltration was found assuming $F(\theta_R) = \theta_R^n$ with $0.8 \le n \le 1$ (Kutílek, 1980, Perroux et al., 1981, and Boulier et al., 1984). In general, errors associated with the uncertainty in $F(\theta_R)$ are less than those owing to our uncertainty in estimating $D(\theta)$. Considering the joint development of (6.102), the procedure should be called the Parlange-Císler-Philip-Knight (PCPK) method' see our introductory remarks and comments to (6.44) and (6.47).

Morel-Seytoux (1982) also provided an acceptable approximation for a constant intensity rainfall. For a variable rainfall his solution is restricted to soils having $K/K_S = \theta_R^n$ and to that initial part of a rainfall when its intensity is increasing with time. His method cannot be applied to the receding part of rainfall inasmuch as the procedure does not consider θ_0 decreasing as water redistributes within the profile. Generally, infiltration stemming from a variable rainfall can only be accommodated by numerical procedures.

The most versatile analytical solution of constant rate infiltration was published by Broadbridge and White (1988). They allowed the soil hydraulic functions to exist in broad limits between those of linear and δ-function soils. The hydraulic functions typical of real soils are expressed by a simple, free parameter C which is easily measured in the field. A detailed description of the procedure involving Kirchoff, Storm, Hopf and Cole and Laplace transformations is beyond the scope of this book. Their strictly analytical solution in parametric form given in Fig. 6.22 is most useful for testing numerical schemes. Moreover, the application of their solution to practical examples contributed to our knowledge of infiltration discussed in the previous section.

Numerical solutions have been reviewed by Vauclin et al. (1979) and van Genuchten (1981). Among improvements provided in the last decade are those of Mls (1982).

6.2.4 Field Infiltration

When field infiltration tests are performed and evaluated, we meet a complex set of effects not fully accounted for by exactly defined infiltration equations. These effects more or less influence the observed data and the applicability of infiltration theory to runoff hydrology, irrigation and other practical domains involving infiltration. Some of these effects will be discussed.

6.2.4.1 Soil Sealing and Crusting

During *ponded infiltration* tests (DBC), the abrupt contact of the soil surface with excess water causes weak aggregates to disintegrate and slake. The migrating smaller particles quickly form a seal on the soil surface within a short period of only a few minutes. The slaked clay particles are gel-like and exhibit a thixotropic behavior not yet fully studied in detail. We expect in the presence of mono-valent cations that this peptization of clays leads to the separation of individual sheets of clay minerals that subsequently reorient into horizontal, parallel configurations that tend to seal the soil surface causing an extremely small value of K_S. Bi-valent cations allow the clay sheets to remain coagulated and in a face-to-edge configuration. Hence, the value of K_S of the seal for bi-valent cations is larger than that for mono-valent cations, but nevertheless orders of magnitude less than that of the original soil. The value of K_S of the seal and its thickness are both time dependent. The formation and quality of this seal are major factors responsible for the difference in infiltration rates between structured and structure-less soils, see Fig. 6.23. Surface sealing is not a rare phenomenon – indeed, it occurs with virtually all arable soils during ponded infiltration.

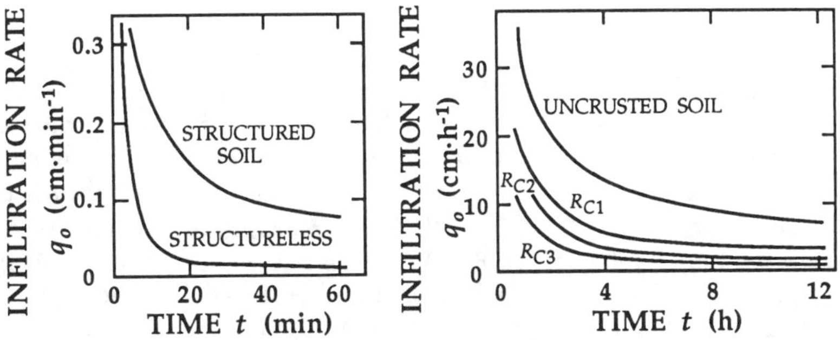

Figure 6.23. Infiltration rate $q_o(t)$ for DBC into a structured and structureless luvisol on loess (left) and $q_o(t)$ for a soil with different resistances (R_{C1} = 1, R_{C2} = 9.1 and R_{C3} = 17 d) on the surface (right) for a loess soil (Hillel and Gardner, 1970).

During *rainfall infiltration*, the slaking of aggregates is enhanced by the kinetic energy of the raindrops. The impact of raindrops upon the soil surface can be compared to a micro-bombardment. The drop after hitting the soil forms a micro-crater with some of the segregated fine soil particles relocating to clog pores and the remainder washed deeper into the soil with the infiltrating water. Inasmuch as the ponding time differs from one point to another, suspended clay particles are transported to small puddles and unevenly deposited across the soil surface. During the subsequent dry period after the rain, the newly formed seal consolidates and forms a crust. With repetitive rainfalls the process of sealing and crusting eventually forms a crust-topped soil. For such a soil crust, McIntyre (1958a and 1958b) defined two layers - a compacted thin layer called a skin and a less dense "washing-in layer". For an originally undisturbed soil having a value of K_S = 36 mm·h^{-1}, he reported a 0.1 mm thick skin having a K_S = 0.018 mm·h^{-1} and a 1.5 to 2.5 mm "washing-in layer" having a value of K_S = 0.115 mm·h^{-1}.

Mualem et al. (1990) defined two types of crusts. *Depositional crusts* are formed by fine particles settling from a suspension reaching a depositional site. The scale of this type of crust is related to the scale of observation. *Structural crusts* are caused by the destruction of soil aggregates exposed to the direct impact of raindrops as we discussed in the paragraph above.

Soil surface seals have relatively large values of bulk density ρ_T. The rate of increase of ρ_T depends upon the kinetic energy of the rain, nature of the soil and its aggregates and initial values of bulk density ρ_{Ti}. The smaller is ρ_{Ti}, the larger is the rate of change of $d\rho_T/dt$. Similar relations hold for K_S and other hydraulic functions of the seal.

Seal formation is more dependent upon rainfall energy than upon cumulative rainfall. Rainfall energy is closely correlated with rain intensity. The kinetic energy of rain which induces seals and crusts ranges from about 0.1 $J \cdot s^{-1} \cdot m^{-2}$ for a rain intensity q_r of about 0.3 $mm \cdot min^{-1}$ to 1.2 $J \cdot s^{-1} \cdot m^{-2}$ for $q_r \approx 2.5$ $mm \cdot min^{-1}$.

Once a seal is developed by a rain event, the physical properties of the seal are usually sustained. Repeated high intensity rainfall or sprinkler irrigation forms deleterious, undesirable crusts in the majority of agricultural soils. For example, below a 5 mm thick surface crust, Passerat de Silans et al. (1989) measured values of $\rho_T = 1.32$ $g \cdot cm^{-3}$ and $K_S = 1.3 \cdot 10^{-4}$ $m \cdot s^{-1}$. Within the crust $\rho_T = 1.45$ $g \cdot cm^{-3}$ and $K_S = 2.8 \cdot 10^{-6}$ $m \cdot s^{-1}$. Similar examples are reported in the literature, e.g. Callebaut et al. (1985). With the nature and extent of soil crusts being highly variable, their behavior falls between the one extreme of manifesting an earlier developed, constant hydraulic resistance and the other extreme of a gradually increasing hydraulic resistance as a seal develops during rainfall on an originally crust-free soil.

6.2.4.2 Infiltration into Crust- and Seal-Topped Soils

The first solution of infiltration into a soil profile having a surface seal of constant hydraulic resistance $R_1 = L_1/K_{S1}$ was proposed by Doležal and Císler (1969) assuming the Green and Ampt approximation, see (6.53). Analogous to (6.63) we obtain

$$\frac{dL_f}{dt}\Delta\theta = \frac{K_{S2}\left(h_o - h_f\right) + K_{S2}\,L_f}{K_{S2}\,R_1 + L_f} \tag{6.103}$$

where index 1 is for the seal, index 2 for the homogeneous soil below the seal and L_1 is the thickness of the seal provided that $L_1 \ll L_f$. After integrating between the limits (0, t) and (0, L_f), and using dimensionless terms (6.65) with dimensionless resistance χ^* where

$$\chi^* = \frac{R_1 K_{S2}}{h_o - h_f} \tag{6.104}$$

we obtain analogous to (6.66)

$$t^* = I^* + \left(\chi^* - 1\right)\ln\left(1 + I^*\right) \tag{6.105}$$

which is graphically demonstrated in Fig. 6.24. The primary disadvantage of this and similar approaches is the error introduced by the simplified conditions $L_1 \ll L_f$ and $\theta_2 = \theta_{S2}$ at the interface. When the wetting front is in the vicinity of the interface between the seal and soil, the soil water content θ_2 immediately below the interface is time dependent, see also Fig. 6.25. A similar approach was later published by Hillel and Gardner (1970).

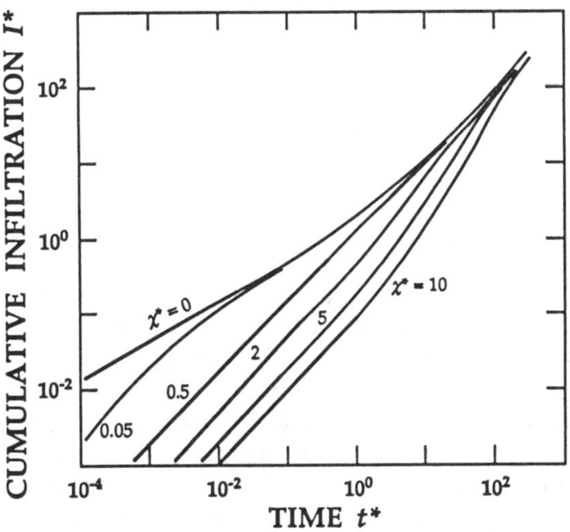

Figure 6.24. Dependence of dimensionless cumulative infiltration I^* upon dimensionless time t^* for DBC when the dimensionless resistance χ^* in (6.104) is defined on the soil surface (Doležal and Císler, 1969).

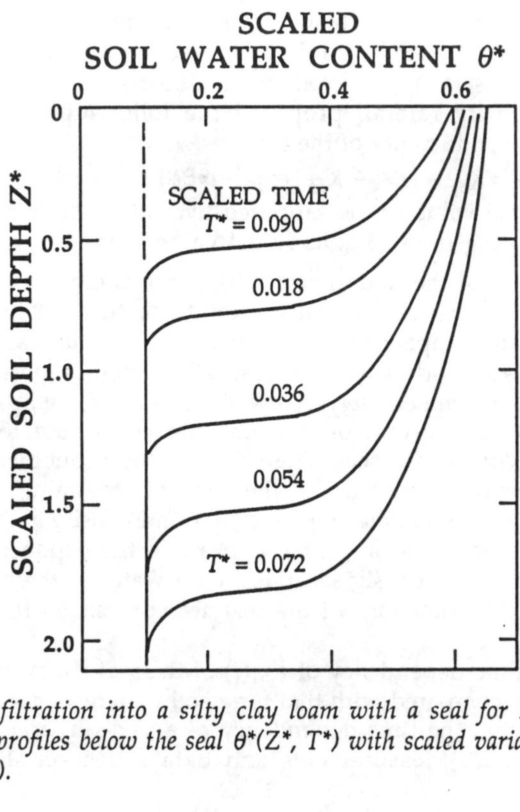

Figure 6.25. Infiltration into a silty clay loam with a seal for DBC. Soil water content profiles below the seal $\theta^*(Z^*, T^*)$ with scaled variables (Kutílek et al., 1991).

When the water capacity in the seal is neglected, the boundary condition at the seal-soil interface can be formulated as a Neuman boundary condition. Neglecting the seal thickness L_1 we write

$$q_o(t) = K(\theta) - D(\theta)\frac{\partial \theta}{\partial z} \qquad z = 0 \qquad t \geq 0 \qquad (6.106)$$

or

$$q_o(t) = -\frac{h(\theta) - h_o}{R} \qquad z = 0 \qquad t \geq 0 \qquad (6.107)$$

Here, the resistance transforms the original DBC to a NBC. The example in Fig. 6.25 (Kutílek et al., 1991) shows infiltration into a soil with a surface seal of resistance R. Water stored in the seal is neglected. The soil water profiles $\theta^*(Z^*,T^*)$ are typical for NBC, compare them with those in Fig. 6.20 or 6.22. We explain more about these scaled variables θ^*, Z^* and T^* in section 8.3.

For rain infiltration (i.e. for a Neuman boundary condition above the seal) a similar procedure leads to $\theta(z, t)$ analogous to that presented in Fig. 6.25. When the rain intensity $q_r > K_{S1}$, where K_{S1} is the saturated hydraulic conductivity of the crust, we obtain the reduced ponding time. The decrease of ponding time t_p with an increase of crust resistance R has a nearly exponential form. When R rises by one order of magnitude, the value of t_p may be reduced by nearly two orders of magnitude (Aboujaoude, 1991).

Up to now, we have considered approximate solutions of infiltration based upon the surface seal resistance being constant in time. However, rain-induced seals as well as their hydraulic functions are time dependent. Edwards and Larson (1969) observing an exponential decrease of K_{S1} (seal) with time during a high intensity rainfall proposed the following formulation of the saturated hydraulic conductivity of the seal

$$K_{S1}(t) = K_{SF} + (K_{So} - K_{SF})\exp(-\alpha E t) \qquad (6.108)$$

where K_{SF} is the final value of the saturated hydraulic conductivity of the seal (usually at the ponding time), K_{So} its initial value at the beginning of the rain (usually taken as K_{S2} of the underlying soil), E the energy of the rain and α a parameter which characterizes the susceptibility of the soil to seal formation including the eventual impact of water quality upon the slaking of the soil. Owing to the large, abrupt decrease in pressure head across the seal, microscopic bubbles of air released immediately below the saturated, compacted seal block the pores of the lower portion of the seal and contribute to a decrease in hydraulic conductivity of the seal. The effect of these bubbles is empirically included in the value of α. Inasmuch as the energy of the rain E is generally not measured, the term E is sometimes replaced by rain intensity q_r (and hence, $\alpha E = \alpha_1 q_r$) or the cumulative value of rainfall appears in the exponent. Inasmuch as the seal is almost instantaneously saturated with water at the beginning of the rainfall, other hydraulic functions of the seal are not needed for the infiltration solution.

Because the time dependency of $K_{S1}(t)$ of the seal shifts the ponding time t_p by only 10 to 20% compared with that for a soil manifesting a seal of constant hydraulic conductivity, the time dependency of a seal's hydraulic conductivity can often be neglected. Measurements and data collection should be aimed

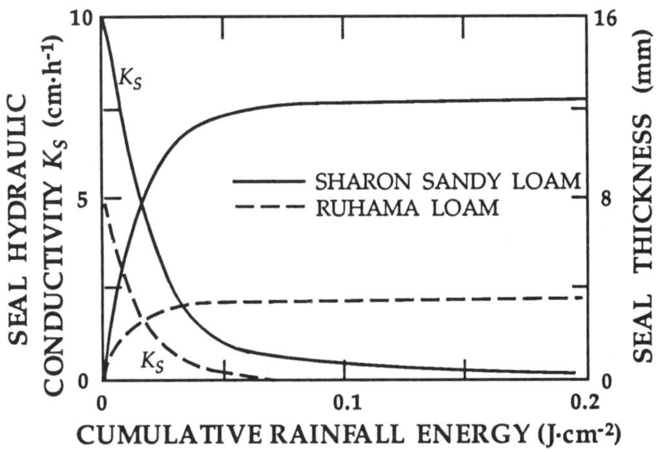

Figure 6.26. Hydraulic conductivity and seal thickness as functions of cumulative rainfall energy (Mualem et al., 1990).

primarily at the determination of K_{SF} in relation to the energy of the rain, see Fig. 6.26.

We denote the time dependency of the hydraulic conductivity of the seal described above as a short-time dependence. This short-time dependence of hydraulic functions of a soil wetted during infiltration remains more intuitive rather than being based upon exact evidence. In addition to a short-time dependence, a well-known large-time dependence of hydraulic functions exists mainly in the topsoil. The time scale of these changes embraces several months, an annual season or years. This large time infiltration variation is heavily dependent upon the plant canopy and the frequency and type of cultivation. Although the effects of cultivation appear to damp out after several months, the effect of root channels is more persistent (Gish and Starr, 1983).

6.2.4.3 Infiltration into Layered Soils

Infiltration is primarily dependent upon the relation of K in each of the layers to gradients of the respective pressure heads h. Let us consider infiltration under a NBC into a simplified profile consisting of only 2 layers: layer 1 on top is texturally finer than that of layer 2, e.g. loam over sand with $K_{S1} \ll K_{S2}$, where the indices 1 and 2 denote each of the layers. Because infiltration into a layered soil profile is dependent upon the relation $\bar{K}(h)$, see earlier Fig. 5.9, the sandy sub layer, even if more permeable at saturation, behaves like a less permeable layer at $h < h_c$. As the wetting front approaches, the hydraulic gradient should increase at the interface of the two layers. The increase is realized through an increase of h at the bottom of layer 1 which is reflected by an increase of θ above the value θ_o at the bottom part of layer 1, see Fig. 6.27. Although $h(z)$ is

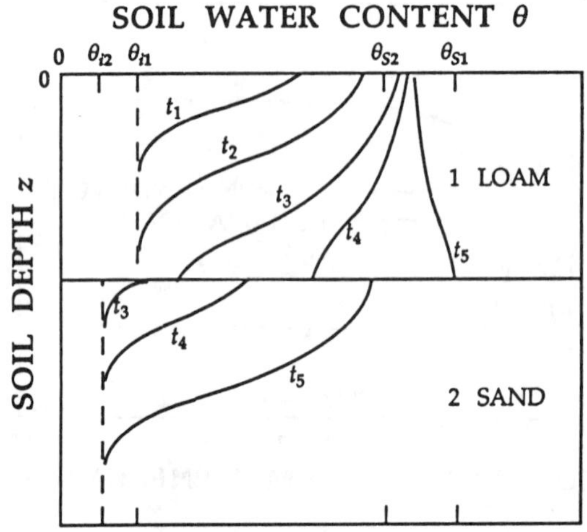

Figure 6.27. Soil water content profiles θ(z, t) for NBC-infiltration into a layered soil. At the interface between the layers, θ may reach θ_S in loam over sandy soil without saturation occurring at the soil surface.

continuous, $\theta(z)$ is discontinuous at the boundary of the layers. If indeed, there were a gravel layer instead of sand below the fine-textured topsoil, the water flux in these macropores and cavities would occur only if $h > 0$ above the interface.

This layering of a loam over a sand can also cause an instability of the wetting front visualized as finger-like wedges below a continuous wetting front during infiltration. The term fingering used by Hill and Parlange (1972) is commonly explained either by hydrodynamic instabilities (Diment et al., 1982) or by descriptive empirical approaches (Hillel and Baker, 1988). When the wetting front arrives at a textural or structural interface, h is too small in the upper layer to allow the entry of water into the coarse pores of the sub layer. As θ increases in the soil above the interface, h increases until it is large enough to allow water to enter clusters of fine pores of the sub layer, thus forming the nucleus of fingers. As the wetting front progresses, the average $\bar{h} > h_c$ with K_2 of the sub layer larger than K_{S1} of the topsoil. Because the topsoil supplies a limited amount of water $q_o \approx K_{S1}$, the sub layer conducts this water only through those originally preferential domains or clusters, and as a result, the fingers conduct the water and increase vertically in size. From the description of this fingering we immediately recognize an imperfection in our quantitative modeling of the soil porous system.

On the other hand, when the topsoil is a layer of loam or sandy soil over a clay sub layer, a zone of $h > 0$ is formed during infiltration governed by a NBC with $q_o < K_{S1}$ (topsoil). The height of this saturated zone in the bottom part of the top layer depends upon q_o and the ratio K_{S1}/K_{S2}. After a large infiltration time the $h(z)$ profile resembles that in the top part of the profile shown in Fig. 6.5. When infiltration starts with a DBC into an unsaturated topsoil underlain

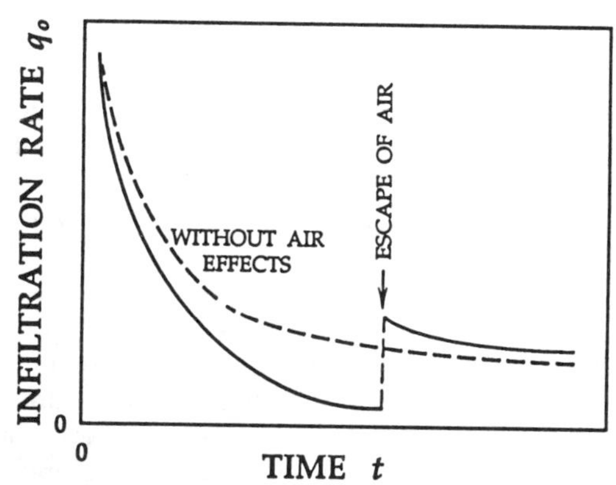

Figure 6.28. The influence of air compression ahead of the wetting front upon the infiltration rate $q_o(t)$.

by a clay subsoil which is either saturated or nearly saturated with water, the approaching wetting front displaces air giving rise to a pore air pressure ahead of the front that increases above atmospheric air pressure. At this time, the continuity of air-filled pores below the front with the atmosphere is interrupted. Hence, the air ahead of the wetting front cannot escape until eventually the increased pressure exceeds the air-entry value h_A of the soil. At that time, the monotonic decrease of q_o with t is interrupted giving rise to a sudden increase of q_o, see Fig. 6.28. On the other hand, if the increased air pressure does not lead to a bubbling of air through the wetted topsoil, the infiltration rate remains substantially decreased. The two-phase formulation of infiltration accounting for the effects of air and the behavior of the air pressure ahead of the wetting front for different values of θ_i has been presented by Morel-Seytoux (1976 and 1983).

6.2.4.4 Further Rainfall Infiltration Effects

Although the area of a rain-simulation plot is usually greater than the REV, local inhomogeneities together with a variable micro-relief oftentimes lead to some phenomena not yet satisfactorily described in theory. As an example, because the ponding time t_p is not well defined experimentally, Rubin and Steinhard (1964) defined three stages:

1. Stage A (absorption and retardation) occurs when the first perceptible retardation of raindrop absorption can be observed.
2. Stage P (puddle formation) occurs when about one-third of the soil surface is covered by puddles of rain water.
3. Stage C (completion of a water mantle) occurs at that instant when the last soil area not yet covered by a layer of free water disappears.

Irregularities of hydraulic properties at the soil surface also lead to the horizontal redistribution of free water on the soil surface which contributes further to a non-uniform soil wetting depth within the infiltrated domain (Clothier and Heiler, 1983). When the surface runoff of water from a small plot is observed, point heterogeneities lead to a suppression of the ponding time indicative of observable runoff. This suppression of t_p by local heterogeneities was confirmed by models of elementary runoff (Vauclin and Vachaud, 1990). A further factor influencing rainfall infiltration and runoff is the existence of preferential flow paths. Their existence renders the quasi-constant infiltration rate q_c dependent upon the rain intensity q_r with the ponding time depressed and practically independent upon q_r, see Fig. 6.29. In such cases observations of infiltration $q_o(t)$ are different than the theoretical function shown in Fig. 6.19 because the micro-catchment area of each of the preferential flow paths is q_r dependent and film flow exists in the majority of the preferential flow paths (Peschke and Kutílek, 1976).

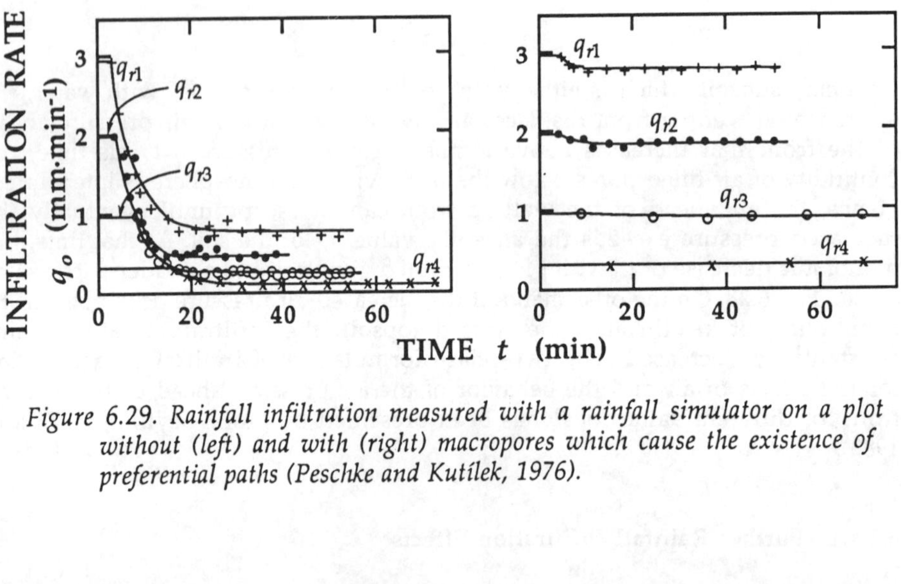

Figure 6.29. Rainfall infiltration measured with a rainfall simulator on a plot without (left) and with (right) macropores which cause the existence of preferential paths (Peschke and Kutílek, 1976).

6.2.5 Infiltration into Seasonally Frozen Soils

Within the season of frost additional factors exist which influence infiltration either directly from snow melt, or from rain combined with snow melt. We have shown in section 5.6.3 that frozen and unfrozen water co-exists at temperatures $T < 0°$ C. Under such a condition the value of the hydraulic conductivity is influenced by the ratio of those two phases of water and by the microscopic and macroscopic characteristic features of ice lenses in the soil, see Fig. 5.22. Generally, the hydraulic conductivity decreases strongly with a decrease in temperature. However, near 0° C the ratio of solid to liquid water is highly variable and under natural dynamic conditions, snow melt infiltration frequently occurs.

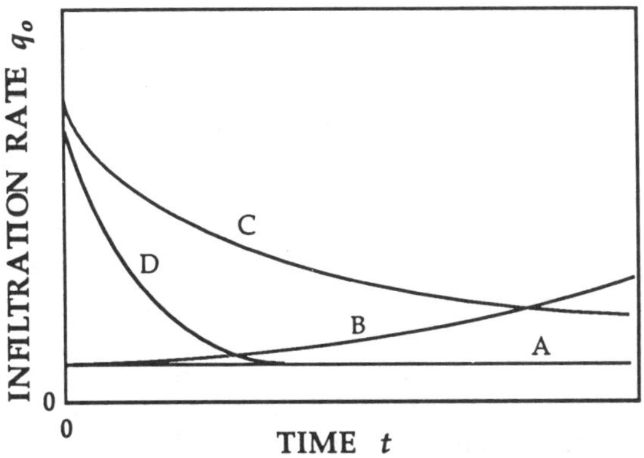

Figure 6.30. Four types of infiltration rates $q_o(t)$ into frozen soils (Gray and Male, 1981).

Because the flux of water is accompanied by a phase change, the simultaneous transfer of both heat and water must be considered. Additionally, the liquid and vapor phases of the water flux have to be treated separately. Water and heat transport processes have to be studied quantitatively in the period before infiltration starts inasmuch as they influence the initial water/ice content in the profile and the eventual occurrence of an impeding ice layer deeper in the profile.

Fig. 6.30 demonstrates the four basic types of infiltration curves $q_o(t)$ according to Gray and Male (1981):

> Curve A is for a soil which is frozen when initially water saturated, or when an impervious ice layer develops on the soil surface during a melt period. The infiltration rate, practically constant with time, is very small and approximately equal to the hydraulic conductivity of the soil frozen at the given temperature.
>
> Curve B is for a frozen soil having a large water content but distinctly below saturation. Some of the melt water penetrating deeper into the soil profile transfers heat and causes ice within soil pores to melt. Progressively, as the soil warms and the ice in the pores melt, the infiltration rate continually increases to values sometimes one-half an order of magnitude greater than that of the initial value of q_o. Eventually, the infiltration rate decreases because the topsoil approaches water saturation and the deeper layers of the frozen profile remain nearly impermeable.
>
> Curve C is for a frozen soil having a small water content and at a temperature near or above freezing. With only the small pores filled with ice which thaws rapidly with a downward movement of infiltrating water, infiltration proceeds in a manner similar to that for a soil in an unfrozen condition.

Curve D is for a frozen soil having a small water content and at a temperature well below freezing when snow at its surface is melting. Water initially infiltrates as in an unfrozen soil but after entering the extremely cold soil, it is frozen everywhere within the soil pores with its movement drastically inhibited. Here, the infiltration rate being very small is described by Curve A.

The curves given in Fig. 6.30 provide only schematical information that must be tempered further by variations which occur that are dependent upon local heat-water-transfer conditions. It should also be recognized that infiltration from snow melt is moderated or controlled by the flux of water through the snow cover. The boundary condition at the soil surface is therefore not easily defined as can be recognized from the model of melt-water infiltration in subfreezing snow (Illangasekare et al., 1990). Although the modeling of melt-water infiltration is feasible and useful for a detailed study and interpretation of observed phenomena, the predictive use of such models remains questionable.

6.3 SOIL WATER REDISTRIBUTION AND DRAINAGE AFTER INFILTRATION

After infiltration ceases, we can detect a gradual decrease of the soil water content even when the soil surface protected by a cover allows no evaporation. The decrease of θ within the originally wetted topsoil is caused by a downward flow of soil water. We distinguish two cases. First, the drier soil below the wetting front is at a great distance from the ground water level, or more generally, we assume that free ground water is absent. In this case, when water draining from the wetted topsoil wets the originally drier subsoil, we speak of *soil water redistribution*. The second case exists when the wetting front is close to the ground water level or reaches the ground water level with water flow at or near steady state conditions, see Fig. 6.1. When infiltration stops, transient conditions allow the excess water from the topsoil to pass directly to the ground water. This process is denoted as *drainage to the ground water level*.

6.3.1 Soil Water Redistribution after Infiltration

When infiltration ceases, a relatively large soil water potential frequently near $h = 0$ exists within the topsoil down to the depth of the wetting front z_f. Below the wetting front, the value of h is very small, and hence, the soil water profile is in a dynamic state rather than one of equilibrium. When we discuss this non-equilibrium process caused by infiltration, we denote the time of cessation of infiltration $t_o = 0$. The non-equilibrium produces a downward water flux within the soil profile without a contribution to the flux from the surface, i.e. at $z = 0$, $q_o = 0$. With the soil between $z = 0$ and $z = z_f$ being drained, water flows below z_f forming a new redistribution wetting front at z_r, see Fig. 6.31. The soil water content profile $\theta(z)$ at t_o is the profile at the end of infiltration. Subsequently, at

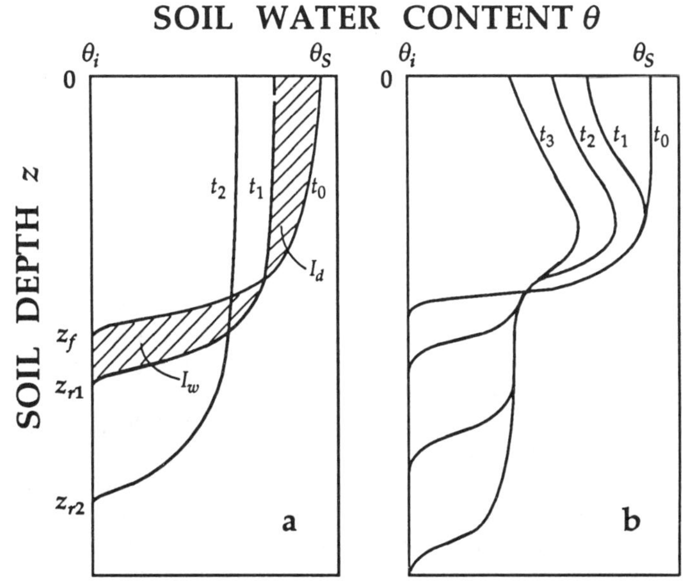

Figure 6.31. Soil water content profiles $\theta(z, t)$ during redistribution after infiltration.

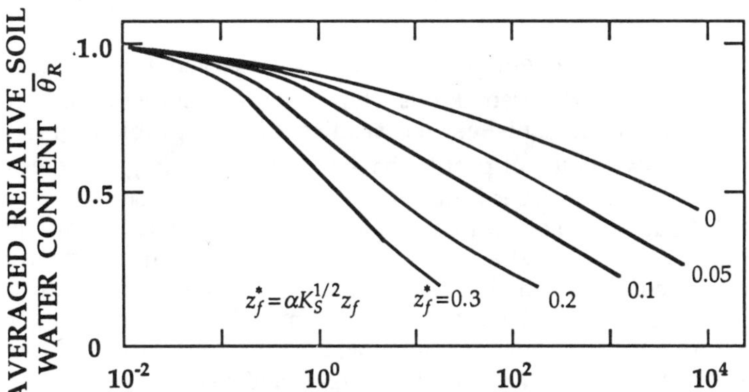

Figure 6.32. Averaged relative soil water content θ_R as dependent upon the depth of the wetting front after infiltration z_f^* and upon time t^* plotted in scaled quantities (Youngs, 1983).

times $t_1, t_2, t_3, \cdots$, redistribution profiles are observed. The cumulative drainage I_d and the cumulative wetting I_w below z_f are equal for the time interval $(0, t_1)$, see the hatched areas in Fig. 6.31a. As the redistribution front moves downward, the mean relative water content $\overline{\theta}_R$ $[= (\theta - \theta_i)/(\theta_S - \theta_i)]$ between $z = 0$ and $z = z_f$

$$\overline{\theta}_R = \frac{1}{z_f}\int_0^{z_f} dz \tag{6.109}$$

decreases with time, see Fig. 6.32. The rate of decrease depends upon the hydraulic conductivity and the depth of the original wetting front from the infiltration. The values of both are included in the scaled variables together with scaling factors α and β in the figure. The decrease of θ_R for the clay is much slower than that for sandy soils. If θ_i is relatively large, the force of gravity has a large influence. We note in Fig. 6.32 that the larger the value of $\Delta\theta$ ($= \theta - \theta_i$), the faster is the redistribution process. The flux density across the redistribution front decreases in time owing to two factors. First, the potential in the wetted topsoil decreases with a simultaneous decrease of the gradient of the total potential dH/dz. Second, the value of the hydraulic conductivity decreases strongly with only slight decreases of θ.

Inasmuch as water redistribution within a soil profile is composed of both drainage and wetting, the entire process is strongly influenced by hysteresis. Hysteresis not only slows down the redistribution process but prevents the soil from subsequently draining to a value of $\theta \leq \theta_i$ inasmuch as for a given h, $\theta_d > \theta_w$ (where the indices d and w again denote drainage and wetting, respectively). Inasmuch as $\theta_d(h_j) > \theta_w(h_j)$, redistribution profile $\theta(z)_{red} > \theta_i(z)$ even at equilibrium. Although a rigorous analytical treatment of redistribution is not feasible owing to hysteresis, approximate solutions of Gardner et al. (1970) confirm most of the above statements. Because analytical procedures are inadequate to predict redistribution in real soils, numerical methods are advisable (Rubin, 1967).

When soil water content profiles $\theta(z)$ are observed for particular time intervals $t_1, t_2, \cdots$, two different kinds of profiles are evident, see Fig. 6.31. Youngs (1983), reviewing his earlier studies of the problem, offered an explanation for the two kinds of profiles based upon the scaling of experimental data and hence, confirmed earlier proposals of Biswas et al. (1966).

In Fig. 6.31a, the soil water profiles during redistribution maintain approximately the same shape of the initial infiltration profile, but progressively become "slimmer" as the soil water content of the topsoil gradually decreases. The soil water content above the sharp wetting front is nearly uniform. This situation develops when the cumulative infiltration is small or when the saturated hydraulic conductivity K_S of the topsoil is relatively small, i.e. that of a fine-textured soil. On the other hand, in Fig. 6.31b, the redistribution profile $\theta(z)$ shows a relative maximum of θ at some depth above the original infiltration wetting front z_f. Below this wetting front persisting during the redistribution at z_f a step-like profile manifesting a small increase in soil water content $\Delta\theta$ advances into the deeper, drier soil. With depth, we note that the value of $\Delta\theta$ ($= \theta_r - \theta_i$) remains relatively constant as redistribution continues. The soil surface desaturated to a greater extent than that portion of the profile near z_f is assumed to be typical for large volumes of cumulative infiltration into a coarse-textured soils.

Although some authors assume for case (a) in Fig. 6.31 that computed values of $\theta(z)$ are obtained when hysteresis is neglected, this type of profile is obtained even with hysteresis when gravity is neglected. In both cases during redistribution , although θ decreases monotonically with time for $z < z_f$, the rate is however different. Below the position of the infiltration front the soil water content at $z > z_f$ first increases in case (a) to a certain maximum and then decreases slightly. In case (b), the soil water content increases monotonically to a constant value.

When a soil profile is layered, the redistribution soil water profiles are similar to NBC infiltration profiles of layered soils with $q_o < K_S$. Except when θ is close to θ_S, a sandy layer may either slow down or even stop redistribution. However, if θ_i was close to θ_S before infiltration and remains large just below the wetting front, a sandy layer in the subsoil may accelerate redistribution. During redistribution, large cavities and macropores will not be filled with water until $h > 0$. When a clay or loam layer is below a sandy soil, it may serve as an impeding layer and cause waterlogging above it.

The practical importance of redistribution is seldom fully appreciated. With the duration of redistribution being much longer than that of infiltration, redistribution plays an important role regarding soil water storage and its availability to plants. It contributes together with evaporation and transpiration to the initial boundary condition for the next infiltration and it contributes directly to the redistribution of solutes within the profile as well as their eventual transport to ground waters.

In addition to the redistribution which occurs after infiltration, redistribution can also occur during rainfall infiltration if the rainfall intensity $q_r(t)$ is strongly time dependent. We distinquish the occurrence of two cases. (i) After ponding time t_p has been reached, redistribution occurs during decreasing portions of $q_r(t)$ at times when q_r drops below K_S, provided that surface runoff is not inhibited. (ii) If $q_r(t)$ is sufficiently small that ponding never occurs, redistribution starts the moment q_r decreases and persists as long as $q_r(t)$ continues to decrease.

If surface sealing develops during rain infiltration, the hydraulic resistance of the soil surface increases. The time-dependent increase of the soil water content θ_o at the seal-soil interface is less pronounced than that of θ_o for a soil without a seal. Eventually, θ_o may even slightly decrease after reaching its maximum during infiltration. In such a case, infiltration is accompanied simultaneously with redistribution and hysteresis.

Detailed solutions of redistribution and its accompanying processes are available only from numerical methods.

6.3.2 Field Capacity

The practical importance of redistribution and its consequences for water stored in the soil profile was recognized early by soil physicists in the last century (Heinrich, 1886). The terms "field capacity", "specific retention" and "retention capacity" were coined to describe the amount of water held in the topsoil after a thorough wetting followed by excessive water draining (now we should say redistribution) to the drier subsoil. The downward movement of water is assumed to stop, or nearly so, within 2 to 3 days. This assumed state of equilibrium (or quasi-equilibrium) is verified in the field by measuring the soil water content as a function of time to ascertain the value of θ when $d\theta/dt = 0$. This field capacity value θ_{FC} was supposed to be a real, universal characteristic constant of a soil. However, when the theory of soil water flow is applied, it can be simply demonstrated that the assumption of universality and constant character of θ_{FC} is not a reality and that the value of θ_{FC} is a rough approximation to be used as a "rule of thumb". The value of θ_{FC} measured in the field for a given soil depends upon θ_i before thoroughly wetting the topsoil and upon the depth to which the soil was wetted before the commencement of redistribution. Because the rate at which $d\theta/dt \rightarrow 0$ is not unique even for a given soil, a rigorous quantification of θ_{FC} is not possible. For a detailed discussion and criticism of the field capacity concept, see Hillel (1971), Baver et al. (1972) and Cassel and Nielsen (1986).

In spite of our theoretical reservations the concept of field capacity is useful in the design of field management schemes for approximating soil water storage. Field capacity θ_{FC} is taken as an upper threshold value of θ within each soil layer such that any water in excess of θ_{FC} quickly drains to the next deeper soil layer. With this concept, the soil profile is considered a vertical sequence of reservoirs with the overflow level for each reservoir representing the value of θ_{FC} for each corresponding layer. During irrigation or rainfall the top reservoir is first filled up to the overflow and if more water flows in, the excess water flows over to fill the next lower reservoir etc. With a judicious selection of the depth of each soil layer, this simple analog of the soil profile is easily modeled and computed.

Inasmuch as field measurements are laborious and time consuming, the value of θ_{FC} is frequently estimated in the laboratory using soil core samples. The traditional European procedure calls for the sample to be first thoroughly saturated and then allowed to drain for a designated time, e.g. for 2 hours on a many layered mat of filter papers or on a layer of dry soil. Terms such as capillary or retention capacity are used to describe the characteristic value of θ derived from specified laboratory procedures (Kopecký, 1914 and Rode, 1952). The American school generally estimates θ_{FC} as the value of θ on the SWRC at 1/3 bar or less. Sometimes, especially for sandy soils, the value of θ corresponding to 0.1 bar defines field capacity. To avoid misunderstanding, the characteristic is called simply the 1/3 bar soil water content with its value recognized as an approximation of field capacity. In general, the European procedure provides a better correlation to field capacity than the 1/3 bar soil water content. However, in both cases, we must keep in mind that these laboratory-determined characteristics are two steps removed from reality. We first assumed that redistribution stops at field capacity. The second step substituted a laboratory procedure for a field test. Therefore, we should be fully aware of the possibility of the laboratory approximation not being a reliable estimate of θ_{FC} or the retention of water in a field soil profile.

6.3.3 Drainage to the Ground Water Table

Here, we differentiate between two conditions. For the first condition, we consider drainage from the soil profile after steady infiltration has ceased. With the ground water table kept at constant depth Z, the profiles $h(z)$ and $\theta(z)$ at $t = 0$ correspond to those in Fig. 6.1 except that now $z = 0$ is at the soil surface and the ordinate z is oriented down positive. Because $\theta(z)$ is greater than the SWRC [$\theta(z) > \theta(h)$] and $dH/dz < 0$, water flows downward to the ground water table. We recognize that the flux density q is time dependent based upon the following discussion. In the topsoil ($0 \le z \le z_c$) where $dh/dz = 0$ and $dH/dz = -1$, $q = K$. If we assume an exponential model (5.39) describes $K(h)$, we obtain

$$\ln q = A + ch \tag{6.110}$$

and for $K(\theta_R)$, (5.34) becomes

$$\ln q = A + n \ln \theta_R \tag{6.111}$$

where $A = \ln K_S$. Values of θ_R decrease rapidly at the beginning of drainage and eventually approach an equilibrium distribution within the profile $\theta_R(z) = \theta_R(h)$, the drainage branch of the SWRC. The position of z_c moves upwards with time. Neglecting the change in θ_R along the curved part of $h(z)$, we obtain

$$\frac{dz_c}{dt} \approx K_S\, \theta_R^{n-1}. \tag{6.112}$$

The rate of change of the flux $q(t)$ leads for the water storage $W = \theta z_c$ to the second derivative of θ with t. The total height of water drained is

$$W_D = \int_0^Z \theta\, dz - \int_{h=-Z}^{0} \theta(h)\, dh \tag{6.113}$$

where $\theta(h)$ is the SWRC.

The second condition of drainage to the ground water table is when water flows from a totally saturated soil profile. Swartzendruber (1969) modified the procedure of Gardner (1962) by linearizing Richards' equation

$$\frac{\partial \theta}{\partial t} = \frac{\partial}{\partial z}\left[K(\theta)\left(\frac{\partial z}{\partial \theta} - \frac{\partial h}{\partial \theta}\right)\frac{\partial \theta}{\partial z}\right] \tag{6.114}$$

using

$$N = -K(\theta)\left(\frac{\partial z}{\partial \theta} - \frac{\partial h}{\partial \theta}\right)\frac{\partial \theta}{\partial z} \tag{6.115}$$

for conditions

$$\theta = \theta_S \qquad 0 \le z \le Z \qquad t = 0 \tag{6.116}$$

$$\theta = \theta_S \qquad z = Z \qquad t > 0 \tag{6.117}$$

$$\frac{\partial \theta}{\partial z} = \frac{\partial \theta}{\partial h} \qquad z = 0 \qquad t \ge 0 \tag{6.118}$$

With the SWRC being simplified to

$$\theta = \theta_S + bh, \tag{6.119}$$

boundary condition (6.118) becomes

$$\frac{\partial \theta}{\partial z} = b \qquad z = 0 \qquad t \ge 0 \tag{6.120}$$

Equation (6.114) now transformed to

$$\frac{\partial \theta}{\partial t} = N \frac{\partial^2 \theta}{\partial z^2} \tag{6.121}$$

is solved using separation of variables. Neglecting all except the first term of the infinite series solution, we obtain

$$\frac{W(t)}{W_c} = 1 - \frac{32}{\pi^3}\exp\left[-\left(\frac{\pi}{2Z}\right)^2 Nt\right]. \tag{6.122}$$

With a weighted diffusivity $\overline{D}$ (Gardner, 1962) the above equation becomes

$$\frac{W(t)}{W_c} = 1 - \frac{8}{\pi^2}\exp\left(-\frac{\pi^2 \overline{D}t}{4Z^2}\right) \tag{6.123}$$

where

$$W_c = Z\theta_S - \int_{h=-Z}^{0} \theta(h)\,dh \tag{6.124}$$

or with (6.119)

$$W_c = Z(2\theta_S - bZ)/2. \tag{6.125}$$

By following procedures of Broadbridge and White (1988), Warrick et al. (1990) extended this solution to more realistic soils of a non-linear character with SWRC characterized by parameter C, see Fig. 6.22. However, the general features of the solution embracing the non-linear character of D do not differ from those of (6.122) or (6.123). In other words, the volume of water draining from the profile decreases exponentially with time and the deeper the ground water table, the slower is the drainage.

Collis-George and Yates (1990), reviewing earlier publications on the role of encapsulated air in drainage, found for $t > 0$ during the first stage of drainage that water leaves the soil without any pores being drained. During this period, drainage occurs as a consequence of the encapsulated air expanding with the value of h decreasing below 0. They designate this early behavior as the first stage of drainage from ponded soils. As we noted earlier, they state that both analytical and numerical solutions of the influence of encapsulated air are virtually impossible with only qualitative descriptions being feasible.

6.4 EVAPORATION FROM A BARE SOIL

Evaporation of water from bare soil is one of the simple processes of water loss from land to the atmosphere. It is not just an academic exercise of theoreticians trying to simplify the more complicated processes of water loss from land surfaces. Such a simple process occurs from plowed soils, from fallow land, from soil between tree and row crops and frequently from agricultural lands during non-vegetative periods. Evaporation involves three events: (i) the transport of water to an evaporating surface located either within the soil profile or on the geographic soil surface, (ii) a phase change from liquid water to vapor water and (iii) the transfer of water vapor from the soil surface to higher elevations within the atmosphere. And, in general, the process also includes the simultaneous transport of matter and heat according to conservative concepts of mass and energy.

In this section, we assume that an energy balance is implicitly included as evaporation occurs from a free water surface, and that the evaporation from a free water surface is identical to that from a saturated soil when both are subjected to equal atmospheric conditions. Theoretically, these two evaporation rates cannot be equal for the same atmospheric conditions because the properties of the two evaporating surfaces are different. The albedo, surface roughness, area of air-water interface, heat capacity and heat conductance each differ and lead to different surface temperatures of the free water and saturated soil. Hence, our assumption regarding the equality of evaporation from a free water surface and from a water saturated soil is merely an approximation which appears acceptable within the limitations of our experimental observations.

We denote the upward water flux density q_o across the soil surface as the evaporation E [LT^{-1}] and conventionally take E as a rate. Similar to the description of infiltration, we distinguish between steady and unsteady evaporation. The intensity of steady or unsteady evaporation is dominated by the evaporativity or potential evaporation E_p. It is quantified here as the evaporation rate from a free water surface. In this section we ignore temperature fields.

6.4.1 Steady Evaporation

Owing to the condition of steady flow ($\partial q / \partial z = 0$), we are solving the transport equation in its simplest form, i.e. the Darcy-Buckingham equation (5.31) for the boundary condition $z = 0$, $h = 0$ and for z positive upward from the ground water table similar to that described in section 6.2.1.1.

6.4.1.1 Homogeneous Soil Profile

For the exponential form of $K(h)$ [see (5.39)], we obtain an equation identical to (6.5) except that $q > 0$ and the sign in the denominator of (6.4) is changed. The solution is (Gardner, 1958)

$$\frac{z}{h} = \frac{1}{ch} \ln\left[\frac{K_S E^{-1} + 1}{K_S E^{-1} + \exp(ch)}\right] + 1 \tag{6.126}$$

where $q_o = E$. We obtain values of $z(h)$ for a given evaporation rate E, or we can compute E if h is known at any elevation z. The maximum possible evaporation rate E_{max} for a ground water table at depth Z is found with the assumption of the most extreme condition of dryness at the soil surface. We integrate (6.4) for $h \rightarrow -\infty$ at $z = Z$ and obtain

$$\frac{K_S}{E_{max}} = \exp(cZ) - 1. \tag{6.127}$$

If E_{max} in this solution is greater than the potential evaporation E_p, $h > -\infty$, and hence, we obtain $E = E_p$ from (6.126) using the appropriate value of h at $z = Z$.

When $K(h)$ is expressed in the hyperbolic form of (5.38), the solution for E_{max} for the boundary conditions above is (Císler, 1969)

$$Z = \frac{\pi b^{1/m}}{m \sin(\pi / m)}\left(\frac{a}{b E_{max}}\right)\left[\frac{a}{b E_{max}} + 1\right]^{(1-m)/m} \tag{6.128}$$

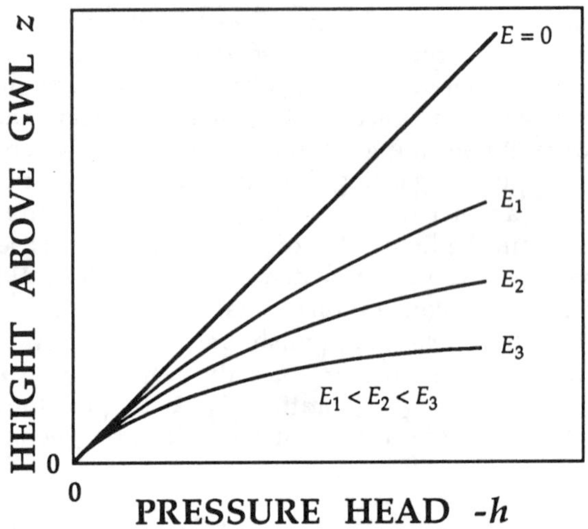

Figure 6.33. Steady evaporation from a soil with ground water level depth z. The distribution of pressure head h with z depends upon the rate of evaporation E.

Numerous solutions available in the literature have been reviewed by Raats and Gardner (1974) and Warrick (1988).

The distribution $h(z)$ within the soil profile during steady evaporation is illustrated in Fig. 6.33. If no evaporation is occurring ($E = 0$), $dh/dz = -1$ and $h = -z$. If $E > 0$, $dh/dz < -1$, and an increase in the evaporation rate is accompanied by a decreased hydraulic gradient dh/dz. Accordingly, $d\theta/dz$ also increases with an increase of E especially near the soil surface, where θ may approach the hygroscopic soil water content θ_H. If the top soil layer manifests a value of $\theta \leq \theta_H$, water also moves as a vapor. In this top layer water transport occurs predominately by diffusion which is a much slower process than liquid flow even for the same hydraulic gradient. Thus, this top layer functions as a strong hydraulic resistance that causes the actual evaporation $E_A \ll$ the potential evaporation E_p.

The real situation is far from the simplifications considered to obtain the solution above. Atmospheric conditions are never constant in time, evaporation rates from a free water surface usually follow a periodicity of 24 hours and the soil-atmosphere system is not isothermic. Nevertheless, steady evaporation solutions provide practical guidelines for some field situations. For example, they can be used to estimate the depth of the ground water table necessary to substantially reduce evaporation, to estimate the contribution from a given water table depth to supply water to plant roots or to estimate the critical depth of the water table to minimize soil salinization. Such guidelines for field soils can be improved by considering some more approximate solutions of steady evaporation from layered soil profiles.

6.4.1.2 Layered Soil Profiles

Steady evaporation from a layered soil profile is formulated similarly to that from a homogeneous profile where the vertical axis is oriented positively upward, i.e. in the direction of flow. The bottom layer (soil 1) is in contact with the ground water table where $z = 0$ and $h = 0$. At the soil surface the soil is assumed to be completely dry ($h \to -\infty$). The simplest case is that of two layers. At the interlayer boundary z_1 (see Fig. 6.34) it is assumed that h is continuous.

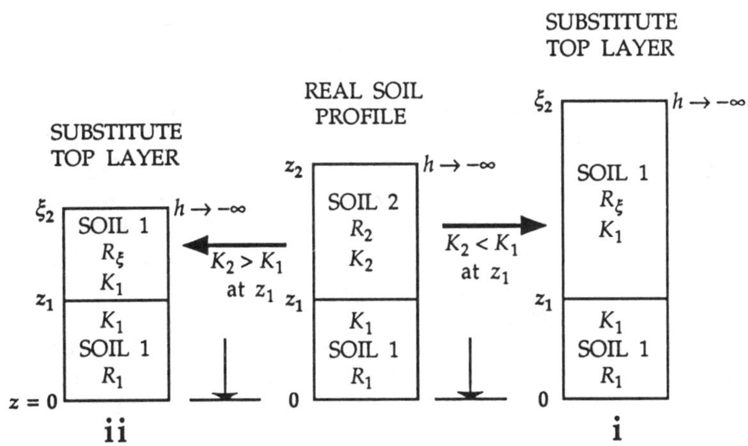

Figure 6.34. Steady evaporation from a layered profile is equal to that from a homogeneous profile of a substitute length ξ offering the same hydraulic resistance.

In the solution, two strategies exist. In the first one we search for the value of h_1 at the interlayer boundary z_1. Evaporation from the top layer is computed for boundaries z_1 and z_2 and for pressure heads h_1 and $-\infty$, respectively. By plotting separate graphs of flux density versus pressure head for the thickness of each layer, Willis (1960) matched the graphs to obtain the evaporation rate and the value of h_1 at z_1. From his graphical iterative procedure he concluded that the evaporation rate from a system of fine-textured soil overlying a coarse soil is only slightly greater than that from a homogeneous, fine-textured soil profile. On the other hand, a large effect on the evaporation rate was observed when the order of the layers was reversed. Willis' conclusions, valid for the conditions he studied, demonstrate the great variation of evaporation rate expected in relation to the hydraulic characteristics, length and relative position of the soil layers. Hadas and Hillel (1972) using the same strategy solved the problem by numerically integrating

$$\int_0^{z_1} dz + \int_{z_1}^{z_2} dz = \int_0^{h(z_1)} \frac{dh}{1+E/K(h)} + \int_{h(z_1)}^{h(z_2)} \frac{dh}{1+E/K(h)}. \tag{6.129}$$

For the second strategy, we consider the resistance of each of the individual layers in a manner similar to that for saturated flow in layered systems, see Fig. 5.2 and (5.9). For a 2-layer profile, indices 1 and 2 represent the bottom and top layers, respectively. Hence, the total resistance $R = (R_1 + R_2)$. Determining the resistance of each layer follows from the dependence $K(z, E)$ which, in turn, follows from $K(h)$ and $h(z, E)$. We distinguish two cases illustrated in Fig. 6.34. (i) This case illustrates a soil with a mulch on its surface or a freshly tilled soil having an increased porosity and an increase in the number of large pores that increase K_S. However, because K/K_S decreases rapidly with decreasing values of h, usually $K_1 > K_2$ at the interface where $h = h_1$. Evaporation from this two-layer profile (case i) will be the same as that from a homogeneous profile consisting only of soil 1 but of a length greater than z_2. Let us denote the elevation of this fictitious substitute surface of the homogeneous profile as ξ. Then $R_\xi = R_2$. Here for case (i), $\xi_2 > z_2$. When ξ_2 is found, the maximum evaporation is simply obtained from (6.127). (ii) This case illustrates a soil with a compacted top layer where K decreases with decreasing h much more slowly than in the subsoil. With the exception of some soil profiles that are water saturated up to nearly the soil surface, usually $K_1(h_1) < K_2(h_1)$. Evaporation from this two-layer profile equals that from a homogeneous profile of soil 1 having a length smaller than z_2. Here, $\xi_2 < z_2$. inasmuch as $R_\xi = R_2$.

For an approximate solution for both cases, we make three assumptions. First, we assume that soils 1 and 2 have an identical functional relationship for $K(h)$ with unique values of parameters that describe each soil. Later we assume that function is (5.39) with K_{S1}, c_1 for soil 1 and K_{S2}, c_2 for soil 2. Second, we assume the inequality of $K_1(h_1)$ and $K_2(h_1)$ properly characterizes the top soil layer as being either less or more permeable in the region $h_1 > h > -\infty$. Third, we assume that h_1 is very low and K changes nearly linearly between h_1 and h_{z2} with $K(h_{z2}) \to 0$.

With the resistance R_2 of layer 2 being first approximated, the substitute resistance R_ξ is found for substitute layer of soil 1. From $R_2 = R_\xi$, we obtain with further approximations (Kutílek and Mls, 1975)

$$\frac{\xi_2 - z_1}{z_2 - z_1} = \frac{K_{S1}}{K_{S2}} \left[\frac{exp(c_1 \xi_2) - exp(c_1 z_1)}{exp(c_1 z_1)[exp(c_1 \xi_2) - 1]} \right]^{\frac{c_1 - c_2}{c_1}} . \tag{6.130}$$

From the value of ξ_2 we can obtain the maximum evaporation rate E_{max} using (6.127) where K_S and c are data of soil 1.

If a soil consists of more than two layers, numerical procedures for the solution are appropriate. In general, when we wish to relate the depth of the water table to evaporation, we should consider the primary role played by the layer having the greatest resistance within the profile. The resistance is that obtained from the unsaturated hydraulic conductivity corresponding to the measured pressure head profile $h(z)$. Resistances estimated from ratios of saturated hydraulic conductivity values K_S for different layers only are obviously not correct. For example, a loamy sand layer in the middle of a loamy profile usually reduces evaporation provided that the layer is considerably above the water table. Or, a sandy layer above a water table and overlain by a loam layer reduces evaporation compared with that of a homogeneous loam profile, provided again that the boundary between the layers is well above the water table.

6.4.2 Unsteady Evaporation

Evaporation from a soil in the absence of a water table is a transient process even for the following most simplified initial and boundary conditions:

$$z = 0 \qquad t = 0 \qquad \theta = \theta_S \tag{6.131}$$

$$z = 0 \qquad t > 0 \qquad q_o = -E \tag{6.132}$$

$$z = L \qquad t = 0 \qquad \theta = \theta_S \tag{6.133}$$

$$z = L \qquad t > 0 \qquad q = 0 \tag{6.134}$$

Here, a soil column of length L initially saturated by water has its soil surface open to evaporation at $t > 0$ and its bottom at $z = L$ closed with no flux across it. Both experiment and theory have shown that three distinct evaporative stages of drying of a soil column exist, see Fig. 6.35 (Kossovitch, 1904, Lemon, 1956, Gardner, 1959, Gardner and Hillel, 1962).

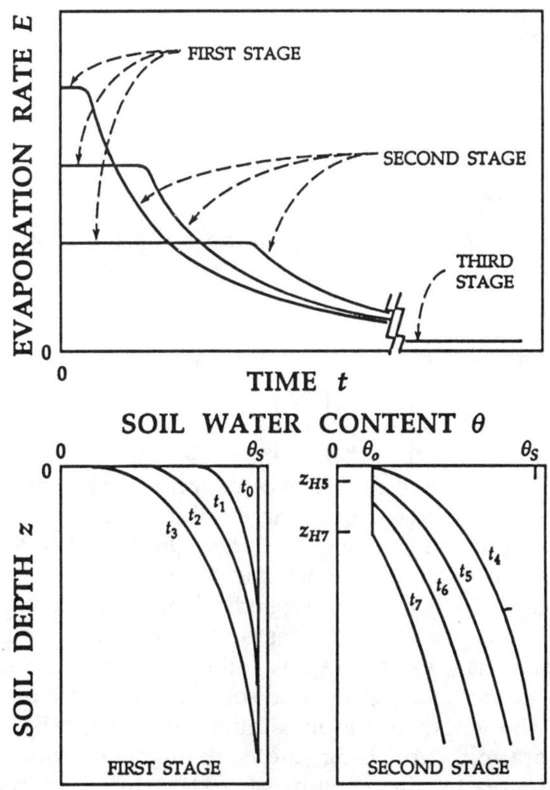

Figure 6.35. Three stages of evaporation from a soil during constant atmospheric conditions. Top: Evaporation rate E as a function of time t for three values of potential evaporation E_p. Bottom: Profiles of $\theta(z)$ for increasing times from t_0 to t_3 in the first stage and from t_4 to t_7 in the second stage of evaporation.

In the initial, first stage evaporation is governed exclusively by external atmospheric conditions and we denote it as potential evaporation E_p. This value of E_p is sustained as the soil water content decreases with time in the topsoil owing to the hydraulic gradient increasing sufficiently to compensate for the concomitant decreasing value of K. Eventually, as still more water is lost from the topsoil, the hydraulic gradient can no longer increase especially when the soil water pressure head approaches a value equivalent to the partial water vapor pressure p/p_o of the adherent atmosphere and $\theta = \theta_o$. At this time when the decrease of K is not compensated by an increase of the hydraulic gradient, the first stage of evaporation ends.

Gardner (1959) derived an approximate description of evaporation during the first stage by neglecting the gravitational force. Assuming water loss by evaporation from a soil profile of depth L is

$$E = -L\frac{\partial \theta}{\partial t} \tag{6.135}$$

and that the diffusivity equation (5.73) with the soil water diffusivity D expressed by (5.74), he obtained after still further approximations, the expression

$$W^* = ln\left(1 + \frac{1}{D^*}\right) \tag{6.136}$$

where

$$W^* = \frac{\beta W}{L} \tag{6.137}$$

and

$$D^* = \frac{2D_o}{E_p \beta L} \tag{6.138}$$

with $D_0 = D(\theta_0)$ and W the water storage in the entire soil profile at the end of the first stage $t = t_1$ being

$$W = \int_0^L \theta(z,t_1)dz. \tag{6.139}$$

Instead of (6.136), Kutílek (1978) provided the exact solution

$$W^* = (D^*+1)^{1/2} ln\left[\frac{(D^*+1)^{1/2}+1}{(D^*+1)^{1/2}-1}\right] - 2. \tag{6.140}$$

The difference between the two solutions can approach 30%, see Fig. 6.36. For a steep $D(\theta)$ relation characterized by a large value of β, the amount of water storage W at the end of the first stage is large. On the other hand for a $D(\theta)$ relation typical of a linear soil, W is substantially reduced.

The duration of the first stage depends upon both atmospheric and soil conditions. If the atmospheric conditions and therefore E_p are the same, the period of the first stage is greater for a clay soil than for a sandy soil, and greater for a structureless soil than for a structured soil. Therefore, the quality of the soil surface influences the length of the maximum evaporation E_p period and the cumulative evaporation. Mulching the soil shortens this first stage of evaporation and increases the amount of water stored in the soil profile. Depending upon the atmospheric conditions, the smaller the value of E_p, the longer is the period of the first stage. And, if the value of E_p is very small, the

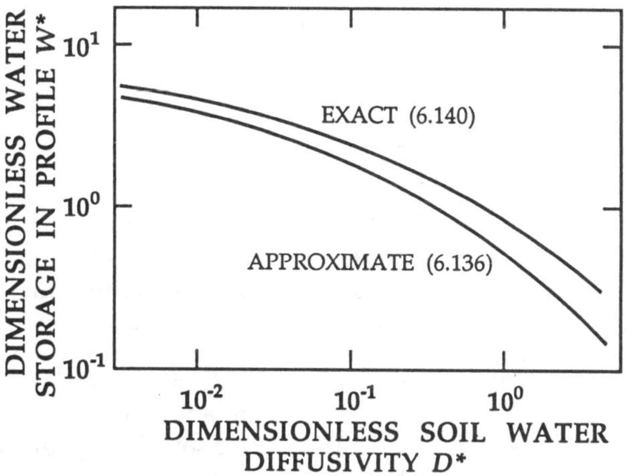

Figure 6.36. The approximate (Gardner, 1959) and the exact (Kutílek, 1977) solution of the first stage of evaporation with $D = D_o exp[\beta(\theta - \theta_o)]$ *plotted in dimensionless terms.*

soil profile is dried more uniformly with depth owing to the small hydraulic gradient required to provide water conduction through the profile.

The second stage of evaporation is characterized by a gradual decrease of the evaporation rate with time with actual evaporation E being less than potential evaporation E_p and the difference between both increasing with time. In addition to atmospheric conditions, evaporation depends upon the rate of transport of water from deeper parts of the profile to the soil surface. Soil water content continues to decrease with time as well as the value of the hydraulic conductivity. Because the increase of the hydraulic gradient is small and often negligible, it approaches a constant value in time, and hence, the rate of water movement to the soil surface layer as well as E decreases. With the soil water content near the soil surface being extremely small and near the value of the hygroscopic θ, the thickness of this dried surface layer increases with time. The progress of this dry front with soil depth is dependent upon $t^{1/2}$ with the water flux through the dry soil surface layer realized as water vapor transport achieved through molecular diffusion. Here, the dry layer acts as a hydraulic resistance and as its thickness increases with time, its resistance also increases.

For an analytical solution of this second stage of evaporation, Gardner (1959) applied solutions for diffusion published by Crank (1957). The description of the boundary conditions was simplified by assuming that the first stage of evaporation did not exist. Hence, it was assumed that $E_p \to \infty$ when the second stage starts at $t = 0$. Subsequently, Gardner uses (5.73) assuming that the influence of gravity is negligible and that $dD/dz = 0$. When D is expressed as a mean weighted diffusivity $\overline{D}$ in (5.80) for boundary conditions

$$\theta = \theta_i \qquad 0 \leq z \leq L \qquad t = 0 \qquad (6.141)$$
$$\theta = \theta_o \qquad z = 0 \qquad t > 0 \qquad (6.142)$$

the evaporation rate after additional approximations is

$$E = (\theta_i - \theta_o)\left(\frac{\overline{D}}{\pi t}\right)^{1/2}. \tag{6.143}$$

And, the cumulative evaporation is

$$E_c = 2(\theta_i - \theta_o)\left(\frac{\overline{D}\, t}{\pi}\right)^{1/2}. \tag{6.144}$$

The dependence of $E(t)$, demonstrated graphically in Fig. 6.35, is only approximately valid for very long soil columns. When the condition of uniform θ with z is not fulfilled, we can assume in accordance with a similar analysis of heat flow (Carslaw and Jaeger, 1958), that soon after the start of evaporation the terms expressing the influence of a non uniform $\theta(z)$ disappear.

In some later attempts to achieve an analytical solution, the term desorptivity [$LT^{-1/2}$] was introduced analogous to sorptivity. However, if $E_p \rightarrow \infty$ is not assumed, the solution for both the first and second stage of evaporation must be treated as a continuous process and without simplifying approximations such as neglecting gravity. Inasmuch as no appropriate analytical solution exists, our discussion here is based upon numerical procedures. Boundary condition (6.142) is replaced by

$$0 < t \leq t_I \qquad z = 0 \qquad q_o = E_p \tag{6.145}$$

$$t > t_I \qquad z = 0 \qquad \theta = \theta_o \tag{6.146}$$

where t_I is the time dividing the first two stages of evaporation, and a zero flux at $z = L$ is still maintained. Numerical solutions as well as experimental data show that E during the second stage depends upon E_p. Evaporation E_c accumulated from the soil during small values of E_p can exceed that during larger values of E_p. Such relations depend primarily upon the hydraulic functions and θ_i of the soil (Jalota and Prihar, 1990). The extent to which evaporation is reduced by mulching not only depends upon the soil hydraulic functions but upon the value of E_p. On relatively fine-textured soils when the value of E_p is small, shallow tillage or mulching greatly reduces evaporation for long periods of time. On the other hand, tillage or mulches do not alter cumulative evaporation from coarse-textured soils for large values of E_p. Between these two extremes a wide range of situations exist (Jalota and Prihar, 1990).

The simplifying assumption that a zero flux condition is maintained at depth L necessitates additional interpretation. When this condition is released, evaporation is accompanied simultaneously by redistribution, and hence, evaporation rates as well as the duration of the first stage are reduced. The redistribution rates are reduced substantially in the presence of evaporation compared with those in the absence of evaporation – the case of zero flux at $z = 0$ considered in section 6.3.

Solutions for evaporation considering isothermal conditions differ markedly from those accounting for non-isothermal conditions (e.g. Hadas, 1975). Errors approaching 20% or more, especially in layered soils are commonly made if temperature fluctuations are ignored.

Jackson et al. (1973) showed that the concept of three stages of evaporation does not strictly hold in field conditions. Diurnal temperature fluctuations and other atmospheric processes induce diurnal fluctuations of evaporativity in the field. During night-time when the air temperature is low, heat flow upward within the topsoil is accompanied by water flow. Additionally, inasmuch as dew causes the value of θ to increase at the soil surface, $E \approx E_p$ at sunrise for a brief time interval followed by a period later on when $E < E_p$. Moreover, cyclic periods of downward heat flow induces coupled downward water flow, especially water vapor to reduce actual evaporation. The overall impact of the temperature induced field effects causes the second (falling) stage of evaporation to begin even before the soil water content has reached hygroscopic values (Jackson et al., 1976).

Soil cracking and other inhomogeneities of field soils can influence evaporation up to 50% in a mild climate. In such cases, quantitative solutions are strongly influenced by thermal fields. For example, within cracks of relatively small depth, water vapor flows downward owing to thermal gradients (Hatano et al., 1988).

The time at which the third stage of evaporation begins is not well defined by the condition $E \to E_{min}$ or when $dE/dt \to 0$. A quantitative measure of the time at which these negligibly small evaporation rates occur within a meaningful fiducial limit is presently not feasible.

In field studies, the depth within the soil profile at which the water flux density is zero is of major importance and relatively easy to measure. It is commonly called the "zero flux plane" z_o. Above that depth, the soil water content decreases owing to evaporation. Below z_o any change of θ is attributed to redistribution. The position of z_o is identical to that depth where $dH/dz = 0$. Here, $H = (h + z)$ with h simply measured using tensiometers, see Fig. 6.37.

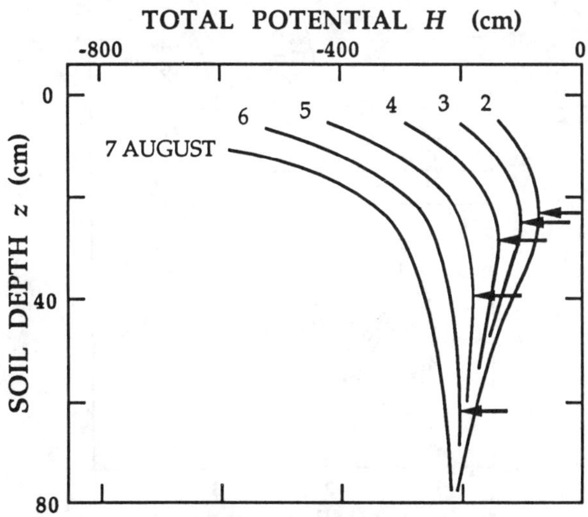

Figure 6.37. Total potential head H(z, t) measured in 1-day intervals during August with a mean evaporation rate E = 3.5 mm·day^{-1} from a soil profile having a silty loam overlying a fine sand. Positions of the zero-flux plane z_o are denoted by arrows (Thony et al., 1979).

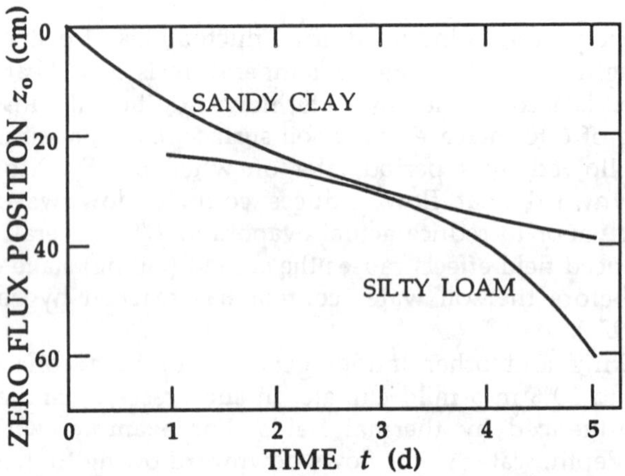

Figure 6.38. The progress of the zero-flux plane position with time in a sandy clay ferrallitic soil (Kalms et al., 1979) and in a silty loam fluvisol (Thony et al., 1979).

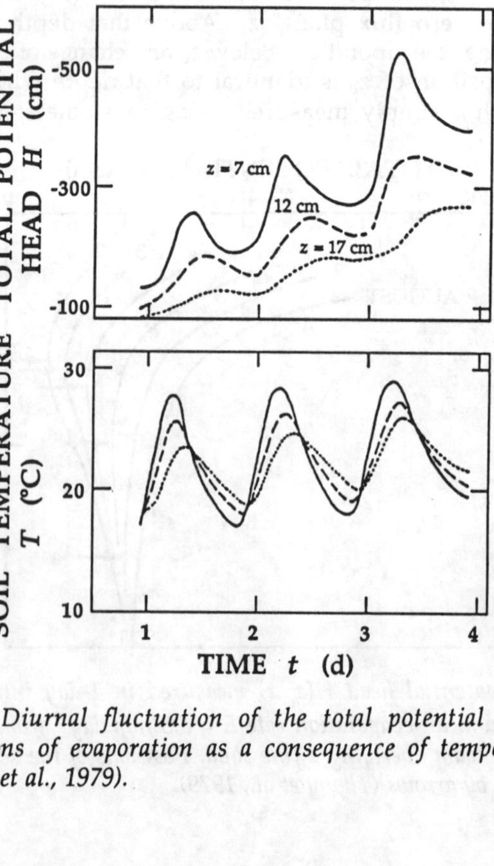

Figure 6.39. Diurnal fluctuation of the total potential head H owing to variations of evaporation as a consequence of temperature variations (Thony et al., 1979).

Although the zero flux plane gradually moves downward, a general relationship for $z_0(t)$ is not unique. Its functional shape depends upon the soil water content distribution before evaporation starts, the soil hydraulic characteristics and the evaporation rate. The position of z_0 as a function of time for two soils, a sandy clay and a silt loam, is given in Fig. 6.38. The data were collected for comparable evaporation rates by the same researchers using the same kind of equipment. Diurnal variations of evaporation dominantly influence the diurnal fluctuations of θ, see Fig. 6.39. The daily fluctuations of H as well as those of soil temperature are clearly evident as the average value of both increase during a 3-day period (Thony et al., 1979).

A better knowledge of the role of non-isothermal conditions is obtained when solutions of an isothermal process are compared with those of non-isothermal evaporation. For the latter case, coupled fluxes and the basic equations were shown in section 5.6.2. Milly (1984) made such a comparison using two types of boundary conditions at the soil surface: (i) a step-like change of θ and T and (ii) a sinusoidal variation of T. He first neglected the thermal water vapor flux component, and next the thermal liquid flux component. The thermal liquid flux is negligibly small in wet soils. When soils having water contents somewhat less than saturation are subjected to the step-like boundary condition, the thermal liquid flux does not have a dominant role near the soil surface. And, it is acceptable to ignore the thermal liquid flux in the prediction of the evaporation. The thermal and isothermal vapor fluxes for diurnal variation of temperature are approximately equal and of opposite direction in wet soil (Jackson et al., 1974). As the soil dries out, the net contribution of vapor transport increases. Errors induced by neglecting the thermal vapor flux are of the same order of magnitude as those for the total flux for short time periods. If the soil surface layer is dry ($h < -10^5$ cm), the thermal vapor flux is responsible for increasing evaporation during the night. On the other hand, neglecting thermal effects for a month introduces an error of only about 1% of the average evaporation rate. Hence, isothermal theory is adequate for estimating cumulative values of evaporation for long time periods. Detailed, non-isothermal models offer estimates for short time predictions but are plagued with uncertainties in the determination of the transport and the coupled coefficients. Recent formulations of coupled heat and mass flow by Passerot de Silans et al. (1989) are sets of equations describing the fluxes, transport coefficients, storage coefficients, aerodynamic resistances and terms for forced and free convection in full complexity. In addition to the classically described fluxes in soil, atmospheric fluxes of water vapor are considered at the same level of physical accuracy. The physics of evaporation into the atmosphere is concisely described by Brutsaert (1982).

6.5 EVAPOTRANSPIRATION

When a plant is rooted in a soil, evaporation E from the soil surface is accompanied by evaporation from the plant, which is commonly denoted as transpiration T_R. Water for transpiration is extracted from the soil by the root system. Macroscopically, we do not describe the influx of water into individual roots, but consider the flux averaged over a representative volume of roots as we did with the flux of water in soil without roots. Although detailed, microscopic studies of a single root-soil interface are highly informative, we keep the description of transpiration here on the same scale as that of evaporation, and hence, the root action is considered macroscopically. Influx into a root hair has its counter part in the description of water flow in an individual soil pore by Navier-Stokes equations. Here we use the macroscopic description of soil water extraction by the root system as a partner in the Darcian formulation of soil water flow. Therefore, the continuity equation (5.62) in its one dimensional form (6.2) is complemented by the extraction term S (Whisler et al., 1968)

$$\frac{\partial\theta}{\partial t} = -\frac{\partial q}{\partial z} - S(z,t). \tag{6.147}$$

The change of soil water content θ in time t is now caused by the change of flux q as well as the extraction of water by roots on a scale of the REV. The dimension of S is $[L^3L^{-3}T^{-1}]$ or, simply $[T^{-1}]$. Transpiration T_R is given by

$$T_R = \int_0^Z \int_0^t S(z,t)\,dt\,dz \tag{6.148}$$

where Z is the depth of the root zone. When (6.147) is combined with the Darcy-Buckingham equation, we obtain

$$\frac{\partial\theta}{\partial t} = \frac{\partial}{\partial z}\left[K(\theta)\frac{\partial H}{\partial z}\right] - S(z,t) \tag{6.149}$$

or some other expression equivalent to (5.64), (5.66) or (5.68).

Because the combination of the two processes (evaporation E and transpiration T_R) is not simply separable, we consider their sum unique and use the term evapotranspiration E_T

$$E_T = E + T_R. \tag{6.150}$$

In section 6.5.2.2 we shall learn that neither $E_T > T_R$ nor $E_T > E$ is a generally applicable condition, even if $E_T > T_R$ and $E_T > E$ occur very frequently.

Evapotranspiration has its deterministic elements in the evaporative demand of the atmosphere and in the transport processes of heat and water vapor from soils and plants through the sublayers which are next to the evaporative surfaces and through plant canopies to the outer atmosphere. The transport processes of heat and water vapor are treated in evapotranspiration principally in two groups of theories. In the first group, the entire plant canopy is considered a single, homogeneous layer characterized by bulk resistances which reflect all molecular transfers through boundary layers as well as turbulent transport. This consideration yields Penman-Monteith equations or their equivalent.

In the second group, the micro climate of the canopy is studied with turbulence within and above the canopy described using multi-layered models. This second approach, not fully covered here owing to its complexity and extensiveness, is considered more as a research topic while the first approach is commonly and extensively used in practice. Quantitative descriptions of individual phenomena and an objective explanation of the dynamics of evapotranspiration can be achieved with the second approach. An appreciation for the second approach is available in review papers (e.g. Thom, 1975, and Raupach, 1988).

We first limit our discussion to transpiration only in order to explain some of its special features, and subsequently in the following section, treat the more complex topic of evapotranspiration.

6.5.1 Transpiration

6.5.1.1 Transport of Water in Plants

Water extracted from soil by the plant root system is partitioned by the plant into that metabolized into plant tissue and that transpired into the atmosphere. The ratio of the mass of water extracted from the soil to that of the dry organic matter of the plant is denoted as the transpiration coefficient (or ratio). The transpiration coefficient for cultivated plants ranges from 300 to 900, and that for trees of the temperate zone from 150 to 700 (Penka, 1985). Water is the main constituent of plant tissues. Although the percentage of water in hydrophytes is about 80 to 85% and that in cultivated plants about 70 to 80%, the water content of individual cells or particular plant parts deviates greatly. The metabolically most active plant parts and young tissues contain about 90% water in relation to dry organic matter, leaves of cultivated plants 70 to 90%, wood of trees about 50 to 60% and seeds as little as 5 to 10% (Penka, 1985, and Slatyer, 1967). The water content in individual parts of the cell usually follows the sequence, cell wall < protoplasm < vacuole. Detailed studies of plant composition and transpiration [e.g. Slatyer (1967 and 1973) and Penka (1985)] lead to the conclusion that about 95% or even more of the water extracted from soil is transpired and flows through the soil-root-stem-leaves-atmosphere system. Water transpired through the tissues of the stem is relatively small. The concept, Soil-Plant-Atmosphere-Continuum (SPAC), describes the water flux through the above system.

According to SPAC, the driving force of the water transport is the gradient of the water potential. Resistances at the soil-root and leaf-atmosphere interfaces and those inside the plant, soil and atmosphere influence the actual fluxes, see Fig. 6.40. The analogy between water flux in the SPAC and current in an electrical resistance network illustrated in the figure was first suggested by Gradmann in 1928 and further elaborated by van den Honert in 1948 according to Molz (1981).

The water potential in plant tissues is defined analogously to that in soils. The total plant water potential is usually partitioned into two components: (i) turgor or turgor pressure p_T identified with a pressure potential and osmotic pressure p_O. Hence, as illustrated in Fig. 6.41,

$$\psi = p_T + p_O. \qquad (6.151)$$

Inasmuch as the cell volume is not constant, its reference value is taken as that volume at zero turgor ($p_T = 0$). An increase of turgor is accompanied by an increase of cell volume and a decrease of p_O to $\psi = 0$ at full turgor. A decrease of turgor owing to dehydration causes the cells to shrink and at a certain threshold

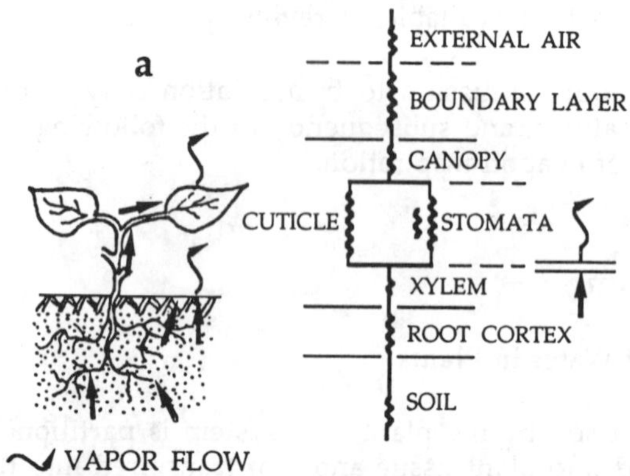

Figure 6.40. a. Schematic demonstration of the concept of soil-plant-atmosphere-continuum (SPAC) with the most important of the hydraulic resistances in the individual parts of SPAC. b. Profiles of the total potential head ψ in the SPAC according to Philip (1957d): 1. normal transpiration, 2. temporary wilting and 3. permanent wilting. Points along the transpiration path are A. a definite distance in the soil away from plant roots, B. surfaces of root hairs and of absorbing epidermal cells, C. the cortex, D. the endodermis, DE. vessels and tracheids in xylem, E. leaf veins, F. mesophyll cells, FG. intercellular space and stomatal cavity, GH. stomatal pore, HJ. laminar sublayer, if present, JK. turbulent boundary layer and KL. the free atmosphere. DPD_W denotes the value of at incipient plasmolysis of root cells.

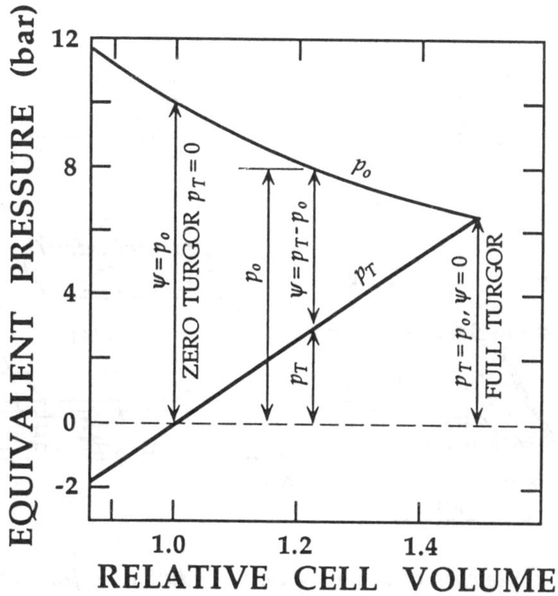

Figure 6.41. Relationship between turgor pressure p_T, osmotic pressure p_O and water potential ψ in an idealized osmotic cell (Höfler 1920, after Slatyer, 1967).

value is macroscopically observable as the wilting of leaves. Complete wilting occurs at $p_T \approx 0$ with p_O occurring in a broad range from 0.5 to 20 MPa (Slatyer, 1967). The total potential ψ at permanent wilting can manifest a broad range of values that depend upon the species and stage of development of the plant, the actual water stress resistance of the plant and the local environmental conditions. This extreme decrease of p_T is accompanied by irreversible morphological changes inside of the cells. Some plant physiologists still denote the absolute value of the water potential of plants as the diffusion pressure deficit $DPD = -\psi$ or $[= (p_O - p_T)]$.

Although the above description of SPAC may appear mechanistically restrictive, regarding processes within and between cells of living organisms, it does not exclude detailed descriptions of metabolic systems which regulate resistances, permeabilities of cell membranes, etc. Resistances, especially those in roots and leaves which vary with transpiration rate, can also be included. A full description of these mechanisms not presented here can be found in the literature on physiology of water regimes in plants.

As a first approximation to a simplified but realistic cross-section of a cell (Fig. 6.42a, b), we show mechanistic models of water transport in series of cells, see Fig. 6.42c, d and e. The first model in the series of cells considers the simple transport from vacuole to vacuole (Fig. 6.42c) as was originally proposed in a quantitative study by Philip (1958a). Although this series is realistic for water

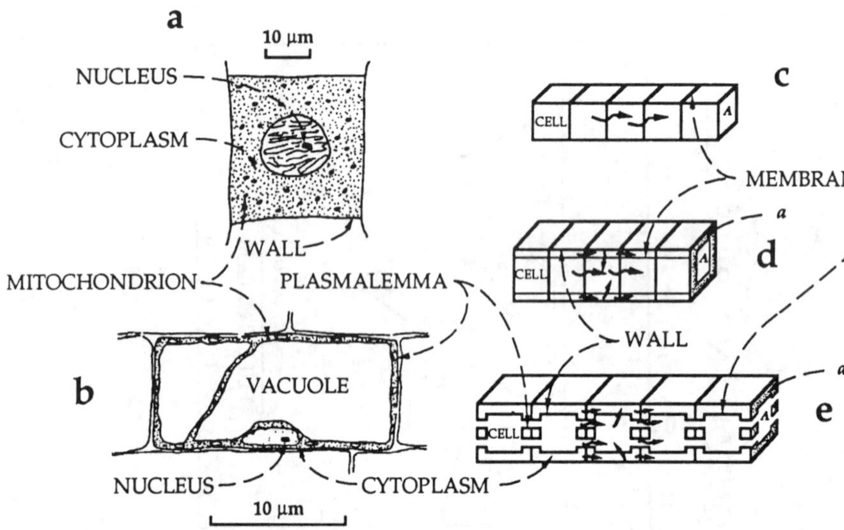

Figure 6.42. Left: Sketches of a. young and b. adult cells from higher plants (Briggs et al., 1961, after Slatyer, 1967). Right: Models of water transport in cells, c. from vacuole to vacuole, d. with contribution of cell walls and e. with contribution of cytoplasm of the cell walls (Molz, 1981).

flow in young cells, it includes a contradiction for adult cells. That is, inasmuch as the cell walls parallel to the direction of flow do not contribute to the total flux, only the perpendicular walls are permeable. Transport in cells is included in the next model given in Fig. 6.42d. In the most complex model, Fig. 6.42e, water transport in cytoplasm realized through plasmodesmata in the walls (cytoplasm pathway) is accompanied by parallel transport in the walls (apoplasm pathway). The main fluxes do not exclude the coexistence of water exchange between the cell wall and the vacuole. Molz (1981) conducted quantitative studies primarily focused on water transport in roots. In spite of the above simplifications, these models provide quantitative insights into the water regime of plants, especially when further developments included the effects of diffusible solutes.

Water enters the plant primarily through the most active part of young roots and root hairs. Hence, if L is the length of the roots, dL/dt is primary information on the proportion of active roots. The cross section of a young root given in Fig. 6.43 allows visualization of the pathway of water entering a root. Water enters the root through root hair cells and through the epidermis of young roots. Its path through the cortex is the location of greatest hydraulic resistance. With this resistance regulated by the plant itself, Molz (1981) concluded from his models that it takes about a half hour for the root cortex to respond fully to a boundary condition change at the endodermis. Thereafter, with quasi-steady fluxes being apparent, water potential gradients in the cortex were greater than those in the neighboring soil. Hence, we should correct the

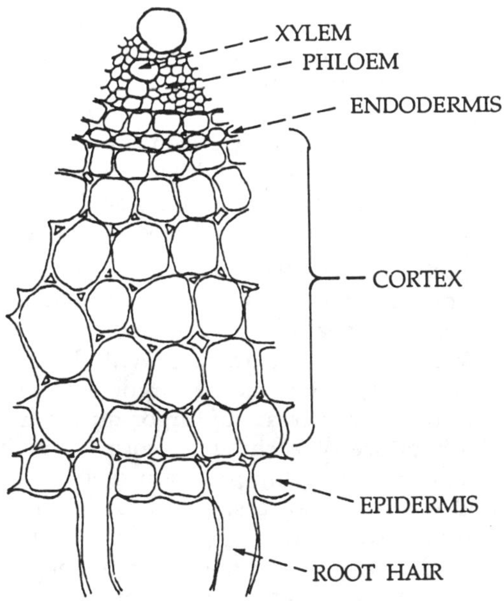

Figure 6.43. Cross section of a young root (Kramer 1959, after Slatyer, 1967).

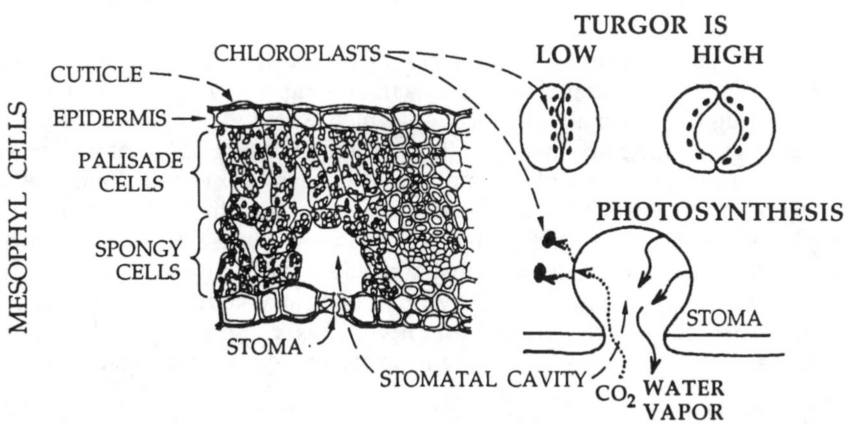

Figure 6.44. Left: Cross section of a leaf (Meyer and Anderson, 1952, after Slatyer, 1967). Right: Closed and opened aperture of stoma and gas exchanges in the open stoma.

original graph of Philip in Fig. 6.40b. Water, after passing the cortex and endodermis, enters the xylem of the root which has a very small hydraulic resistance. Aging of a root is accompanied by cutinization and suberization with cork eventually completely replacing cortex in the roots of woody species. The hydraulic resistance of those aging tissues drastically increases.

From the root xylem water enters the stem xylem where it easily and continuously flows at a very small resistance. Xylem can be considered the main "river bed" of water transport in plants.

The dominant locality of transpiration is the leaf. A schematic cross section is in Fig. 6.44. The surface of a leaf is constituted by an epidermis having its outer cell walls thickened and lined with cutin and further covered by a "crusty" often waxy cuticle. This epidermal outer surface has a large hydraulic resistance. In the epidermis, micro holes or pores called stomata exist at a density of 50 to 500 mm^{-2} (Slatyer, 1967). Stomata occur on both surfaces of the leaf of cultivated plants, but they may occur more commonly on the lower surfaces of tree leaves. The shape, opening and closure of each stoma is regulated by its two kidney-shaped guard cells. With the turgor in the leaf cells transferred to the guard cells, a decrease of turgor deforms the guard cells to initiate closure of the stoma. An increase of turgor functions an opposite way to open the stoma. The aperture of each stoma forms a pathway for diffusive transfer of water vapor. However, the function of the guard cells including stomatal opening and closure is not a simple process that reacts only to cell turgor and simply to plant water stress, see Zeiger (1983). Sensors below each stoma react to concentration of CO_2 in the air, radiation, temperature etc. For example, the presence of toxic gases in the air causes stomatal closure. Potassium ion concentration and proton transport across the membrane of guard cells play a dominant role regarding the dynamics of opening and closing of stomatal apertures (Raschke,1975). The proton transport is accompanied further by biochemical reactions.

The actual transformation of water from the liquid to the vapor phase takes place within two domains of the leaf, see Fig. 6.40a. The first domain is that of the outer epidermal wall through which water vapor flows through its cuticle with great difficulty owing to its resistance r_c being constant and of very large value. The upper part of the mesophyll cells is the second domain where water vapor flows through the air-filled space below and through stomata to the atmosphere. Inasmuch as the resistance r_S is determined by the area of the stomatal aperture, its value is not constant. Water vapor moves readily through stomata regardless of them being fully open or nearly completely closed. Even when they are closed, the flux through them is roughly equal to the flux through the cuticle. The total resistance to vapor transport by the leaf r_L is

$$\frac{1}{r_L} = \frac{1}{r_S} + \frac{1}{r_c} \qquad (6.152)$$

where r_S is the stomatal resistance and r_c the cuticle resistance which does not change substantially. The primary self-regulating mechanism of transpiration by a plant is realized through r_S. With a gradual closing of the stomata, r_S increases and transpiration decreases. Transpiration does not stop however, even if the stomata close completely. And, if the transpiration flux exceeds the influx into the roots, the plant tissues loose water and turgor. If the loss of turgor continues beyond a critical threshold value, wilting occurs. Wilting can occur even if the soil is fully moist when the evaporative demand of the atmosphere is large and the influx to the roots is lower than the transpiration loss. Often during a hot summer midday, plants having large leaf areas (e.g., the sugar beet) temporarily wilt. Subsequently, in the evening hours when the extreme air conditions cease and the transpiration rate decreases, turgor in the leaf tissues returns. When extreme air conditions prevail or when the soil remains relatively dry for several days, the loss of turgor in cells accompanied by irreversible changes in the plant finally causes permanent wilting. Stomata apertures also provide pathways into the plant for CO_2 needed for photosynthesis and the growth of plant tissues. Hence, a correlation between transpiration and crop yields exist and is frequently documented. Some authors claim that the priority for receiving CO_2 necessary for photosynthesis indeed regulates stomatal aperture.

6.5.1.2 Potential and Actual Transpiration

Analogous to potential evaporation E_P, potential transpiration T_{RP} is defined as the loss of water from plant tissues to the atmosphere according to the evaporative demands of the atmosphere with the stomata fully opened. Additionally, it is understood that water movement from the soil to the plant roots does not limit the process. With atmospheric conditions including the energy source and radiation controlling the phase change of water, energy regulates the process for a given stomatal density. The term unstressed transpiration is also used for T_{RP}.

Let us now assume that the soil water content θ suddenly decreases. Concomitantly, values of both h and unsaturated hydraulic conductivity K drastically decrease. In such a case, the evaporative flux is maintained by the gradient of the water potential ψ increasing – the cell water potential continually decreases and is accompanied by a loss of cell turgor. Inasmuch as the regulating mechanism of stomatal dynamics is not uniquely controlled by the water potential, we can not precisely define the critical value of guard cell turgor at which the stomata start to close. This critical value also depends upon plant type, its variety, its susceptibility to water stress and upon the local environmental conditions – quality and intensity of light, CO_2 concentration and surface temperature of leaves. We must also remember that transpiration has an important cooling effect on the plant. If we hold all conditions constant except the evaporative demand of the atmosphere, the critical value of turgor pressure h_D when stomata start to close depends upon the value of the potential transpiration T_{RP}, see Fig. 6.45. After this critical value is reached, the decreasing cross sectional area of the stomata causes a rise of the stomatal resistance r_S and hence, a rise of the leaf resistance. With the transpiration rate being reduced, the actual transpiration $T_{RA} < T_{RP}$. With further extraction of soil water, the unsaturated hydraulic conductivity decreases substantially and the compensating increased soil water potential gradient causes a drastic decrease of plant water potential and a sufficient decrease of turgor to initiate complete closure of stomatal apertures. The greatly reduced transpiration rate allows the cell turgor in the leaf to decrease to such an extent that wilting is clearly evident, see the range of wilting in Fig. 6.45. The whole process depends upon the value of T_{RP}, see Fig. 6.45. Close to saturation ($\theta/\theta_S \rightarrow 1$), when redox conditions in the soil reduce the root activity of cultivated plants, $T_{RA} < T_{RP}$ in the range RO. For

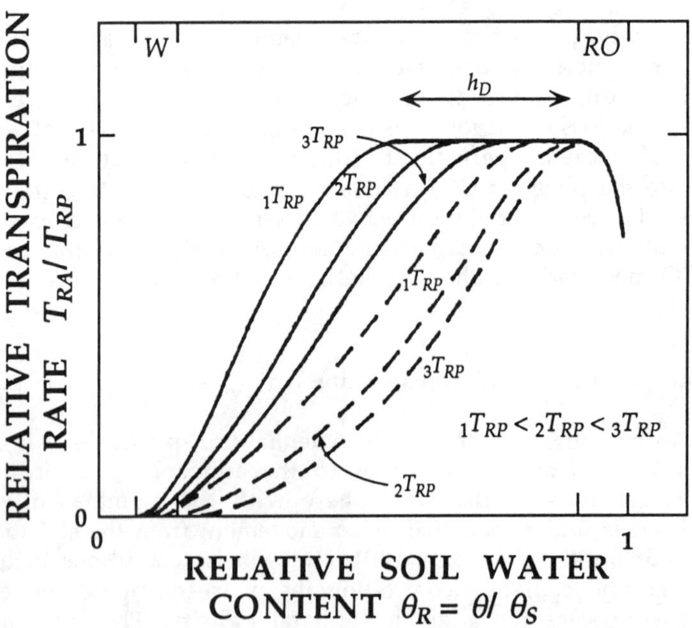

Figure 6.45. The dependence of actual transpiration rate T_{RA} upon relative soil water content θ/θ_S and upon the rate of potential transpiration T_{RP}. The full curves are for dense rooting while the broken curves are for half density of roots.

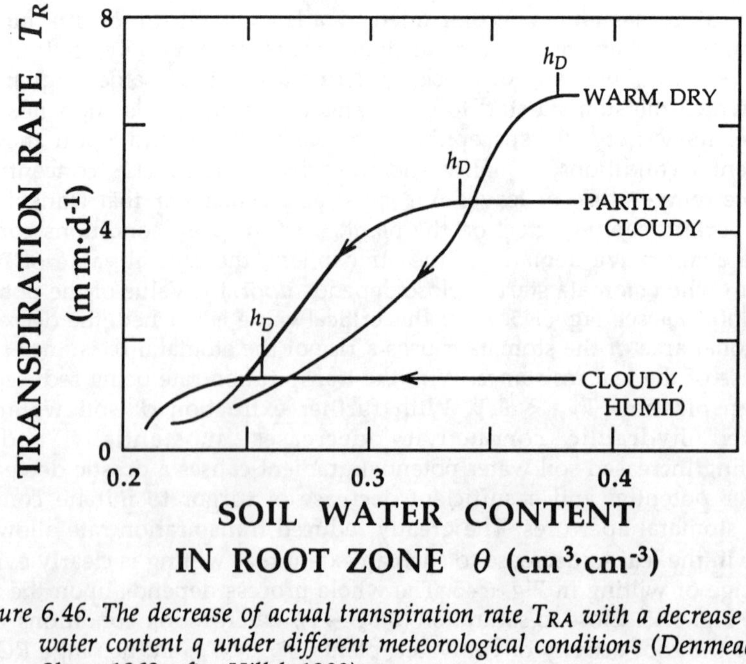

Figure 6.46. The decrease of actual transpiration rate T_{RA} with a decrease of soil water content q under different meteorological conditions (Denmead and Shaw, 1962, after Hillel, 1980).

a range of θ/θ_S very often between field capacity and θ corresponding to h_D in the range of -500 to about -3000 hPa, $T_{RA} = T_{RP}$. If the earlier-mentioned factors are kept constant, the value of h_D associated with decreased availability of soil water depends upon the evaporative demand of the atmosphere characterized by T_{RP} in Fig. 6.45. From model studies of Federer (1982) whose results appear in Fig. 6.45, we recognize that the density of the root system is not negligible. The shift of h_D and the dependence of T_{RA} upon meteorological conditions demonstrated by Denmead and Shaw (1962) are illustrated in Fig. 6.46.

Up to now we have dealt only with average values of transpiration. However, as was already mentioned earlier in the section on evaporation, the evaporative conditions of the atmosphere manifest large diurnal amplitudes. A detailed analysis of factors which contribute to diurnal fluctuations can be found in Brutsaert (1982). Such atmosphere characteristics occurring during the day and night cause fluctuations of leaf water potential, see Fig. 6.47. In the root, the amplitude of the water potential is smaller than that in the leaf and occurs later manifesting a lag or shift of phase. In the root zone, diurnal amplitudes of soil water potential driven by atmospheric demand are negligible, except perhaps at shallow soil depths where large daily temperature fluctuations occur.

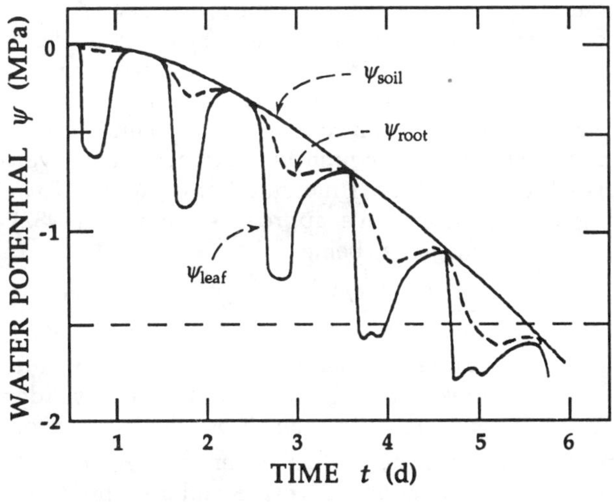

Figure 6.47. Scheme of diurnal changes in leaf and root water potential ψ together with soil water potential in a sequence of days without precipitation (Slatyer, 1967).

Having now used the microscopic approach to describe water flow within the plant and its path within the SPAC, we now return to (6.148) to macroscopic formulations frequently used to model transpiration in soil hydrology or plant production. Two different kinds of extraction terms S can be grouped together. In the first group, the value of potential transpiration is related to the maximum value of soil water extraction S_{max} by roots extending within the profile to a root zone depth z_r by $T_{RP}/z_r = S_{max}$. This relation is subsequently reduced schematically in a fashion similar to that in Fig. 6.45, or by some deterministic function of root length or related plant parameter. Using a formulation belonging to this first group, Feddes et al. (1978) suggested:

$$S = 0 \qquad 0 \geq h > h_1 \qquad (6.153a)$$

$$S = S_{max} \qquad h_1 \geq h > h_D \qquad (6.153b)$$

$$S = S_{max}\left(\frac{h - h_W}{h_D - h_W}\right) \qquad h_D \geq h > h_W \qquad (6.153c)$$

$$S = 0 \qquad h_W \geq h \qquad (6.153d)$$

Here, inasmuch as S_{max} is assumed constant, any change of S is caused by a decrease of θ beyond the limits of h shown above. The threshold value h_1 is related to the gas diffusion coefficient of the soil to avoid anaerobiosis. The corresponding θ/θ_S varies roughly between 0.85 and 0.95 and depends upon both soil texture and soil structure. The broad range of possible values of h_D was shown in Fig. 6.45. Pressure head of wilting h_W will be discussed in the next section. Later on, instead of S_{max} being constant with depth, S_{max} was assumed to decrease linearly with depth (Prassad, 1988) or exponentially with depth (Novák, 1987). Inasmuch as S_{max} depends upon the development of the root system, we suggest that it is entirely appropriate to describe $S_{max}(z, t)$.

In the second group, the resistance to water flow is considered at the soil-root interface or within the roots. The water flux in the plant is computed and related to T_{RP} by an iterative procedure. Inasmuch as direct measurement of resistances of plant water flow and xylem water potentials is difficult, plant hydraulic characteristics are frequently obtained from values of plant parameters adjusted to match directly measured or computed T_{RP}.

Combining both of the above approaches, Novák (1987) derived and experimentally confirmed $S_{max}(z)$ as being

$$S_{max}(z) = T_{RP}\left[\frac{\alpha\, exp(-\alpha z/z_r)}{1 - exp(-\alpha)}\right] \qquad (6.154)$$

where α is an empirical constant usually of order 10^0 for well-developed root systems and leaf canopies. The above relation, nearly identical to the distribution patterns of active roots in the soil profile, is valid for $h > h_D$ when transpiration reaches its potential value. When $h < h_D$, formulations (6.153) apply. Instead of the linear relation between S and h within the range $h_D \leq h \leq h_W$, a smooth curve is sometimes introduced even though it does not necessarily lead to greater accuracy. Generally, when $h < h_D$, $S(z)$ reflects roughly $h(z)$. Consequently, the soil depth manifesting h_{max} is the position of the greatest value of $S(z)$.

In the above discussion we have shown that transpiration as well as $S(z, t)$ are strongly influenced by stomatal resistance. If formulated as functionally dependent upon leaf water potential and solar flux, stomatal resistance can be modeled to an external control of transpiration. A review of this type of modeling is given by Lynn and Carlson (1990).

6.5.1.3 Wilting Point

We have already shown that if the rate of soil water extraction by plant roots is insufficient to meet a reduced transpiration rate, the plant is wilting. If the plant does not regenerate its turgor and biological activity after a rewetting of the soil, we define such θ as the soil water content of wilting. Resuming the discussion of flow of water within plants, we recognize that the soil water content of wilting is not the same for different plants. Even if we restrict our consideration to a particular cultivar, the soil water content of wilting depends upon the developmental stage of the plant, its water stress susceptibility and the evaporative conditions of the atmosphere. Moreover, the composition and concentration of the soil solution plays a role. An increased soil solution concentration increases the soil water content of wilting through a decrease of the total soil water potential. The increased value of K and D owing to an increased soil solution concentration may also influence the value of the soil water content of wilting provided that there are no other solution effects linked to soil structure and transport across cell membranes.

Because the soil water content of wilting separates soil water into two categories (one useful and available to plants and one that is not) practitioners introduced the concept of the wilting point θ_W especially for the production of cultivated crops. The "point" is indeed nothing more than an average of soil water content of wilting values which are closely correlated with the permanent wilting percentage.

Wilting point is now defined in soil science as the soil water content at a soil water potential of -1.5 MPa (or equivalently as the logarithm of the pressure head expressed in cm, pF = 4.18). Inasmuch as this value is usually reached as the soil becomes progressively drier, the wilting point is that θ which is related to the pressure head on the drainage branch of the soil water retention curve. The pressure plate method is commonly used to measure θ_W at -1.5 MPa. Because soil structure does not significantly influence the value of θ_W, soil samples need not have to be the volume of a REV. Desiccator methods were previously employed to estimate θ_W using a relative partial pressure of water vapor either above a saturated solution of K_2SO_4 or above 3% H_2SO_4 at a temperature of 20°C.

With the concept of continuous water flow in SPAC accepted, it would appear more appropriate to relate the soil water content of wilting to the dynamic characteristics of soil water quantified by values of the unsaturated hydraulic conductivity K or soil water diffusivity D. However, measuring soil water potentials in the vicinity of the root system is much easier than measuring K and D to ascertain critical flux conditions. Assessing critical flux conditions offers a real challenge to our understanding of SPAC. For example, if we assume that $D = 10^{-3}$ $cm^2 \cdot min^{-1}$ is a critical minimal value for avoiding wilting of plants, this value is never attained in the whole range of soil water contents in confined vertisols having small values of EC with ESP > 15% (Kutílek, 1973).

The definition of wilting point based upon biological experiments requires a standardized observation of the permanent wilting of an indicator plant (e.g. barley).

Soil texture can be used as an indicator of the wilting point. In 1924, Solnař suggested that θ_W = (silt + clay)/2.4, where (silt + clay) is the percentage of particles < 0.01 mm (Kutílek, 1978).

6.5.2 Potential Evapotranspiration

Just as we defined potential evaporation and potential transpiration, we assume an unlimited supply of evaporating surfaces with water in the definition of potential evapotranspiration. However, before proceeding, we must first clear up some specific terminology.

The reference value denoted as the potential evapotranspiration E_{TP} is often taken to be the evaporation from a plant stand of densely covered short cut grass freely supplied with water. Penman (1963) and his colleagues (see e.g. Monteith, 1981) have used the term potential transpiration inasmuch as evaporation from the soil is truly negligible for a densely populated grass. We note that Penman's (1948) original derivation described evaporation from a free water surface or any surface freely covered by water (e.g. a plant canopy immediately following rainfall). On the other hand, Thornthwaite (1948) coined the term potential evapotranspiration. The terminology is further confused by some authors who prefer to speak of evaporation from vegetation. Inasmuch as the soil surface is frequently not completely shadowed by a plant canopy, we prefer the term potential evapotranspiration. An unambiguous definition of E_{TP} in practice is only achieved by reporting the method of its computation or the procedure by which it was measured.

6.5.2.1 Computational Methods for Estimating E_{TP}

Provided that steady state fluxes and conditions are maintained, evaporation from any surface can be obtained from three basic equations using the so-called turbulent diffusion theory. From latent heat transport from the surface with roughness for water vapor z_o to the atmospheric surface layer position z_a, the evaporation rate E [kg·m^{-2}·s^{-1}] is

$$E = \rho_A D_t (e_o - e_a), \tag{6.155}$$

the turbulent flux of heat Q_A [W·m^{-2}] is

$$Q_A = \rho_A c_p D_t (T_o - T_a) \tag{6.156}$$

and the partitioning of the net radiation R [W·m^{-2}] at the evaporative surface is

$$R = Q_S + Q_A + \chi E.. \tag{6.157}$$

These equations express the similarity between turbulent heat and water vapor transport where D_t is the turbulent transfer coefficient [m·s^{-1}] which is assumed to be identical for latent and sensible heat transfer, χ the latent heat of vaporization [J·kg^{-1}], Q_S the flux of heat into the soil [W·m^{-2}], T_o [°C] the temperature of the surface at the scalar roughness for sensible heat z_{oh}, T_a [°C] the temperature of the atmospheric surface layer at z_a, c_p the specific heat of air at constant pressure [J·kg^{-1}·K^{-1}], ρ_A the density of air [kg·m^{-3}] and e_o and e_a [kg·kg^{-1}] the specific humidity at z_o and z_a , respectively.

Values of T_o and e_o are not routinely measured at standard micro meteorological stations owing to the difficulty of their observation. For the purpose of estimating E_{TP}, their measurement is eliminated by the following procedure. The functional dependence of the saturated vapor pressures at the surface $e^*_{So}(T)$ and $e^*_{Sa}(T)$ are known. Using a Taylor series expansion leads to the major approximation of Penman (1948)

$$e^*_{So} = e^*_{Sa} + \left(T_o - T_a\right)\phi_a \tag{6.158}$$

where e^*_{So} is the air saturated vapor pressure at the temperature of the surface at z_o, e^*_{Sa} the air saturated vapor pressure at the temperature of the surface z_a and ϕ_a is the derivative $d\,e^*_S/dT$ at T_a. The specific humidity e is related to the vapor pressure e^* by $e = 0.622\ e^*/(R_d T_a \rho_A)$ where R_d is the Universal gas constant for air (R_d = 287.04 J·kg^{-1}·K^{-1}). Substitution of (6.157) into (6.155) and combining the result with (6.156) and (6.157) yields an equation for the potential evapotranspiration (Novák and Hurtalová, 1987)

$$E_{TP} = \frac{\phi_a\left(R - Q_S\right) + \rho_A c_p D_t\left(e^*_{Sa} - e^*_a\right)}{c_p + \chi\,\phi_a}. \tag{6.159}$$

Note that $(e^*_{Sa} - e^*_a)$ is equivalent to the saturation deficit at elevation z_a. Equation (6.159) is comparable to Penman's (1948) equation which is traditionally written as

$$\chi\,E_{TP} = \frac{\phi_a}{\phi_a + \gamma} H_Q + \frac{\gamma}{\phi_a + \gamma}\,\chi\,E_A \tag{6.160}$$

where E_A is the "drying power of the air" (i.e. a function of wind velocity and saturation deficit of the air) and γ the psychometric constant [Pa·K^{-1}] in (6.160) defined by

$$\gamma = \frac{c_p\,p_A M_A}{\chi M_V} \tag{6.161}$$

where p_A is the air pressure [Pa] and M_V and M_A are values of the molecular mass of water vapor and air, respectively. $H_Q = (R - Q_S)$ comprises the radiative flux minus the soil heat flux as well as heat storage in the vegetation and photo synthetically used energy according to Penman which are often neglected.

The net radiation R is either measured or computed according to standard methods (FAO, 1977). Although the integral of E_{TP} usually yields a negligible value of the heat flux into the soil Q_S for a day, its diurnal values should be estimated for periods shorter than a day. A major problem is estimating the turbulent transfer coefficient D_t. And, the difficulty is not normally overcome by using Penman's empirical definition of $E_A = f(\mu)d$ where $f(\mu)$ is a wind velocity function and d the saturation deficit ($e^*_{Sa} - e^*_a$) [Pa].

In Penman's original equation and in subsequent modifications, the use of a wind velocity function is probably appropriate in order to obtain E_{TP} for integral values of a day or even somewhat longer periods. On the other hand, for diurnal (say hourly) computations the variation of atmospheric stability plays an important role. Expressions of D_t which account for atmospheric stability to solve the problem are discussed in detail by Brutsaert (1982). When atmospheric stability plays a role, we have

$$\chi E_P = w(R - Q_S) + \chi(1-w)E_A \tag{6.162}$$

where $w = \phi_a/(\phi_a + \gamma)$ on a diurnal basis and the drying power function E_A may be written in terms of atmospheric Monin-Obukhov (1954) similarity theory,

$$E_A = k u_* \rho(e_{Sa} - e_a)\left[\ln\left(\frac{z_a - d_{ov}}{z_{ov}}\right) - \psi_v\left(\frac{z - d_{ov}}{L}\right)\right]^{-1} \tag{6.163}$$

where von Karman's constant $k = 0.4$, $u_* = (\tau_o/\rho)^{1/2}$ is the frictional velocity, τ_o the surface shear stress, d_{ov} the displacement height for water vapor and e_a and e_{Sa} are the specific humidity of the air and the saturation specific humidity at the air temperature, respectively (Brutsaert, 1982, and Katul and Parlange, 1992). The stability correction function ψ_v depends upon $(z - d_{ov})/L$ with L being the Obukhov length defined by

$$L = \frac{-u_*^3}{kg\left[H_v/(\rho c_p T_a)\right]} \tag{6.164}$$

where $H_v = (Q_A + 0.61\, T_a c_p E_p)$ is the specific flux of the virtual sensible heat at the surface. The value of u_* is obtained from the mean horizontal wind speed v defined by the surface layer model of Monin-Obukhov as

$$v = \frac{u_*}{k}\left[\ln\left(\frac{z - d_o}{z_o}\right) - \psi_m\left(\frac{z - d_o}{L}\right)\right] \tag{6.165}$$

where d_o is the momentum displacement height, z_o the surface roughness and ψ_m the momentum stability correction function. For unstable conditions ($L < 0$),

$$\psi_m = \ln\left[\frac{(1+x)^2(1+x^2)}{(1+x_o)^2(1+x_o^2)}\right] - 2\tan^{-1}x + 2\tan^{-1}x_o \tag{6.166}$$

and

$$\psi_v = 2\ln\left(\frac{1+x^2}{2}\right) \tag{6.167}$$

where $x = (1 - 16y)^{1/4}$, $y = (z - d_o)/L$ and $x_o = (1 - 16z_o/L)^{1/4}$. For stable conditions ($L > 0$),

$$\psi_v = \psi_m = -5\ln\left(\frac{z - d_o}{z_o}\right). \tag{6.168}$$

The scalar roughness z_{ov} can be substantially smaller than z_o. Some suggestions are given by Brutsaert (1982) for possible parameterizations of z_{ov}. An iterative scheme is required to actually solve for E_{TP}. With values of ψ_v and ψ_m set equal to zero for the first iteration, (6.165) is solved for a first estimate of u_* and (6.163) for E_A. Next a first estimate of L in (6.164) is made using E_p from (6.162) and the surface energy balance to obtain Q_A. Next, the stability correction functions are included in (6.165) and (6.163). This system of calculations are repeated until the value of E_p converges. Usually convergence is achieved in only a few iterations. Katul and Parlange (1992) and Parlange and Katul (1992) tested this formulation of Penman's equation for a wet bare soil and found it to be remarkably accurate during the day even under conditions with strong, dry winds passing over a wet land surface such that $Q_A < 0$ (i.e. $L > 0$) and $\chi E_p > (R_M - Q_S)$. In these extreme conditions, sometimes called an oasis problem or effect as found in arid regions (e.g. western United States or Australia), it is crucial to account for atmospheric stability to ultimately ascertain the total daily evapotranspiration.

At least two other approaches should be mentioned here regarding evapotranspiration equations. Thornthwaite (1948) proposed an empirical equation based upon air temperatures and an air temperature dependent empirical heat index. The equation provides integral monthly values of E_{TP} and its use is restricted to mild climatic zones. Efforts to obtain daily data from the equation appear to be useless. A second approach is that of Priestley and Taylor (1972). They proposed a simplified version of the Penman equation to calculate potential evapotranspiration

$$E_{TP} = \alpha\left[\frac{\phi_a(R-Q_S)}{\gamma+\phi_a}\right] \tag{6.169}$$

for extended wet surface conditions with symbols used in (6.159) and (6.160) such that $Q_A > 0$. The empirical coefficient α falls between 1 and $(\phi_a + \gamma)/\phi_a$ and is best estimated according to Priestley and Taylor with a value of 1.26. Inasmuch as this model will break down when there is strong local advection as described above, prudence should be exercised.

6.5.2.2 Structure of Evapotranspiration

The ratio of the two components of evapotranspiration, transpiration T_R and evaporation from soil E is called the structure of evapotranspiration (Budagovskij, 1969). This ratio has a diurnal rhythm – during the day the plant canopy is intensively transpiring, and during the night transpiration practically ceases while evaporation from the soil plays a more important role. During the season, the structure of E_{TP} depends upon the leaf area index ω which is the ratio of the total area of leaves and other green parts of plants related to the reference area of the soil. Methods for measuring ω are described in the literature on photosynthesis. The value of ω is also seasonally dependent and usually has a range $0 \le \omega < 10$. An example of the seasonal evolution of $\omega(t)$ of wheat is given in Fig. 6.48. We should note that ω of wheat reaches very high values compared to those of other cultivated plants, especially row crops. Quantifying the canopy density with a value of ω is important in order to convert the evapotranspiration from a densely populated crop with complete canopy cover to that from a frequently observed incomplete canopy cover. Ritchie (1972) found that the ratio of radiation R_a above a canopy to that at the soil surface R_S below the canopy has the exponential form

$$\frac{R_a}{R_S} = \exp(-\alpha\,\omega) \tag{6.170}$$

with $\alpha \doteq 0.398$. Novák (1981) applying the procedures of Budagovskij (1969) based upon a set of equations describing the transport of water vapor and heat in the canopy obtained similar relationships and distinguished the contributions

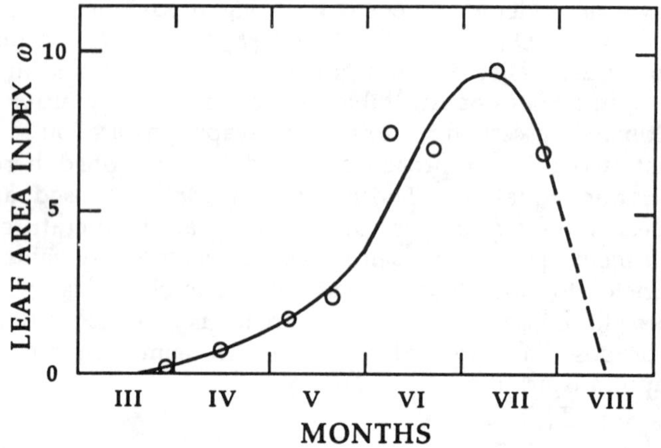

Figure 6.48. The seasonal dependence of leaf area index ω measured for wheat in Slovakia (Novák, 1981).

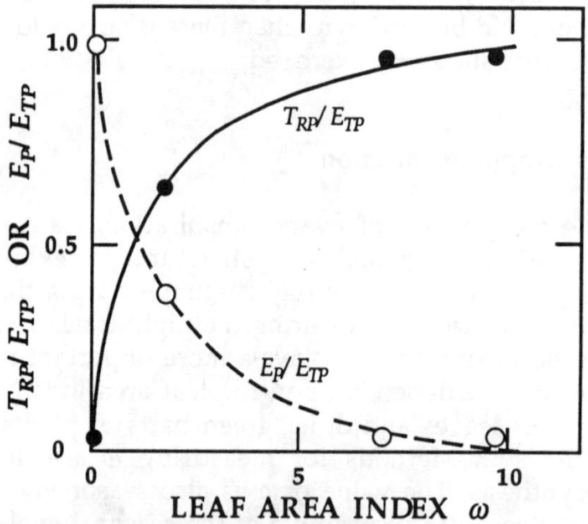

Figure 6.49. The ratio T_{RP}/E_{TP} and E_P/E_{TP} as related to leaf area index ω in Fig. 6.48. T_{RP} is potential transpiration, E_{TP} potential evapotranspiration and E_P potential evaporation from the soil only (Novák, 1981).

of transpiration T_{RP} and evaporation from soil E_P (see Fig. 6.49) as

$$T_{RP} = E_{TP}\exp(-\alpha\,\omega) \quad (6.171)$$

and

$$E_P = E_{TP}\left[1-\exp(-\alpha\,\omega)\right]. \quad (6.172)$$

We note in Fig. 6.50 that the ratio T_{RP}/E_{TP} is closely related to ω for various plants. Moreover, the constant value of the coefficient α is probably not

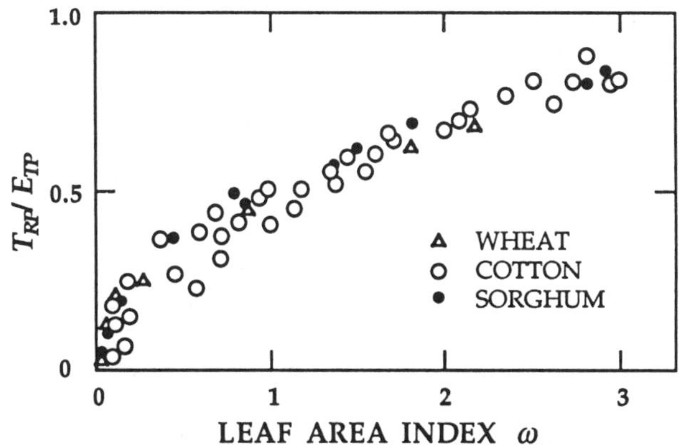

Figure 6.50. The dependence of T_{RP}/E_{TP} upon leaf area index of various plants. Data collected from the literature by (Novák, 1981). T_{RP} is potential transpiration and E_{TP} potential evapotranspiration.

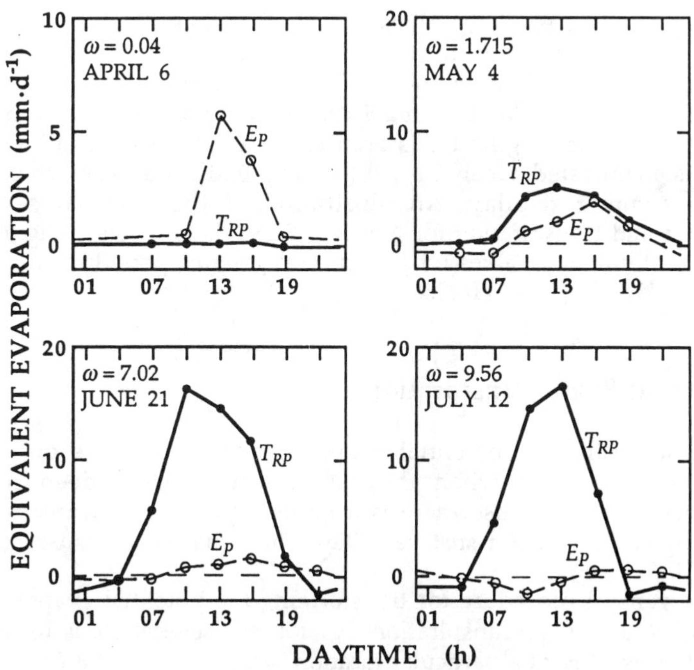

Figure 6.51. The diurnal variation of separated potential terms – E_P is soil evaporation and T_{RP} is transpiration (Novák, 1981).

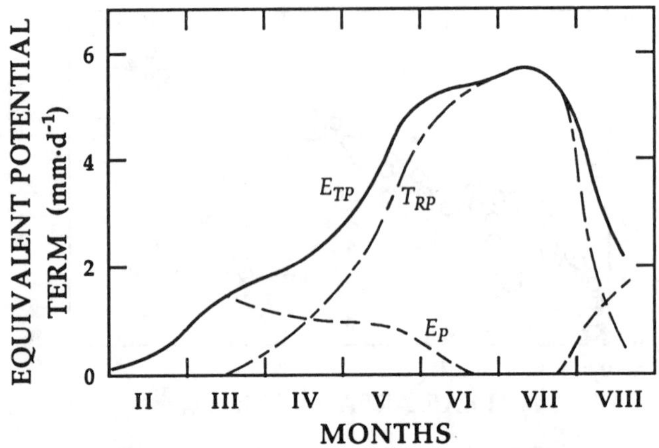

Figure 6.52. The seasonal behavior of separated potential terms for a chernozem on loess in Slavakia. E_P is soil evaporation and T_{RP} is transpiration separated from evapotranspiration E_{TP} for wheat (Novák, 1981). All terms represent potential values.

universally applicable. Daily values of E, T_R and E_T strongly depend upon the plant's ontogenesis stage. With an increase of ω the daily sum of E_P decreases and the maximum peak shifts to the afternoon hours [phase shift of the quasi-sinusoidal curve $E_P(t)$]. At about $\omega \geq 1$ condensation of water vapor occurs at the soil surface and the length of this occurrence is extended as ω increases. This increase is manifested when $E_P < 0$ (see Fig. 6.51) and even when $E_{TP} < T_{RP}$ during a summer midday. An illustration of the seasonal separation of evaporation and transpiration for wheat in Slovakia is given in Fig. 6.52. In this case of wheat, integral values for the entire vegetative period manifest values of $T_{RP}/E_{TP} = 0.78$ and $E_P/E_{TP} = 0.22$.

6.5.3 Actual Evapotranspiration E_{TA}

Let us first recall that potential evapotranspiration occurs only when the surfaces are wet, and when they are not wet, we indeed have actual evapotranspiration, because there is a limited quantity of water reaching the evaporating soil and plant surfaces. The term actual evapotranspiration E_{TA} is introduced, $E_{TA} \leq E_{TP}$.

The general procedure for transforming the potential evapotranspiration E_{TP} to the actual evapotranspiration E_{TA} follows several steps to theoretically compute the real E_{TA} at a particular location.

1. E_{TP} is computed for the reference short green grass with wet surfaces and adequate water supply without advection.
2. This reference E_{TP} is transformed using specific site related values influenced by slope, land use and albedo.

3. The site related E_{TP} is adjusted to the particular type of vegetation including its stage of development.
4. The value of E_{TA} is obtained by reducing the value of the site related E_{TP} to account for the limited supply of water derived from the soil profile. This reduction is achieved empirically by applying relations like (6.152) or by introducing actual resistances which reduce the flux through the SPAC system.

When steps 2 and 3 are omitted, we have the development presented by Monteith with the final result often being named the Penman-Monteith equation. Here, with the stomatal resistance r_S being introduced, the Penman-Monteith equation for E_{TA} is

$$E_{TA} = \frac{\frac{\phi_a}{\gamma}\left(\frac{H_Q}{\chi}\right) + \frac{\rho_A c_p d}{\chi \gamma r_A}}{\phi_a + \gamma\left(1 + \frac{r_S}{r_A}\right)} \qquad (6.173)$$

where r_A [s·m^{-1}] is the aerodynamic resistance against the turbulent transfer of water vapor from the evaporating surface to the atmosphere. Its values are in the range $10 \le r_A \le 300$. This equation describes the actual evapotranspiration of a particular reference plot owing to an extension of Penman's equation by Monteith (1981). It represents the entire canopy as a "big leaf" (Lynn and Carlson, 1990) and is still in agreement with the integral single layer concept of Penman.

Deterministic models of stomatal resistance (see Lynn and Carlson, 1990) can also be applied for estimating E_{TA}. Alternatively, soil water models which empirically describe water absorption by plant roots expressed by (6.153) allow calculations of E_{TA} or E_{TA}/E_{TP}.

The bulk stomatal (or surface) resistance r_S in (6.173) is generally considered a "fitting" parameter in land surface simulation models or a diagnostic water stress index of plant canopies (e.g. Brutsaert, 1982). Recently, many complicated models of r_S have been proposed for drying land surfaces, especially in the context of atmospheric modeling, but for practitioners in the field, r_S is generally unknown. Because it is so difficult in practice to obtain a field scale measure of r_S to assess the aridity of a region, we mention here models based upon ideas of Bouchet (1963) concerning potential evaporation and the drying of the land surface. In these approaches (e.g. Brutsaert and Stricker, 1979, and Morton, 1983) the aridity of the region is assessed from atmospheric measurements. Bouchet basically suggested that as an initially wet surface dries, the decrease in actual evaporation corresponds to an equivalent increase in potential evaporation. When the land surface is wet

$$E(actual) = E_P(potential) = E_{P_o}(surface\,is\,wet). \qquad (6.174)$$

As the water supply to the surface becomes limiting ($E < E_{P_o}$) for the same amount of energy available for evaporation,

$$E - E_{P_o} = q_1 \qquad (6.175)$$

such that q_1 becomes available and increases E_P. Bouchet proposed that

$$E_P = q_1 + E_{P_o} \qquad (6.176)$$

in order that the complementary relationship

$$E + E_p = 2E_{p_o} \quad (6.177)$$

assures that q_1 does not alter the available energy and that no external energy suddenly enters the region.

Brutsaert and Stricker (1979), adopting the Penman equation for potential evaporation E_P and the Priestley-Taylor equation for E_{p_o}, solved (6.177) for E. The primary restriction of this model is that the Priestley-Taylor model is only valid when there is no local advection. Parlange and Katul (1992a and b) showed that in order to conserve energy whenever $Q_{A_p} < 0$ ($Q_{A_p} = R - Q_S - \chi E_P$) the Priestley-Taylor formulation of E_{p_o} must be adjusted by taking $\chi E_{p_o} = [\chi E_{p_o}(\text{Priestley-Taylor}) + |Q_{A_p}|]$. Parlange and Katul tested this model using 20-minute time step measurements and found it to be extremely accurate, though more work is needed before it becomes more widely accepted and verified. The advantage, of course, is that r_S does not have to be specified.

Inasmuch as all models or estimates have to be first verified against directly measured data, we shall briefly discuss the measuring techniques of evapotranspiration in the field. The methods can be grouped into three classes: (i) soil water balance, (ii) plant measurement and (iii) micro meteorological methods.

6.5.3.1 Soil Water Balance

The basic water balance of an elementary area where horizontal fluxes are negligible is

$$E_{TA} = I + Q_V + \Delta W \quad (6.178)$$

where I is the cumulative infiltration from irrigation and rain equal to precipitation if surface runoff is eliminated, Q_V the vertical flux at the base of the soil profile (when the groundwater table is very deep, $Q_V < 0$ and usually represents the vertical drainage component) and ΔW the difference in water storage for a given period of time Δt $(t_2 - t_1)$ from which E_{TA} is evaluated. Measuring the water content distribution within the soil profile $\theta(z, t)$ at times t_1 and t_2 allows the magnitude of ΔW to be ascertained using

$$\Delta W = \int_0^Z \theta(z,t_1)dz - \int_0^Z \theta(z,t_2)dz \quad (6.179)$$

where Z is the depth of the bottom of the soil profile where Q_V is evaluated. The principle of the zero flux plane Z_z is used if $Z_z > z_r$ where z_r is the rooting depth. The determination of the zero flux plane (where $Q_V = 0$) was previously explained and illustrated in Fig. 6.37. If tensiometers are installed and the soil hydraulic functions known, the flux across the plane at Z can be computed.

In spite of the relative simplicity of the soil water balance method, soil heterogeneity, described in Chapter 8, plays an important role in the strategy of selecting observation sites (Vachaud et al., 1983) and defining field scale hydraulic properties based upon evaporation measurements (Parlange et al., 1992 and 1993).

In order to integrate the diverse distribution of vertical and potentially horizontal fluxes occurring at the microscale, lysimeters are frequently used to measure Q_V at the bottom of the soil profile. The term indicates that the measurement of soil leaching was the original aim of lysimeters. Nowadays, the term identifies a container filled with soil covered with natural or cultivated vegetation. The water percolating through the soil is collected either gravimetrically or through a suction plate which is maintains a negative soil water pressure head h identical to that in the field next to the lysimeter. Because gravitational drainage from a lysimeter of finite length differs from that in the field and disturbs the natural development of $h(z)$, the depth of the lysimeter should be as great as possible. To achieve the same internal structure of the porous system as in the natural soil, it is desirable to place a monolithic block of field soil into the container. Lysimeters containing soil whose natural structure has been disturbed have limited applicability.

Lysimeters are either weighable or non weighable. Weighable lysimeters provide information about the change of water storage W for any time period even in the absence of percolating water, while non weighable lysimeters yield only the water percolating from the soil column. Instead of directly measuring the weight of a lysimeter, weight changes are sometimes determined from fluid displacements in a larger surrounding container in which the lysimeter is "floating". These floating lysimeters can be instrumented to observe the force required to keep them centered within the larger container within which they float. These observations provide additional information regarding the shear stress of the wind as it passes through and over the vegetative canopy.

6.5.3.2 Plant Surface Measurements

These measurements neglect the soil water evaporation component of evapotranspiration. In forests, the sap flow in the tree trunk is measured. The time required for either a heat pulse or a radioactive tracer pulse to travel along the trunk is measured by an appropriate technique. A similar approach is also applied to trace the flux of water within the xylem of cultivated plants.

The porometer method is more fully developed. With the flux of water measured on individual leaves, the stomatal resistance r_S is obtained from an appropriate calibration. The canopy resistance r_c ($r_c = r_S/\omega$) is calculated from r_S and the leaf area index ω. Models of evapotranspiration include values of r_c as directly measured information.

Integral values of evapotranspiration are also estimated from remote sensing. Values of E_{TA} are obtained from a surface energy model which has inputs of measured surface temperatures based upon infrared (IR) images. The sensible heat flux Q_A can be calculated from Monin-Obukhov similarity using

$$T_S - T_a = \frac{Q_A}{k u_* \rho c_p}\left\{\ln\left[\frac{z-d_{oh}}{z_{oh}}\right] - \psi_h\left[\frac{z-d_{oh}}{L}\right]\right\} \qquad (6.180)$$

where T_S is obtained from IR remote sensing of surface temperature, T_a is taken in the surface layer of the atmosphere, z_{oh} is the heat roughness length, d_{oh} the sensible heat displacement height (often taken equal to d_o) and ψ_h the sensible heat stability correction function taken equal to the value of ψ_v defined earlier. Actual evapotranspiration is solved once more from the energy balance ($\chi E = R - Q_S - Q_A$) in an iterative scheme where once again ψ_m and ψ_h are initially set equal to zero, E is estimated from the energy balance and a first estimate of L can then be obtained. The system is repeated with the iterative values of ψ_m and ψ_v until the value of E converges.

Alternatively, temperatures of a reference plot (where E_{TP} exists) with an optimal water supply are monitored and used for evaluating E_{TA} of a vegetative cover under natural conditions for which temperatures are monitored.

6.5.3.3 Micrometeorological Methods

These methods are used to estimate the value of E_{TA} from directly measured atmospheric data above the land surface.

In eddy covariance methods, the use of fast response sensors permits measured vertical wind speed fluctuations w' to be correlated with measured temperature and humidity fluctuations T' and q', respectively. Hence, average values of the flux of water vapor during short time intervals (e.g. one hour or less) are obtained where $E = \rho\, \overline{w'q'}$ and $Q_A = \rho c_p\, \overline{w'T'}$. The fluctuations for the measurement time period (e.g. T') are defined by $T' = T - \overline{T}$, where $\overline{T}$ is the mean value over the time period of interest and the condition $\overline{T'} = 0$ holds. For further study of eddy covariance methods, see Stull (1988) or Qarratt (1992).

The Bowen ratio β ($= Q_A/\chi E$) is applied in order to partition the energy budget and from it, E_{TA} is estimated. Note the Bowen ratio is the ratio of the sensible and latent heat fluxes. It is obtained by measuring temperature and the partial vapor pressure at two heights together with the atmospheric pressure assuming the transport of heat and water vapor are identical.

Aerodynamic methods are based upon Monin and Obukhov (1954) similarity theory which provides a good description of the mean boundary layer surface layer flow (in the turbulence sense). The main theory is given here for completeness. The flux of momentum, sensible heat and water vapor can be given by

$$v = \frac{u_*}{k}\left[\ln\left(\frac{z-d_o}{z_o}\right) - \psi_m\left(\frac{z-d_o}{L}\right)\right] \tag{6.181}$$

$$T_r - T = \frac{Q_A}{k u_* \rho c_p}\left\{\ln\left[\frac{z-d_o}{z_r-d_o}\right] - \psi_h\left[\frac{z-d_{oh}}{L}\right] + \psi_h\left[\frac{z_r-d_{oh}}{L}\right]\right\} \tag{6.182}$$

and

$$q_r - q = \frac{E}{k u_* \rho}\left\{\ln\left[\frac{z-d_o}{z_r-d_o}\right] - \psi_v\left[\frac{z-d_{ov}}{L}\right] + \psi_r\left[\frac{z_r-d_{ov}}{L}\right]\right\} \tag{6.183}$$

where the subscript r refers to a reference height near the surface and d_{oh} is the displacement height for heat. These three equations can be iteratively solved simultaneously when atmospheric surface layer measurements of wind speed, temperature and humidity are available.

As it follows from this brief presentation, the majority of difficulties in the estimation of E_{TA} is in fact the direct measurement of water vapor fluxes into the atmosphere is not simple in practice and that reliable flux meters have not yet been developed. However, this is not only the case for atmospheric fluxes but for the direct measurement of soil water fluxes as well as we discuss in the next chapter.

PROBLEMS

1. Derive the solution $z(h)$ of steady infiltration into a homogeneous soil profile if $K(h)$ is described by (5.38) for $m = 1$ and for $m = 2$. Follow a procedure analogous to that in the derivation of (6.5). Hint: Insert $A = b - a/q$ into the denominator of the integral.
2. Derive the implicit equation for h_l in a crust-topped profile (Fig. 6.4) if $K(h)$ is described by (5.39) and the crust (layer 2) is fully saturated.
3. What is the least value of L_1 in Fig. 6.4 when the value of $dh/dz = 0$ just below the crust. L_1 is the thickness of subsoil 1 between the crust and the ground water level GWL. Is the solution also valid for a homogeneous soil without a crust with L_1 the depth of the GWL and $q_o < K_{S1}$?
4. Modify (6.12) for soil 1 with air-entry value h_{A1} when you solve steady infiltration into a crust-topped profile.
5. Assuming steady infiltration, what is the elevation of the water level in each of the two piezometers – one inserted in the soil profile at the top and the other at the bottom of the zone of saturation in Fig. 6.5?
6. Determine the soil water content $\theta(x)$ at 2-cm intervals along a soil column during horizontal infiltration at time t_2 = 100 min. based upon the following measurements of $\theta(x)$ at time t_1 = 25 min.:

x (cm)	0	2	4	6	8	10	12	14
θ ($cm^3 \cdot cm^{-3}$)	0.507	0.505	0.497	0.470	0.408	0.312	0.185	0.082

   Values of θ_S and the initial soil water content θ_i are 0.507 and 0.082, respectively.
7. Using the Green and Ampt approximation, derive the value of m in the relation $A = mK_S$ where A is the parameter in (6.67).
8. For Green and Ampt's approximate solution of vertical infiltration, derive the validity of
$$\lim_{t\to\infty} q_o(t) = K_S.$$
9. For a constant rainfall rate, derive the value of the ponding time t_p for infiltration when infiltration for DBC is described by Green and Ampt's approximation. Proceed analogously to the derivation of (6.93).
10. Derive the value of h_f in Green and Ampt's approximation (6.53) when you use: a. (5.38) or b. (5.39) for $K(h)$.
11. Is $(\theta_{FC})_1$ < or > $(\theta_{FC})_2$ if field capacity is measured on the same soil once at θ_{i1}, second time at θ_{i2} and $\theta_{i1} > \theta_{i2}$?
12. Sketch the change of soil water content θ versus time for the two redistribution processes (a. and b. demonstrated in Fig. 6.31) when θ is measured at $z_{r1} > z_f$.

13. With the GWL being 180 cm below the soil surface of two different soil columns, find the difference in maximum evaporation rate E_{max} from the two columns. Column A is a homogeneous profile of a loamy soil characterized by (5.39) with $c = 0.02\ \text{cm}^{-1}$ and $K_S = 20\ \text{cm·d}^{-1}$. Column B is a profile of the same loamy soil having an interlayer of sand [characterized by (5.39) with $c = 0.08\ \text{cm}^{-1}$ and $K_S = 200\ \text{cm·d}^{-1}$] extending from 80 to 110 cm below the soil surface.
14. Estimate the actual evapotranspiration rate at $\theta/\theta_S = 0.5$ if potential evapotranspiration is 5 mm·d^{-1} for which the critical pressure head of decreased availability h_D (Fig. 6.45) is related to $\theta/\theta_S = 0.7$ and the wilting point is at $\theta/\theta_S = 0.35$. Assume linear relationships.

7 ESTIMATING SOIL HYDRAULIC FUNCTIONS

Similar to the measurement of saturated hydraulic conductivity K_S, we prefer to estimate the hydraulic functions $K(\theta)$, $K(h)$ or $D(\theta)$ from field experiments if we intend to use them in solutions of field problems. We revoke the great difference between laboratory- and field-measured saturated hydraulic conductivity K_S, even if a core sample is used in the laboratory. This difference is slightly reduced when we measure K in the neighborhood of $h < 0$ inasmuch as the macropores are empty and do not contribute to the flux. If indeed, a REV is sampled, the smaller the value of h, the smaller is the difference between K determined on the core sample and that measured in the field. In addition to the reliability of the data, a decision to analyze a sample in the laboratory leads to a practical consideration. That is, in structured soils, inasmuch as a REV is at least 10^3 cm^3, sampling and laboratory measurements are more difficult and time consuming than field investigations. Additionally, the overburden pressure which naturally occurs in the field is difficult if not impossible to reliably achieve in the laboratory. And, with the field length continuity of large capillary pores usually disturbed or destroyed as we curtail the length of soil cores, our preference for field methods is reasonable.

Here, after briefly discussing laboratory procedures for core samples, we describe field methods in more detail. The measurement of $h(\theta)$ is still preferentially performed in the laboratory on core samples in order that we define the hysteretic history of the soil water retention curve (SWRC) more exactly. Reliable SWRC measured in the field with tensiometers and an appropriate soil water content meter are difficult to ascertain owing to the experimental data being scattered around the fitted curve and an underestimation of θ for a given value of h when compared with laboratory-determined drainage curves (Bork and Diekkrüger, 1990). The latter difficulty is caused by hysteretic effects while the former is complicated by soil heterogeneity. The method for determining the SWRC described in Chapter 4 is generally valid for field investigations. The particular method used to obtain the SWRC should be selected and modified according to the nature of the field problem to be solved.

Disturbed, repacked samples cannot be substituted for soil core samples intended to characterize field hydraulic properties. Data obtained on repacked samples of "core size" are virtually useless in field applications even in relatively homogeneous sandy soils, see Clothier and White (1982). We do not imply that experiments performed on repacked samples and columns are of no use. Experiments with repacked samples are frequently and profitably used for experimental verification of theories and especially approximate solutions. Nevertheless, such results cannot be directly transferred to quantify soil water behavior in the field.

Procedures for estimating soil hydraulic functions can be grouped into three classes: (i) Steady or quasi-steady flow experiments, described by simple Darcy-Buckingham equations, performed with properly selected boundary conditions. (ii) Unsteady water flow experiments carried out under restrictive conditions and with boundary conditions selected to allow the application of analytical, semi-analytical or approximate solutions. Typically, some sort of a specific hydraulic conductivity function is prescribed. (iii) Unsteady water flow experiments performed without restrictive conditions that are analyzed with parameter estimation procedures. The reliability of the hydraulic conductivity functions is best tested against the results from an additional, simply formulated and performed field flux experimental test. These additional data are compared with theoretical values from the equation containing the estimated soil hydraulic conductivity function.

7.1 LABORATORY METHODS

Here we concentrate upon principles without going into detail on technique, equipment or algorithms of evaluation. These details are available (e.g. Klute, 1986, Topp et al., 1992).

7.1.1 Steady State Method

The steady state method for determining $K(h)$ or $K(\theta)$ can be performed with an apparatus constructed according to the principle described in Fig. 5.8 (Biggar and Nielsen, 1961). Note that the wall of the core sample must allow an exchange of air while θ changes and the flow is unsteady. After steady state flow ($q_{inflow} = q_{outflow}$) is obtained, evaporation through the wall must be avoided. Quantitative measurements of both inflow and outflow are usually achieved volumetrically. According to the Darcy-Buckingham equation (5.30), $K(h)$ is related to the mean h between the inflow and outflow ends of the sample. The corresponding value of θ can be estimated from the initial soil water content and the difference between volumetric water inflows and outflows during the experiment. Because porous, semi-permeable plate assemblies contribute to the hydraulic resistance of the entire system, a tensiometer is installed just behind the inflow plate and a second one just in front of the outflow plate. The quality of the water used should match that in the local field condition and should be recirculated through the core when necessary. Instead of using simple, hanging water columns, air pressure (like that in the pressure plate apparatus) can be used to control the degree of unsaturation. The steady state method, laborious and time-consuming, is described in detail by Klute and Dirksen (1986).

7.1.2. Transient methods

Transient methods are generally less time consuming and more simply performed, but less accurate compared with steady state methods. There are two kinds of transient methods commonly used in the laboratory: (i) infiltration methods and (ii) pressure outflow methods.

7.1.2.1 Infiltration Methods

Recognizing from (6.41) that $q_0 \rightarrow K_o$ as $t \rightarrow \infty$ for a homogeneous soil profile initially at soil water content θ_i subjected to Dirichlet's boundary condition (6.14b), Davidson et al. (1963) utilized a highly permeable glass bead plate in the laboratory to impose values of $h_o < 0$ at the soil surface of a vertical column to ascertain values of $K(\theta)$. By measuring q_0 versus time in this infiltration method for each imposed value of h_0 until the infiltration rate was essentially constant for two different soils, they suggested a relatively simple laboratory method to measure $K(\theta)$. The convenience of the laboratory allows verification of a constant value of q_0 throughout the upper portion of the wetted profile even though the pressure head drop across the plate may not be nil. Youngs (1964), apparently the first to verify the method using slate dust as a porous material, provided incentive for others to develop the crust-topped soil method now used in the field and discussed in section 7.2.1.

7.1.2.2 Pressure Outflow Methods

In the pressure outflow method a soil core sample placed in a pressure plate apparatus is brought to an equilibrium state between θ and an initially applied air pressure. Subsequently, the pressure is abruptly raised by Δp and the outflow of water is recorded. The pressure steps are purposely kept small in order that D is considered constant over the interval $\Delta\theta$ caused by Δp. From the solution of the diffusion equation (5.73), neglecting the gravitational field and considering only the first term of the series solution, Gardner (1956) obtained

$$ln(V_o - V_t) = ln\left(\frac{8V_o}{\pi^2}\right) - \frac{\pi^2 Dt}{4L^2} \tag{7.1}$$

and

$$K = \frac{DV_o \rho_W g}{V_S \Delta p} \tag{7.2}$$

where V_o is the total outflow volume of water, V_t the outflow volume at time t, L the length of the sample, V_S the bulk soil volume and Δp the air pressure increase [Pa]. The computed D is related to the mean $\overline{\theta}$ in the interval of the applied pressure increase Δp.

In order to increase the efficiency of the procedure, Doering (1965) proposed the *one-step outflow method* based on Gardner's solution (1962) without the restriction of D being constant. The method is also less sensitive to the hydraulic resistance of the plate which several investigators attempted to correct in the previous method. The measuring principle is the same as that of the pressure outflow method except that Δp is substantially greater. The greater is Δp, the more reliable are the results. Based upon numerical simulations, the value of Δp should not be less than 100 kPa. Neglecting the gravitational field, the approximate solution is

$$D = -\frac{4L^2}{\pi^2(\theta_t - \theta_f)} \frac{d\theta_t}{dt} \tag{7.3}$$

where θ_t is the average value of θ integrated over the entire soil sample and θ_f is the final value. The transformation of outflow to the change of θ of the sample is simple. For a measured set of values of V_t the value of D is easily computed and related to appropriate values of θ. With still further analytical development, Muazu et al. (1990) obtained

$$D(\overline{\theta}) = 0.27\,L^2\left(\frac{d^2 V/dt^2}{dV/dt}\right) \tag{7.4}$$

where V the cumulative outflow and

$$\overline{\theta} = \frac{1}{L}\int_0^L \theta\,dx$$

are both evaluated at time t.

All equations show that the time required to reach equilibrium is dependent upon L^2. Because this requirement restricts our choice of the volume of the sample, the criterion of a REV is rarely met. Although some slight improvements of the analytical development can be found in the literature, the primary simplifications remain. For example, Kool et al. (1985) tried to overcome the inaccuracy of the analytical approximation by using an inverse solution for parameter estimation. Nonetheless, restrictions owing to sample size prevailed.

The soil water diffusivity function $D(\theta)$ could be measured on sufficiently long core samples with application of absorption theory, see Fig. 6.11. However, the sampling of 20 or 30 cm long cylinders is difficult without soil compaction. A further difficulty also arises – the measurement of soil water content $\theta(x)$ of those samples taken from the field. Therefore, Dirksen (1975) proposed ascertaining $D(\theta)$ from a set of sorptivities $S(\theta_o)$ where $\theta = \theta_o, x = 0$ and $t \geq 0$. Values of θ_o were governed by the values of soil water pressure head $h_o < 0$ maintained at the inflow end ($x = 0$) of each soil column. Sorptivities were evaluated with (6.35) using the flux density $q_o(t)$ measured at the inflow end of each column. Searching for $D[S(\theta_o)]$, Dirksen combined the simple relation between a constant diffusivity and S with the equation (5.79) for a weighted mean diffusivity $\overline{D}$ to obtain

$$D(\theta_o) = \frac{\pi S^2(\theta_o)}{4(\Delta\theta)^2}\left[\frac{\Delta\theta}{0.725}\frac{d(\log S^2)}{d\theta_o} - 0.198\right] \tag{7.5}$$

where $\Delta\theta = \theta_o - \theta_i$ and θ_i is the initial soil water content kept constant for the entire series of soil columns. The graph of log S^2 versus θ_o is constructed by a polynomial in order to obtain simple readings of the derivatives.

A one step desorption method, conceptually opposite to the above absorption methods, is the *hot air method* (Arya et al., 1975). In addition to the existing difficulties mentioned above, this method neglects the coupled process of heat and water transport induced by the hot air used to enhance evaporation at the end of the soil column. Although according to the authors the soil surface is heated only to 50° C by the stream of hot air, the temperature of the soil surface has been reported to increase to 90° C. Hence, the method is presently not recommended without further improvement

7.1.2.3. Centrifuge Methods

Centrifugal methods offer an opportunity to utilize a body force other than that of gravity. Briggs and McLane (1907) used a centrifugal method to measure the "moisture equivalent" of soils. They introduced the method to estimate the "water holding capacity" of soils using a quantitatively known force "sufficient in magnitude to remove the water held in the larger capillary spaces". The magnitude of the force is easily controlled by choosing the radius and speed of rotation.

Consider a small soil sample initially water-saturated and at rest (we neglect the gravitational field) with the soil water matric potential ϕ_w being everywhere zero. Let the sample be rotated about a radius r at a constant angular velocity ω within a plane parallel to the earth's surface. Its tangential velocity $v = \omega r$ and the its acceleration directed toward the center of rotation equals $v^2 r^{-1}$ (Russell and Richards, 1938). Hence, the centripetal force per mass is $\omega^2 r$. With the sample located between $r = r_1$ and $r = r_2$, we integrate the force along the sample length to obtain $\phi_w(r)$ from

$$\int_{\phi_w(r)}^{0} d\phi_w = \omega^2 \int_r^{r_2} r dr \tag{7.6}$$

where $\phi_w(r_2) = 0$. Within the sample being centrifuged the matric potential ϕ_w eventually becomes invariant with time. Hence,

$$\phi_w = -\frac{\omega^2}{2}\left(r_2^2 - r^2\right). \tag{7.7}$$

Notice that $\phi_w(r)$ is a parabola denoting that the soil becomes drier moving inward along the radius toward the center of rotation. Also note that there is only one force acting on the water and that is the centripetal force stemming from the interaction between the soil matrix and the water. Briggs and McLane, being aware of the confusion between centripetal and "centrifugal" forces made the following footnote in their 1907 paper: "There is, of course, no actual force exerted in a centrifuge machine which tends to pull the water out of the soil. What actually occurs in the machine is just the opposite; that is to say, the soil is pulled away from the water. There is an acceleration of the soil directed inward toward the axis, which leaves the water behind. The result is, of course, the same as if the water were subject to an equal acceleration outward." Hence, when water no longer escapes the rotating sample, the water remaining in the soil is being subjected to an unbalanced force identical to that being received by the soil. That force is the negative derivative of $\phi_w(r)$ in (7.7). The soil receives the force mechanically from the centrifuge while the water receives the force owing to the interaction between the soil matrix and the water.

The moisture equivalent methodology proposed by Briggs and McLane was modified (e.g. Veihmeyer et al., 1924) and remained for many years a standard laboratory procedure for estimating field capacity. Integrating (7.7) and dividing by g and the sample length ($r_2 - r_1$), we find the mean value of soil water pressure head $\bar{h}$ is

$$\bar{h} = -\frac{\omega^2}{6g}(2r_2 + r_1)(r_2 - r_1). \tag{7.8}$$

By trial and error, it was found that a sample length of 1 cm rotated for 30 min at an angular velocity yielding a force of 1000g at r_2 = 15 cm gave satisfactory estimates of θ_{FC}.

Using a centrifuge to establish known distributions of $\phi_w(r)$, Alemi et al. (1976) and Paningbatan (1980) measured $K(\theta)$ on repacked and field soil core samples. With one of their techniques using the method of moments, they gravimetrically measured the redistribution of water within the soil after it had been initially brought to a known $\phi_w(r)$. They also provided another transient technique, a variation of the pressure outflow method, whereby the volume of water draining out of a sample as a function of time is used to calculate $K(\theta)$ when the speed of centrifugation is suddenly increased. The first technique is versatile because porous plates are not required, water does not have to be added or removed from the soil and measurements are made simply with an analytical balance. For the second technique, unlike the pressure outflow methods, the porous plate needs only to support the soil and need not remain water saturated.

More recently, Nimmo et al. (1987) and Conca and Wright (1992) used a steady state centrifugation method to measure steady state values of hydraulic conductivity. Water is supplied at a constant rate by means of a constant, ultra flow pump or by a water reservoir and a porous ceramic plate at the inflow surface of the sample. A constant value of ϕ_w is maintained at the outflow surface of the sample by means of a free-water surface kept slightly above the bottom of the ceramic plate holding the sample. The steady state flux density of water is calculated from volumes of water accurately measured with calibrated reservoirs or strobe lights.

Under a centripetal acceleration when water is moving through a soil, water flows owing to both the matric potential gradient and the inertial effect of the centripetal acceleration ("centrifugal force"). Hence, the steady-state water flux density is described by a modified Buckingham-Darcy equation

$$q = -\frac{K(\phi)}{g}\left(\frac{d\phi_w}{dr} - \omega^2 r\right). \tag{7.9}$$

For large values of ω, $\omega^2 r \gg d\phi_w / dr$ and the value of ϕ_w is nearly constant over the length of the sample. For such conditions with each constant angular velocity established, the value of the derivative is neglected and K for each particular, nearly constant value of ϕ_w is simply calculated by dividing q by $\omega^2 r g^{-1}$. [Note that when q = 0, the parenthetic difference is zero – the starting point above for the derivation of (7.7).] For values of ω for which the matric potential gradient cannot be neglected, further development is required to be able to measure $\phi_w(r)$ *in situ* during centrifugation.

Although centrifuge methods will undoubtedly be improved in the future by incorporating *in situ* measurements of both $h(r,t)$ and $\theta(r,t)$ during centrifugation, their greatest limitation is that of soil deformation, particularly for surface soils having fragile soil structure and small bulk densities.

We note that most all the laboratory methods for estimating hydraulic functions are continually being improved by using new techniques to better control boundary conditions, measuring h and θ at one or multiple locations within the sample as a function of time and exploring improved formulations and estimations of soil water parameters with inverse methods (e.g. Malicki et al., 1992; Wendroth et al., 1993; Dirksen and Matula, 1994; Eching et al., 1994).

7.2 FIELD METHODS

Although many of the laboratory disadvantages can be avoided in the field, field measuring techniques are often less accurate and more expensive. Moreover, it is often difficult to rigorously establish boundary conditions in the field required in the theoretical procedure. In a typical field location, soil water boundary conditions are never invariant – they fluctuate in space and time owing to local weather and regional subsurface conditions modified by irrigation and other cultural activities. When dealing with field measuring methods, it is advisable to first study the daily and annual variations of $\theta(z)$ and $h(z)$. Two examples follow.

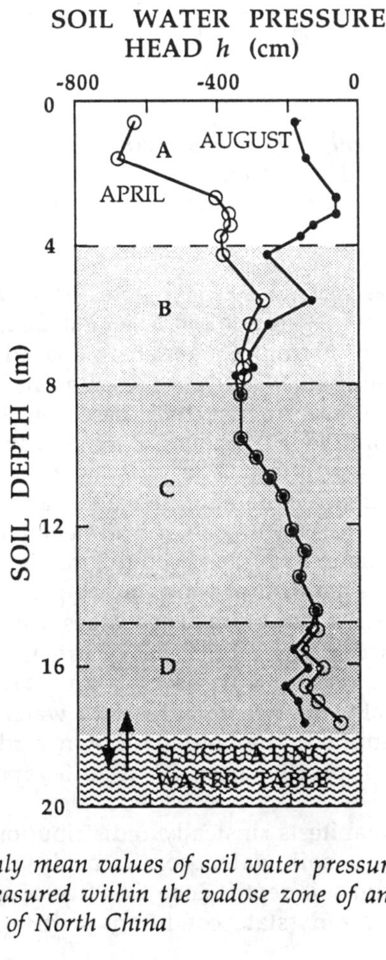

Figure 7.1. Monthly mean values of soil water pressure head h for two months of 1985 measured within the vadose zone of an irrigated soil in a semi-arid region of North China

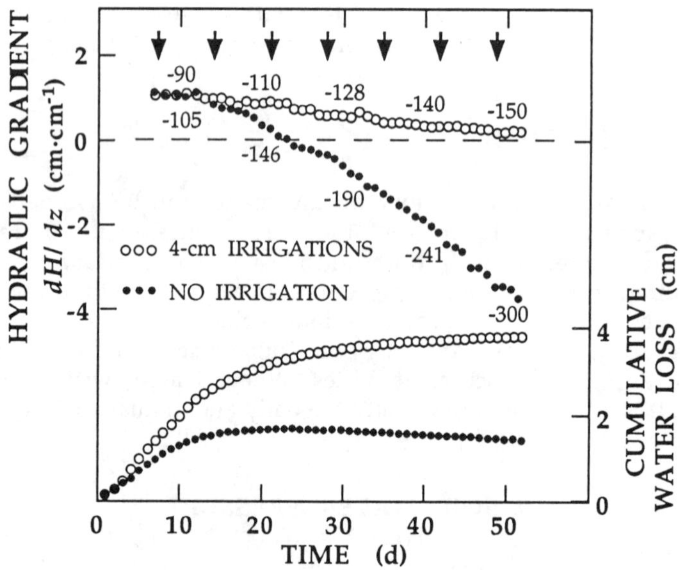

Figure 7.2. Hydraulic gradient, soil water pressure head and cumulative water lost from a 1.8-m soil profile cropped to ryegrass during the summer without any rainfall under two irrigation regimes (LaRue et al., 1968).

Fig. 7.1 shows annual variations of soil water pressure head h measured within the vadose zone of an irrigated soil in a semi-arid region of North China (Yang et al., 1992). Values of h from 65 unevenly spaced tensiometers were recorded every 10 d beginning in 1985. Data presented in Fig. 7.1 are monthly mean values of $h(z)$ for April and August, 1985. Based on the fluctuations of h, the vadose zone is divided into four sub zones. Zone A is strongly influenced by infiltration and evapotranspiration. The soil water potential fluctuates significantly under the influence of rainfall, irrigation and evapotranspiration. The hydraulic gradient manifests both positive and negative values throughout this zone that usually extends to a depth of about 2 m. Fig. 7.2 shows values of h, dH/dz and the cumulative amount of water moving past the 1.8-m depth of a soil uniformly cropped with rye grass. The values were observed for two irrigation regimes established after an initial heavy irrigation. Applying a 4-cm irrigation once a week allowed values of h and dH/dz at the 1.8-m depth to diminish gradually for two months with water draining from the profile. With no water applied, the initially downward flux reversed its direction up into the soil profile after 22 d of evapotranspiration and redistribution.

Zone B in Fig. 7.1 manifests unsteady redistribution with values of h decreasing with depth and soil water continuously moving downward. Seasonal effects decrease with soil depth. Zone C is free of seasonal effects with water moving under near steady-state conditions with a hydraulic gradient close to unity. In Fig. 7.1, the standard deviation of h for the entire year throughout this zone is less than 4 cm. Zone D is characterized by seasonal water table fluctuations. The average daily flux density just above the water table during 1985 was 0.27 $mm \cdot d^{-1}$.

Depending upon the depth of the water table, each of the four zones described above may or may not be apparent. In arid regions, especially when the water table is at great depth, none of them may be readily apparent. Tyler et al. (1994) measured $h(z)$ as well as the relative isotope abundances of ^2H, ^{18}O and ^{36}Cl in 300-m deep boreholes to determine current and past soil water flux densities throughout the vadose zone in an arid environment of Nevada (annual precipitation < 12.5 $cm \cdot y^{-1}$). The distribution of ^{36}Cl indicated that water last moved into the greater depths of the vadose zone beginning 120,000 years ago. Local paleoclimatic indicators clearly show cooler temperatures and much greater effective precipitation at that time – it was the end of penultimate glaciation. Fig. 7.3 shows the soil water pressure head distribution $h(z)$ from the soil surface to the water table (~ 250 m). These recently measured values of h scattered about the broken 45° line indicate an equilibrium condition with $dH/dz = 0$. Indeed, isotope abundances of ^2H and ^{18}O revealed that recharge of water into the vadose zone has not occurred during the past 10,000 - 15,000 y.

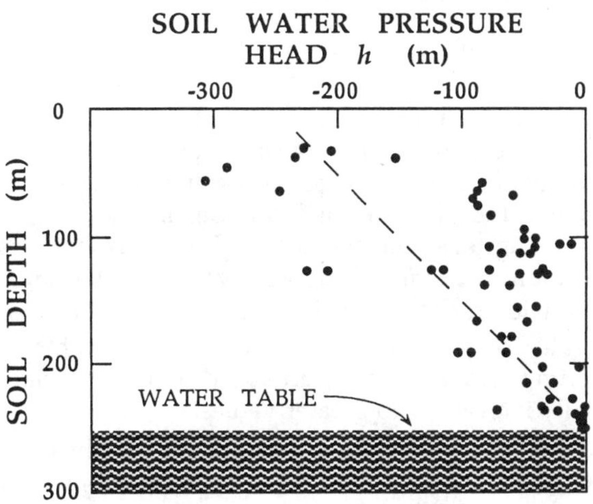

Figure 7.3. Soil water pressure head distribution measured within the deep boring of the vadose zone of an arid environment. The broken line indicates no flow (dH/dz = 0).

We utilize here the experience gained earlier in the laboratory with soil cores and repacked soil columns, and we differentiate between steady and unsteady flow methods. In one group of methods we restrict ourselves to search for a function descriptive of the hydraulic conductivity (e.g. a power or an exponential function). In another group of methods where we seek the hydraulic conductivity without restricting its functional behavior, the results are closer to reality but require more labor and a sophisticated analysis. Because the quality of water can drastically alter the magnitude of soil hydraulic functions (see sections 5.3.2 and 5.3.4) especially in arid and semiarid zones, field investigations should always be conducted with a water source indigenous to the area.

7.2.1 Steady Flow Methods

Although these methods are based upon solutions for steady state infiltration, we rarely achieve upper and lower boundary conditions within a soil profile that satisfy the definition of steady flow. Hence, following from our previous discussion of unsteady infiltration, we assume that quasi-steady conditions are eventually reached after a sufficiently long time even for non-steady infiltration. With this assumption being the only deviation from the theoretical requirements, we consider these steady state field methods as being highly accurate provided that the fluxes are those for a sufficiently large enough area to minimize the errors owing to the divergence of flow paths. Moreover, tensiometers are ideally suited to observe whether or not steady conditions ($\partial h/\partial t = 0$) actually prevail within the profile.

The *sprinkler method* or sprinkler infiltrometer method requires an arrangement of droppers or nozzles supplying the soil surface with a uniform inflow of water such that the flux density at the soil surface q_o is a constant (Peterson and Bubenzer, 1986). We have shown in section 6.2.3 that both θ_o and h_o increase at the soil surface until a value of h_o is reached that allows $dH/dz = 1$, and hence, $q_o = K(h_o)$. The measured value of q_o is that of K commensurate with the tensiometer reading h_o or soil moisture reading θ_o. For each value of q_o, a confirmation of a steady state condition should be achieved using tensiometers to verify that $dh/dt = 0$ rather than the less accurate verification of $d\theta/dt = 0$. Because values of the functions $K(h)$ and $K(\theta)$ are sought for a wide range of h or θ, we initiate the sprinkler method with the least possible value of q_o which is usually about 1 $mm \cdot h^{-1}$. After reaching steady flow for the imposed value of q_o and reading the associated values of h and θ, we sequentially increase q_o to greater values.

The method has two difficulties that are not easily overcome. (i) The application of water must be uniform with respect to rate as well as with respect to its distribution over the entire measured area. A closely spaced network of sprinkler nozzles is required, and in addition, each nozzle delivering identical rates of flow for each and every value of q_o. (ii) The continual sealing of the soil surface during the experiment must be avoided by minimizing the kinetic energy and size of the water drops. Additionally, if distinct layering is present, an accurate measure of the pressure head distribution $h(z)$ is a necessity. Sets of vertical, closely spaced tensiometers provide the best estimate of $h(z)$. Horizontally spaced tensiometers can be used to identify the divergence of flow that may occur in a layered profile. While carrying out the method, the soil surface has to be protected from both rain infiltration and evaporation for considerable periods of time. The method is somewhat laborious and time consuming with the results affected by diurnal temperature fluctuations.

The *crust-topped soil method* relies upon a reduction of the flux density owing to a steep hydraulic gradient across the less permeable crust of the topsoil, see section 6.2.1.2 and Fig. 6.4. Placing the cup of a tensiometer just below the crust enables us to verify that steady conditions are reached. This observation of h together with the control and measurement of a constant flux at the soil surface provides the evidence that $dq_0/dt = 0$ and $dh/dt = 0$. When steady state is reached, $q_0 = K$ and the value of K is related to that of h measured with the tensiometer below the crust. If θ is also simultaneously measured, we obtain $K(\theta)$. Using a laboratory-determined SWRC for recalibration of field-measured h versus θ is not advised even if the wetting branch of the SWRC is used.

Normally, we begin the method with a least value of q_0 using a crust of large resistance. The crust is usually formed from a simple mixture of our field soil and clay – the more clay, the greater the resistance. The flux is derived from cumulative infiltration simply measured as in the case of a ponded double-ring infiltrometer. The resistance of the crust reduces the opportunity for lateral flow.

In more sophisticated equipment, porous ceramic or metallic plates of various resistances are placed on the soil surface inside of the single ring. In order to avoid a lateral component of flow, the soil is sometimes dug out around the cylinder of the infiltration ring with the cylindrical soil column lined with an appropriate impermeable material. Such a procedure is generally not very efficient. The crust-topped soil method may lead to unexpected phenomena. For example, if the resistance of the plate is sufficient to cause a

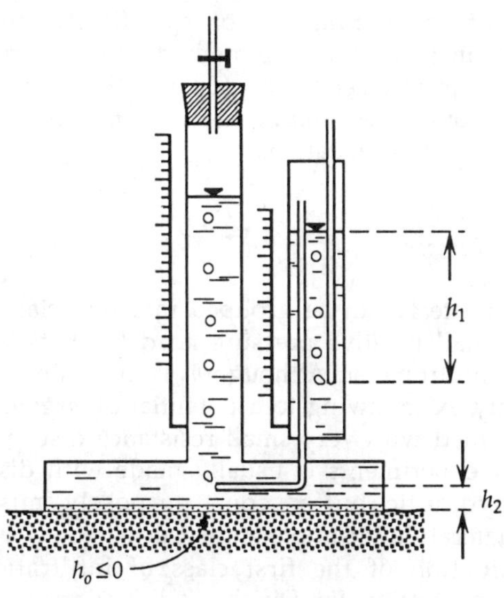

Figure 7.4. Disc permeameter by Perroux and White (1988).

great decrease in the hydraulic head, air excluded from water at the plate-soil interface can effectively plug the pores of the plate rendering the resistance of the plate to be unstable and not constant. Moreover, with the lesser permeable crust overlying the more permeable profile, instability of the wetting front accompanied by fingers within the profile can lead to erroneous estimates of the hydraulic conductivity. The negative pressure at the soil surface is not necessarily only achieved by a hydraulic resistance imposed at the soil surface. an alternative is the application of a controlled constant negative pressure at the soil surface with water infiltrating according to Dirichlet's boundary condition; $h = h_0 < 0$, $z = 0$ and $t \geq 0$. Infiltration is evaluated as unsteady flow, or we monitor the quasi-steady infiltration following the condition dI/dt is a constant and apply the theory of three dimensional steady infiltration. The apparatus used to impose the boundary condition $h_0 < 0$ is called a tension infiltrometer or disk permeameter, see Fig. 7.4. In the next section, we shall describe the apparatus and the evaluation procedure for both the unsteady and steady flow conditions.

7.2.2 Restrictive Unsteady Flow Methods

Among the unsteady flows, infiltration and redistribution of soil water belong to the processes which are relatively simple to execute and monitor in the field. Nowadays, they are the basis of most unsteady field methods.

In the majority of the methods described next, we prescribe the type of function for $K(\theta)$, $K(h)$ or $D(\theta)$ that enters into the mathematical derivation of the evaluation procedure. Therefore, the procedure restricts its search to yield numerical values of the parameters of specifically prescribed functional relationships. In one instance only, the method allows us to evaluate measured data to a single value or to a series of $K_j(\theta_j)$ or $K_j(h_j)$ data. These data are then used as matching factors for estimating $K(\theta)$ or $K(h)$ from the SWRC, see (5.49), (5.51) and (5.52) or analogous equations.

7.2.2.1 Infiltration Tests

Restrictive infiltration tests can be grouped into two classes. (i) Infiltration is performed with a small positive pressure head h_0 at the soil surface, usually with a double ring infiltrometer. Although h_0 is maintained at a small, constant value, an uncertainty exists owing to a potential divergence of flow paths. (ii) Infiltration is performed with very small resistance discs placed upon the soil surface. A series of experiments is usually made with discs of various radii. With the outer hydraulic buffer zone purposely missing, a correction accounting for the lateral flow component is calculated.

For the evaluation of the first class of infiltration tests, equations describing ponded infiltration for Dirichlet's boundary condition are applied, see section 6.2.2. When the test is performed, the cumulative infiltration I versus time t is recorded. A Mariotte flask is commonly used to supply known amounts of water and to keep the water ponded at a constant level. Falling head infiltrometers with h_o changing (e.g. from 15 to 2 cm) should be avoided. The first reading after flooding the surface involves an error composed of two components: (1) The soil surface is not a plane, and (2) the required h_o is not immediately reached at time $t = 0$. This latter error can be substantially reduced by optimization of the first reading. If a pit is excavated, the procedure above can also be performed on the steps of the pit to estimate values of K_S and S of individual layers and horizons of the soil profile. The test should end when the wetting front reaches the next deeper layer or horizon.

The cumulative infiltration measurements are generally best analyzed with one of the three-parameter equations (6.69) to (6.73) or with Parlange-Haverkamp's equations (6.45) and (6.46). Although the three-parameter equations (6.69) to (6.73) are theoretically sound, each of them is very sensitive to local field conditions, e.g. to variations of $\theta_i(z)$, to even the slightest degree of soil layering, to instability of the porous system, etc. These conditions and disturbances in the field sometimes culminate into non acceptable results that yield $K_S < 0$. Philip's algebraic equation (6.67) is relatively reliable for an evaluation of sorptivity S from the early stage of infiltration provided that the ponding does not invalidate the boundary conditions assumed for the solution of Richards' equation. Transformation of A to K_S could be frequently misleading, owing to the unknown value of m in the relationship $A = mK_S$, see our discussion of (6.68).

When the soil layers are thin and infiltration lasts only for a short time dominated by sorptivity, Talsma (1969) proposed estimating K_S from two parallel infiltration tests, each run with different values of h_o. One test would be run with h_{o1} corresponding to S_1, and the other test with h_{o2} corresponding to S_2. Using the equation of Green and Ampt (6.62), we obtain

$$K_S = \frac{S_1^2 - S_2^2}{2\Delta h \Delta\theta} \tag{7.10}$$

where $\Delta h = (h_{o1} - h_{o2})$ if $h_{o1} > h_{o2}$. Because soils are strongly non uniform (see Chapter 8), the pair of tests should be sufficiently repeated to obtain data amenable to a statistical evaluation. Otherwise, two principal errors are included : (i) The error of evaluating S according to Green and Ampt and (ii) the error owing to soil heterogeneity. When we assume a probable error of about 10^1 %, it is advantageous to evaluate K_S from early readings of $I(t)$ according to some of the three-parameter equations – (6.69) to (6.73).

The idea of pairing the infiltration tests at various values of h_o (primarily for $h_o < 0$) is a commonly held proposal of the Australian school. Unhappily, Green and Ampt's theorem is frequently hidden somewhere in the theoretical development. Hence, the approaches lack compatibility with reality because most field soils relative to $D(\theta)$ are frequently more close to the linear soil hypothesis than that of the δ-function (Clothier and White, 1981). Being aware of this type of restriction, White and Perroux (1989) also presented solutions of Burger's linear soil model for infiltration with $h_o < 0$. Analogous to their proposal for unsaturated conditions, the right side of (7.10) should be multiplied by a correction factor $M \approx 0.9$ recognizing that theoretically $\pi/4 \le M \le 1$.

If a soil is homogeneous with depth, we are able to take measurements for times approaching quasi-steady conditions. The time of reaching quasi-steady flow can be considered as roughly equal to t_{grav} of Philip (1957a) where for $t > t_{grav}$ the gravitational force dominates the infiltration process. From

$$t_{grav} = \frac{S^2}{(K_S - K_i)^2} \tag{7.11}$$

an estimate of K_i related to the initial soil water content θ_i can be obtained.

Macropores, frequently in the topsoil, are able to conduct water from the ponded soil surface to greater soil depths to such a degree as to render infiltration data to such an extent that the correspondingly evaluated sorptivity no longer represents the real sorptivity of the soil matrix as defined by (6.32) or, pragmatically, in (6.35), inasmuch as the conditions for the validity of Richards' equation were violated. Because water frequently flows in macropores oriented in directions other than vertical, even the value of q_o for quasi-steady conditions is just an estimate of K_S for the entire profile. Conceptually, q_o is the sum of K_S of the matrix and K_S of the macropore system. There is a distinct discontinuity at $h = 0$ between K_S of the matrix and K_S of the entire soil (including the macropore system). For ponded conditions $[h = (0 + \delta h)]$, measured values of K_S are frequently an order of magnitude greater than those of K_S measured without ponding $[h = (0 - \delta h)]$, see Fig. 5.12a.

Because of the behavior of the macropore system, another class of infiltration tests was developed that utilizes a disc having a very small hydraulic resistance which establishes and maintains a negative value of h at the soil surface, see Fig. 7.4. The resistance of the disc or plate results in a loss of hydraulic head h_c across the plate which is usually negligibly small. When the plate is properly selected and the value of h_o is only -1 to -2 cm, we obtain an acceptable approximation of K_S for the soil matrix which excludes the macropore system.

The crucial point of the inverse solution of infiltration is the de convolution of soil water diffusivity from the integral in the sorptivity formula. In one of the first developments of the problem, Dirksen (1975) explicitly obtained

$$D(\theta) = \pi\left[4(1+\gamma)(\theta_o - \theta_i)^{\gamma}\right]^{-1} \frac{d\left[S^2/(\theta_o - \theta_i)^{1-\gamma}\right]}{d\theta_o} \tag{7.12}$$

where the empirical coefficient γ ranges from 0.50 to 0.67. Note that the use of (7.12) requires the derivative of $S(\theta_o)$, and hence, to obtain a sorptivity value from just one experiment will not suffice. A series of infiltration tests involving negative pressure heads maintained at the soil surface z_o are required to obtain values of S for different values of θ_o. For a value of $\gamma = 0.5$, (7.12) is subsequently mildly modified for soils manifesting very steep functions of $D(\theta)$. The mass balance integral method involving the flux concentration relation was intensively exploited for estimating $D(\theta)$ and $K(\theta)$ from the sorptivity (White and Perroux, 1987, and 1989). However, neither of these theoretical proposals has found a broader application in the field.

Dirksen has already shown the advantage of experimentally determining $S(\theta_o)$ for ascertaining hydraulic functions of soils. Perroux and White (1988) constructed a convenient disc permeameter (sometimes called a sorptimeter or tension infiltrometer) to measure $S(\theta_o)$, see Fig. 7.4. The apparatus uses a double Mariotte bottle which is connected to a porous membrane or disk inside an infiltration ring driven into the soil. The Mariotte bottles produce a constant negative pressure applied to the infiltration surface via the membrane or disk. One Mariotte bottle supplies water to maintain the constant negative h_o while the second Mariotte bottle allows readings of cumulative infiltration from which we obtain the flux q_o, see Elrick and Reynolds (1992) for details. To obtain a value of S, Clothier and Smettem (1990) utilized Wooding's (1968) analysis of steady flow with a constant flux density q_o in Wooding's approximation

$$q_o(h_o) = K(h_o) + \frac{4\phi_o^*}{\pi r} \tag{7.13}$$

where $q_o(h_o)$ is the quasi-steady flux density at the pressure head h_o of the permeameter and ϕ_o^* the matric flux potential (Gardner, 1958) derived from (5.6) and (5.70)

$$\phi_o^* = \int_{h_i}^{h_o} K(h)dh = \int_{\theta_i}^{\theta_o} D(\theta)d\theta. \tag{7.14}$$

ϕ_o^* is related to the macroscopic length λ_c by $\phi_o^* = K_o\lambda_c$ if K_i is neglected and $\lambda = c^{-1}$ in (5.39). With this dependence of K upon h_o clearly understood and with $D(\theta)$ exponentially defined according to (5.74), the value of $S(\theta_o)$ is calculated from

$$S(\theta_o) = \left[\frac{\phi_o^*(\theta_o-\theta_i)}{b}\right]^{1/2} \tag{7.15}$$

where

$$b = D_o\left[\frac{\exp(\beta)-1}{\beta}\right], \tag{7.16}$$

see Thony et al. (1991). For $0 < \beta < \infty$, the range is $0.5 < b < \pi/4$ (Philip, 1969a). When disc permeameters of two different radii (r_1 and r_2) are used for a given $h_o < 0$, we proceed to non-restrictive methods where $K(h)$ is not prescribed as a specific mathematical function but with individual values of $K(h_o)$ being obtained for each series of measurements for particular values of h_o:

$$K(h_o) = \frac{q_{o1}r_1 - q_{o2}r_2}{r_1 - r_2} \tag{7.17}$$

and

$$\phi_o^* = \frac{\pi(q_{o1}-q_{o2})}{4}\left(\frac{1}{r_1} - \frac{1}{r_2}\right)^{-1}. \tag{7.18}$$

In order to minimize the errors owing to soil heterogeneity, the test with each of the two different sized permeameters should be sufficiently repeated to provide adequate statistical evaluation. Sorptivity S_o can be evaluated from early time infiltration according to (6.33) which is $I_o \approx S_o t^{1/2}$ where I_o is the cumulative infiltration for the imposed value of h_o. The limiting time during which the influence of gravity and the presence of 3-dimensional flow can be neglected is discussed by Thony et al. (1991). An alternative procedure has been proposed by Vauclin et al. (1991). If a series of replicated infiltration tests is performed with discs having three different radii, the interception of q_o versus

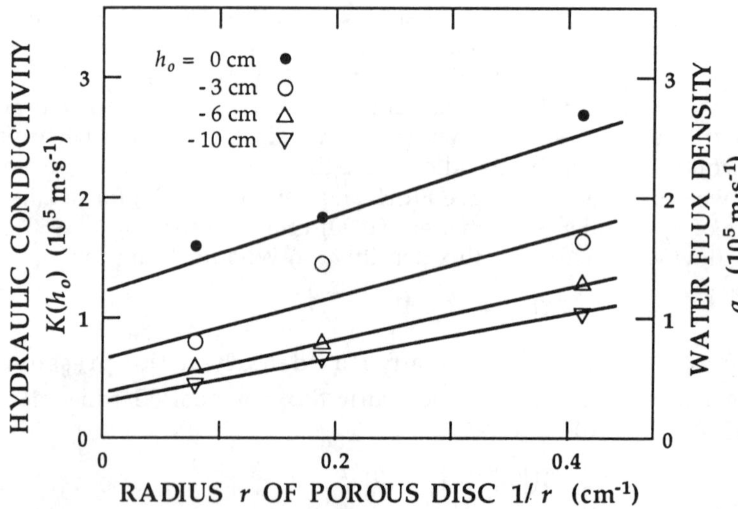

Figure 7.5. Evaluation of soil water flux densities obtained by disc permeameters of various radii r and at various pressure heads h_0 (Vauclin et al., 1991).

r^{-1} provides a value of K related to the value of h_o, see Fig. 7.5. The recent models for disc infiltrometers are discussed in the comparative study of Warrick (1992).

7.2.2.2 Redistribution Tests

The hydraulic properties of soil layers deeper within the profile rely on our knowledge of redistribution of water within the profile and deep drainage from the soil profile after excessive wetting instead of soil water movement during infiltration. In all methods associated with redistribution and drainage, we face the critical problem that the flux is not directly measurable within the soil. The flux is ascertained from the change in soil water storage within soil layers of thicknesses dependent upon the soil water meter resolution and the type of installation of the measuring sensors rather than the naturally occurring genetic structure of the soil profile. Similarly, gradients of H obtained from the direct measurement of h with tensiometers are not necessarily exactly associated with the genetically derived soil layer thicknesses. Inasmuch as a lower limit of measurement of $h \approx$ -600 to -700 cm is achieved through tensiometric observations, measurements of K by redistribution tests cannot be obtained for relatively dry soil water contents. In addition to the restrictive nature of the instruments, the water content of the subsoil during redistribution frequently does not decrease markedly from values close to saturation.

Experiments are best performed during the season when the soil is relatively dry or at least not close to saturation and without vegetation. Within a small, level plot, a set of tensiometers is installed reaching to either equidistant depths throughout the entire profile or to those depths associated with genetically distinct horizons. Triplicate tensiometers at each depth is advisable in order to ascertain if one of the tensiometers is malfunctioning. Access tubes for measuring θ as well as tensiometers should be installed to avoid disturbances that cause leakage along the access tubes or tensiometers.

The plot is usually ponded with water as quickly as possible to wet the profile by infiltration, and if the area of the plot allows it, readings of cumulative infiltration versus time are useful. After a sufficient amount of water has infiltrated to fully wet the upper portion of the profile, the source of water to the plot is discontinued and the time for the initiation of redistribution is recorded when infiltration ceases. Under practical field conditions, owing to the fact that the small plot is never perfectly level, cessation of infiltration is the time when one-half of the plot is covered with water. The tensiometers and the soil moisture meter are read at a frequency to adequately record the continual change of h and θ, respectively. Initially, readings should be taken every few minutes, and as the redistribution slows, once a day or even longer suffices. The impact of evaporation and non isothermal conditions on the redistribution process is avoided by covering the plot with a plastic sheet and an insulating mulch material. Such a covering guarantees that $q_o = 0$ at $z = 0$, and that all changes of soil water content within the soil without a vegetative canopy are attributed to the redistribution process.

The amount of water stored W_t at time t in the profile between the soil surface and depth z is obtained from measurements of $\theta(z, t)$

$$W_t = \int_0^z \theta(z,t)dz. \tag{7.19}$$

During the time interval $\Delta t = (t_2 - t_1)$, the water loss from the soil layer (0, z) is calculated from (7.19) as $\Delta W = (W_{t2} - W_{t1})$. The average flux density during this time interval is

$$\bar{q}(t) = -\frac{\Delta W}{\Delta t}. \tag{7.20}$$

The Darcy-Buckingham equation in finite difference form is

$$q = -K(\bar{\theta})\frac{\Delta H}{\Delta z} \tag{7.21}$$

with $\bar{\theta}$ related to mean $\bar{h}$ of Δh in ΔH. Substituting (7.20) into (7.21) and utilizing our measured values of $\theta(z, t)$ and $h(z, t)$ for an admissibly short time interval Δt, we obtain

$$K(\bar{\theta}) = -\left(\frac{\Delta W}{\Delta t}\right)\left(\frac{\Delta H}{\Delta z}\right)^{-1}. \tag{7.22}$$

The equations above are essentially those of Richards et al. (1956) when they modified the laboratory technique of Richards and Weeks (1953) to calculate the hydraulic conductivity of the surface layers of a field soil from redistribution data following an irrigation. The technique, readily adopted by others (e.g. Nielsen et al., 1964, and Rose et al., 1965), was named the *instantaneous profile method* (Watson, 1966) and is now commonly practiced.

Integrating Richards' equation, we have

$$\int_0^z \frac{\partial \theta}{\partial t} dz = \left[K(\theta)\frac{\partial H}{\partial z}\right]_{z=z} - \left[K(\theta)\frac{\partial H}{\partial z}\right]_{z=0} \tag{7.23}$$

and because $q_o = 0$ at $z = 0$, the last term of (7.23) vanishes. Following the suggestion of Gardner (1962), Davidson et al. (1969) provided the first successful measurements of $K(\theta)$ under the assumption of a unit hydraulic gradient in the drained portion of the profile [$\partial H/\partial z = -1$ in the second term of (7.23)] which were further elaborated starting with Hillel et al. (1972) and Nielsen et al. (1973).

Recognizing that the first term of (7.23) is

$$\int_0^z \frac{\partial \theta}{\partial t} dz = z\frac{\partial \overline{\theta}}{\partial t} \tag{7.24}$$

where $\overline{\theta}$ is the mean soil water content between depths 0 and z, we have

$$z\frac{\partial \overline{\theta}}{\partial t} = \left[K(\theta)\frac{\partial H}{\partial z}\right]_{z=z} . \tag{7.25}$$

With the mean soil water content $\overline{\theta}$ accurately described (Simmons et al., 1979, and Libardi et al., 1980) by

$$\overline{\theta} = a\theta + b \tag{7.26}$$

and assuming an exponential relation for $K(\theta)$

$$K = K_o \exp[\beta(\theta - \theta_o)], \tag{7.27}$$

we obtain for a unit hydraulic gradient (Libardi et al., 1980)

$$az\frac{\partial \theta}{\partial t} = -K_o \exp[\beta(\theta - \theta_o)] \tag{7.28}$$

where θ_0 and K_0 are taken at the end of quasi-steady infiltration (i.e. at $t = 0$ during our redistribution experiment). Note that all parameters of (7.28) pertain to depth z. Integrating (7.28) and neglecting higher order terms in the logarithmic expansion (Libardi et al., 1980) gives

$$\theta_o - \theta = \frac{1}{\beta}\ln t + \frac{1}{\beta}\ln\left(\frac{\beta K_o}{az}\right). \tag{7.29}$$

The slope and intercept (at $t = 1$) of a graph of measured values of (θ_o - θ) versus ln t yields values of $1/\beta$ and K_o, respectively. Taking averages $\overline{\theta}$ and $\overline{\theta}_o$ for the soil layer from z = 0 to z = z, eliminates the necessity of obtaining the value of a. Prescribing that $K(\theta)$ should be an exponential function (7.27) maintains the restrictive nature of the method.

Using an empirical relation for $W(t)$ provides a convenient method for coping with the variability of observations of θ. Inasmuch as $q = -dW/dt$, if we assume the empirical relation

$$W = a\ln t + b, \tag{7.30}$$

we have simply

$$q = -at^{-1}. \tag{7.31}$$

Or, if we assume (Chong et al., 1981)

$$W = At^B, \tag{7.32}$$

then

$$q = -ABt^{B-1}. \tag{7.33}$$

The above expressions can be applied to (7.22) with further developments and simplifications.

If we do not assume an empirical function for $K(\theta)$, the assumption of a unit hydraulic gradient allows us to estimate $K(\theta)$ using (7.22) without measuring h

$$K(\theta)_{z=z} = -\frac{\partial W}{\partial t}. \tag{7.34}$$

On the other hand, if we fit the experimental data to one of the empirical functions $W(t)$ given above [e.g. (7.32)], we have

$$K(\overline{\theta}) = A^{1/B} B W^{(B-1)/B} \tag{7.35}$$

where K is related to $\overline{\theta} = W/z$. Similarly, Chong et al. (1981) obtained for additional readings of $h(t)$ measured at depth z

$$|h| = m t^{n} \tag{7.36}$$

which provides the better estimate

$$K(h)_{z=z} = z A B m^{-(b-1)/n} |h|^{(b-1)/n}. \tag{7.37}$$

Similarly, $D(\theta)$ can be obtained from field readings of $\theta(z, t)$ and $h(z, t)$. Detailed descriptions of measuring methods are available (Green et al., 1986) as well as computer-aided programs (Vachaud et al., 1990b)

In general, the accuracy of field-measured estimates of hydraulic properties will increase as the number of simplifying assumptions decreases. Moreover, the accuracy increases as the number of directly measured data sets increases. Monitoring both $\theta(z, t)$ and $h(z, t)$ during redistribution is recommended. Because of field heterogeneity and the nature of the measurements that manifest differences in resolution of θ and h, it is always advisable to verify the hydraulic functions before they are entered into models. The verification should be done by a simply defined and easily performed field experiment, e.g. ponded infiltration before redistribution starts.

7.2.3 Unrestrictive Unsteady Flow Methods

If no restrictions are imposed regarding the initial condition, and the boundary conditions are allowed to be non trivial, a transient flux experiment is easily performed. Although we speak of non-triviality, we still insist on an exact description of those conditions.

7.2.3.1. Inverse Solutions

When the flux density is measured in an experiment and/or the flux-controlled variables such as $\theta(z, t)$ and $h(z, t)$ are also measured, a general inverse solution can be sought by simulating the flow process numerically. We start the numerical procedure with guessed values of parameters that quantify the searched functions which have a prescribed analytical form. Simulations are reported with improved parameter estimates until the simulated results match the observations. Optimization procedures separate the error in the search of parameters between the two searched functions $h(\theta)$ and $K(\theta)$, see e.g. Kool and Parker (1988).

Because the inverse problem requires sufficient sets of information to guarantee a unique solution (Hornung, 1983), a single set of observations does not provide unique values for the parameters defining the sought functions. For example, measuring only cumulative infiltration $I(t)$ from an infiltration experiment does not yield a unique function $K(\theta)$. The numerical method of solution should be tested with regard to its stability. Unstable solutions react to slight variations of input data to yield disproportionately large perturbations in the solution. Kool and Parker (1988) determined $K(\theta)$ and $h(\theta)$ from flow events during ponded infiltration. Their sensitivity analyses indicated that observations during ponded infiltration should be made near the position of the wetting front. Note that the sensitivity of the parameters is not unique inasmuch as it includes consideration of the observational errors.

The inverse procedure is hastened if we start with guessed values of the parameters obtained from approximate solutions of the problem when the initial or boundary conditions are averaged. Or, the procedure for field data can be initially supplemented by laboratory data which are subsequently revised in the optimization procedure (Šír et al., 1988).

7.2.3.2 State-Space Solutions

Emerging state space formulations now can combine a deterministically based physical equation with a statistically based observation equation to simultaneously examine i. alternative formulations of the differential equation to describe the soil process, ii. alternative functions for soil parameters contained in the equation, iii. alternative frequencies of spatial and temporal measurements to match the theoretical considerations, and iv. the adequacy of accepting different levels of uncertainty always inherent in observation instruments and their calibration including human errors. For example, the description of infiltration and drainage in unsaturated soils is generally complicated by uncertainties in estimating $K(\theta)$.

Katul et al. (1993) formulated state space equations to estimate $K(\theta)$ for a pedon. For the details of our brief presentation here, see their original publications (Parlange et al., 1992, and Katul et al., 1993). After the soil surface of a small field plot had been ponded with water for 10 d, the final infiltration rate was 0.06 $cm \cdot min^{-1}$. Subsequently, the soil surface was covered with a plastic sheet to impose a no-flux surface boundary condition at $z = 0$. One neutron probe access tube and 3 tensiometers per depth located 15 and 30 cm below the soil surface were used to monitor the stored water in the surface 22.5 cm and the hydraulic gradient at 22.5 cm, respectively. Neutron probe and tensiometric observations were recorded every 6 hours for the first day and every 12 h afterwards for a total period of 43 d. The instrumentation noise of the neutron probe was considered a white noise and found to be much smaller than the calibration noise. The standard error of estimate of the calibration curve was equivalent to a soil water content of 0.021 $cm^3 \cdot cm^{-3}$, and was considered to be the standard deviation of the neutron probe calibration.

An optimal nonlinear filter provides hydraulic conductivity function parameters that minimize the mean square error of a differential equation describing the redistribution of soil water content relative to measured soil water content. A simple scheme accounts for both measurement and system uncertainty for processing and updating the time sequence of measurements of soil water content θ and hydraulic gradient dH/dz. The method presumes that the simplifying assumptions for the soil water transport equation result in a random noise to an initial value differential equation characterizing the water stored within the soil profile. The hydraulic conductivity parameters stemming from discrete stored water and hydraulic gradient measurements can then be determined iteratively from a sequence of prediction-updating steps to maximize a defined objective function.

Simplifying Richards' equation for one-dimensional soil water flow to an equation in terms of stored water W in a soil profile of depth b, we have

$$\frac{\partial W}{\partial t} = -K(W)\left(\frac{\partial H}{\partial z}\right)_{z=b} \tag{7.38}$$

where $K(W)$ is the hydraulic conductivity function. In the derivation of (7.38), the following assumptions were made: (i) horizontal transport is neglected, (ii) the depth-averaged soil water content θ^* is described by (7.26) and (iii) thermal and salinity effects on $K(W)$ are neglected. Here we assume that $K(\theta)$ is of the form $A\exp(B\theta)$, but the results can be generalized for any analytic form of the function. In state-space formulation, these assumptions generate a stochastic noise component in the description of time variation of stored water.

In state-space notation, (7.38) may be written as

$$dX(t) = -A\exp\left[BX(t)\right]\left(\frac{\partial H}{\partial z}\right)dt + \varepsilon_s(t)dt. \tag{7.39}$$

The state noise $\varepsilon_s(t)$ results from the various simplifications invoked in the derivation of the soil water transport equation. It is assumed that (7.38) describes the main dynamical characteristics of soil water flow so that a zero mean state noise assumption is appropriate. The product $\varepsilon_s dt$ defines a Wiener increment dI_t of the Wiener process (Gardiner, 1990; Gardner, 1990) with zero-mean and covariance function defined by $Q[\delta(t - s)]dt$. Q is the variance per unit time of the state noise process for which we treat as an unknown to find, and δ is the Dirac delta function.

A corresponding discrete observation equation at time t_k (k = 0, 1, 2,$\cdots$) is

$$Z_m(t_k) = X(t_k) + v_m(t_k) \tag{7.40}$$

where $X(t)$ is now a stochastic state variable representing the stored water W, and $Z_m(t_k)$ is the measured stored water (between z = 0 and b) at time t_k. The stored soil water is measured by a neutron probe with measurement noise $v_m(t_k)$ at time t_k, which is approximated by a zero-mean Gaussian distribution.

The noise components in the observation and state equations permit separate accounting for measurement errors and physical model errors (Gelb et al., 1974). Given the random nature of the additive noises, the state variable $X(t)$ and the neutron probe water storage observations $Z(t_k)$ become random variables in time. Determining the evolution of the mean behavior of $X(t)$ in

(7.39) leads to

$$\frac{d<X(t)>}{dt} = -A\exp[B<X(t)>]\left[\frac{dH}{dz}\right]_{z=b}. \qquad (7.41)$$

Equation (7.41) approximates the dynamic evolution of the mean behavior of the stored water in time between $z = 0$ and $z = b$. Other expressions for the hydraulic conductivity function could have been used [see Wendroth et al. (1993)].

The variance $P(t)$ given by

$$P(t) = \langle [X(t) - <X(t)>]^2 \rangle \qquad (7.42)$$

is predicted with

$$\frac{dP(t)}{dt} = -2\left\{\left(\frac{dH}{dz}\right)AB\exp[B<X(t)>]\right\}P(t) + Q \qquad (7.43)$$

where Q is the spectral density function or error variance per unit time (Gelb, 1974).

In order to integrate the extended Kalman filter equations for continuous dynamic (7.41) and discrete observations (7.43) for $<X(t)>$ and $P(t)$, the coefficients A, B, and Q as well as the initial conditions for the mean $<X(t_o)>$ and the variance $P(t_o)$ need to be known. Once the initial conditions are specified, (7.41) and (7.43) can be simultaneously integrated to yield a prediction of $<X(t)>$ and $P(t)$ at time t_k when an observation $Z(t_k)$ is available. At time t_k predictions of $<X(t)>$ and $P(t)$, denoted by $<X(t_k)>^-$ and $P(t_k)^-$, respectively, are corrected by employing the information contained in the neutron probe measurement $Z(t_k)$. Because of the measurement error variance R in $Z(t_k)$, the predictions and measurements have to be weighted according to their respective variances in the updating steps. The solution requires equations for the updated mean $<X(t_k)>^+$, the updated variance $P(t_k)^+$ and K_g, the Kalman Gain

$$K_g = \frac{P(t_k)^-}{P(t_k)^- + R} \qquad (7.44)$$

which defines how to weigh the two sources of information – predictions and observations.

The structure of the nonlinear extended Kalman filter is shown in Fig. 7.6. For determining the hydraulic conductivity, values of $<X(0)>$, $P(0)$, A, B, R, and Q are needed. The initial mean $<X(0)>$ was taken as the water storage between $z = 0$ and b immediately after ponding, and the observation error variance R was determined from the neutron probe calibration curve. The hydraulic conductivity parameters A and B as well as Q and $P(0)$ were determined following an iterative multivariate optimization scheme. Values of the initial mean $<X(0)>$ and R were (0.478 x 22.5 cm) and $(0.021 \times 22.5\ \text{cm})^2$.

The raw data for θ^* and the hydraulic gradient dH/dz are given in Fig. 7.7 for the 43-d redistribution period. The estimated values of A, B, Q, and $P(0)$ were 1.244×10^{-8} cm·d^{-1}, 1.625 cm^{-1}, $2.413 \cdot 10^{-2}$ cm^2·d^{-1} and $1.504 \cdot 10^{-2}$ cm^2, respectively.

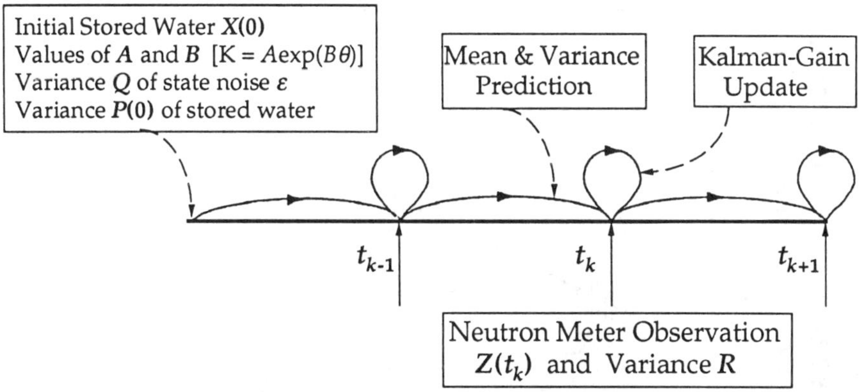

Figure 7.6. Structure of the prediction-updating scheme for the extended Kalman filter.

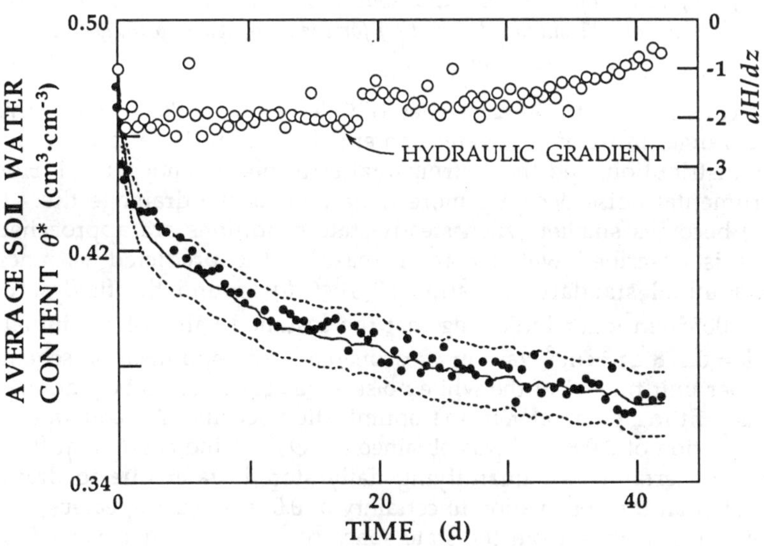

Figure 7.7. Measured values of average soil water content θ^* and hydraulic gradient dH/dz, and predicted values of θ^* (solid line) including ±1 standard deviation (broken lines).

With these final estimates of A, B, Q, and $P(0)$, the predicted average soil water contents and their standard deviations are also shown in Fig. 7.7.

The variation of the standard deviation in time indicates that the proposed dynamic equation performs better during rapid drainage (Fig. 7.7),

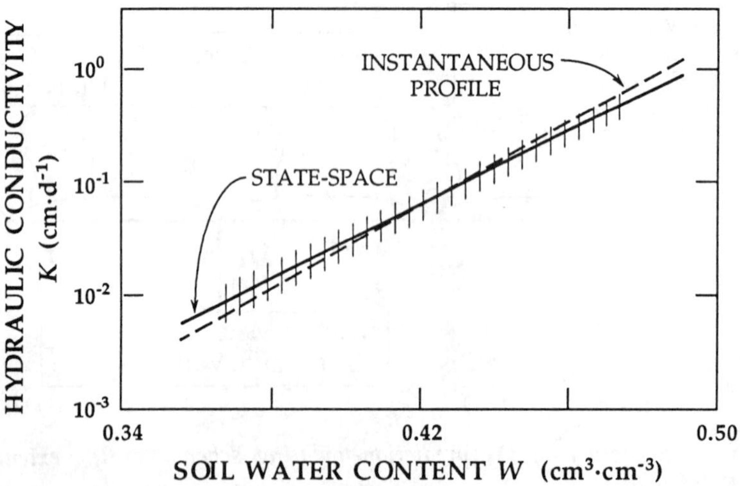

Figure 7.8. Hydraulic conductivity functions determined by the state space formulation and the instantaneous profile method (7.22). The vertical lines denote ±1 standard deviation for the state space formulation.

which is consistent with other studies (Flühler et al., 1976). Also in this near saturated range, most of the water content fluctuation is caused by drainage during redistribution, and the instrumental error noise is not critical. However, the instrumental noise becomes more noticeable as the drainage flux at $z = b$ (22.5 cm) becomes smaller when steady-state conditions are approached (Fig. 7.7). This is described well by an increase in the predicted variance. The computed initial standard deviation $[P(0)^{1/2}/b]$ around the field saturation value is 0.0055 $cm^3 \cdot cm^{-3}$ indicating that the assumed value of the initial mean $<X(0)>/b$ = 0.478 $cm^3 \cdot cm^{-3}$ was very reasonable in the optimization scheme. The variance per unit time Q of the white noise ε was assumed to be stationary and treated as a fitting parameter in the optimization scheme. An equivalent water content variation of 0.0069 d^{-1} was obtained for $Q^{1/2}/b$ indicating that the model prediction uncertainty for quantifying daily stored water due to drainage is 0.0069. Note that the observation uncertainty of 0.021 is much greater .

The $K(\theta)$ relation from the state space solution and that using (7.22) are shown in Fig. 7.8. Unlike current field methods which provide $K(\theta)$ relations in the mean from arbitrarily smoothed field observations, this example of state space using Kalman filtering provides mean estimates of $K(\theta)$ and their variances based on direct raw measurements. It also allowed a quantitative assessment of the validity of equation (7.38) on a daily basis, based on the state variance per unit time Q. For a daily time step, the isothermal and nonhysteretic assumptions assumed for this transient field experiment were acceptable

For the state-space solution above, the original partial differential equation of Richards must be reduced to an ordinary one. This reduction is achieved by adding more information, e.g., providing a functional behavior of transfer coefficients, neglecting second order or less important features of the processes, and deterministically defining magnitudes of more complicated initial and boundary conditions. With these and other approximations, the simplified ordinary differential equation is certainly not exact, but on the other hand, no equation is absolutely exact, particularly when it is used to describe soil hydrology. The advantage of the method over most all other field methods is that we acknowledge that the equation used is definitely approximate and contains a model error. Indeed, we explicitly solve for the model variance, and through its examination, ascertain the impact of our simplifications, and potentially identify improvements for a more realistic equation.

The state-space approach includes an observation error which can be treated as a known, measured quantity as was done above with the neutron probe calibration, or alternatively, treated as an unknown for which a solution is found in the numerical scheme. The magnitude of a known observation error allows a reconsideration of the state variable in the equation or an improvement in instrumentation or calibration. On the other hand as we discuss in the next chapter, treating the observation error as an unknown, its behavior in space and time can be related to spatial and temporal correlation lengths that may manifest themselves within the domain of the field being studied.

Regardless of the instrument or field sampling technique used to observe a soil physical process, most measurements represent some sort of depth-averaged value. Hence, it is not only convenient but more correct to use equations that describe depth-averaged phenomena. A large array of combinations of state and observation variables together with different functions for parameter estimation provides attractive opportunities to enhance the maturation of our field technology in soil hydrology.

PROBLEMS

1. When sorptivity $S(\theta_i)$ was determined from the ponded infiltration test, derive the explicit form $D(\theta)$ from (6.51). Assume an exponential form of $D(\theta)$, see (5.75) and take $D(\theta_R) = \theta_R$.
2. Derive a transformed (7.3) when the volumetric soil water content θ replaces the volume of outflow V.
3. In the pressure outflow method a constant reading of the outflow volume was reached after 22 h for a soil sample having a height of 3 cm. For an identical soil having a height of 6 cm, how long will it take to reach equilibrium for the same Δp?
4. What is the value of the soil macroscopic length λ_c when c is related to the c in (5.39) if the simulated rainfall rate q_o = 6 mm·d^{-1}, K_S = 150 cm·d^{-1} and the tensiometer reading was h = -120 cm? Assume that steady state infiltration was reached.

5. Evaluate the unsaturated hydraulic conductivity K from the following disc permeameter data:

h (cm)	r (cm)	q_o (cm·min^{-1})
-2	15	$7.2 \cdot 10^{-2}$
-2	5	$14.5 \cdot 10^{-2}$
-10	15	$1.8 \cdot 10^{-2}$
-10	5	$2.8 \cdot 10^{-2}$

6. Evaluate the unsaturated hydraulic conductivity K by the instantaneous profile method. During redistribution, we have measured θ and h on a 1-day time interval at depths z_1 = 20 cm and z_2 = 40 cm.

day	depth	θ (cm^3·cm^{-3})	h (cm)
first	z_1	0.345	-122
first	z_2	0.358	-131
second	z_1	0.307	-142
second	z_2	0.329	-150

7. If a unit hydraulic gradient exists for a long time during redistribution after infiltration from the soil surface to depth z , show that (7.32) is obtained if it is assumed that $K(\theta)=K_o\left(\theta \cdot \theta_o^{-1}\right)^n$ and (7.26) is satisfied. The value of A turns out to be a function of a, n, z, θ_o and $\overline{\theta}_o$ while that of B is a function of only n.
8. Starting with (7.21) and assuming that a unit hydraulic gradient exists during redistribution after infiltration from the soil surface to depth z and that the depth-averaged SWRC $\overline{\theta}(h)$ (from the soil surface to depth z) is identical to the SWRC $\theta(h)$ at depth z, show that the soil water diffusivity $D(h)$ at depth z could be estimated with readings from only one tensiometer at depth z. In other words,

$$D(h) = z\frac{dh}{dt}.$$

9. Derive an expression similar to (7.25) that accounts for a constant evaporative flux E at the soil surface.
10. Does the rotation of the earth have any influence on the redistribution of water within a 2-m deep soil profile? An answer can gained by neglecting its huge mass, using the radius of the earth ($r = 6.4 \cdot 10^8$ cm) in (7.7) and recognizing that it revolves on its axis once a day. Calculate $h(r)$ for a rotating column of soil 100 cm long.
11. With the measured observations of $H(z)$ given in Fig. 6.37 and assuming an upward flux E = 3.5 mm·d^{-1}, compute and draw $K(h)$ for $-400 < h < -100$ cm. Consider the soil homogeneous for $10 \leq z \leq 30$ cm.
12. A soil core sample of height L = 3 cm and volume V = 67 cm^3 taken from a loamy sand fluvisol was analyzed with the one step outflow method in the laboratory. Initially at h = -10 cm (θ = 0.374 cm^3·cm^{-3}), the sample was subjected to a pressure head step Δp = -190 cm. The following pairs of values (Q_i,t_i) for the cumulative outflow Q (ml) at time t (min) were recorded: (0.55, 0.5), (1.55, 1), (2.10, 1.5), (2.75, 2.5), (3.10, 3), (3.85, 4), (4.35, 5),

(4.65, 6), (5.20, 7), (5.55, 8), (5.80, 9), (6.10, 10), (6.40, 11), (7.00, 15), (7.85, 25), (8.50, 36), (8.80, 45), (8.95, 50), (9.05, 55), (9.35, 65), (9.50, 75), (9.70, 85), (9.80, 95), (10.05, 115), (10.25, 135), (10.40, 155), (10.65, 215), (10.80, 265), (10.90, 305), (11.00, 385), (11.15, 445) and (11.15, 555). Using the Gardner-Doering procedure (7.3) and that of Muazu et al. (7.4), calculate $D(\theta)$. Compare both results.

13. Using the results from the previous problem, compute $K(\theta)$ for $K_S = 0.09$ cm·min^{-1} and for $\alpha = 0.03$ cm^{-1}, $n = 1.45$, $\theta_S = 0.385$ cm^3·cm^{-3} and $\theta_r = 0.105$ cm^3·cm^{-3} in (4.43).

8 FIELD SOIL HETEROGENEITY

Through their genetic development, soils are spatially variable natural bodies. Variations of parent material and vegetation across the landscape from which soils are derived influence the variability of soils even at relatively short distances. Hence, all soil properties, be they ascertained from visual observations of soil profile morphology in the field or those determined analytically in the laboratory or the field, change from one location to another. The heterogeneity of many soils within present-day mild climatic zones reflect variations of Holocene climate as well as cryoturbate features of Pleistocene glacial - interglacial periods of the Quaternary. Soil variability increases with parent material according to the sequence loess, till fluvial deposits, tectonic rocks, moraines, cryoturbates and gravity disturbates. Pedoturbation like gilgai increases variability over short distances within vertisol profiles, while soils of polygenetic character vary more in space than monogenetic soils.

We have already shown in Chapter 2 that homogeneity in soils does not exist unless we consider the concept of a representative elementary volume REV. And, if our scale of observation is less than that of the REV, the system is heterogeneous. In Chapter 2 we described REV for porosity. Other soil properties (physical, chemical and microbiological) can be treated in a similar manner. Because each soil property manifests its own REV, a value of an REV common for all kinds of observations does not exist. Generally, values of REV for capacity or static parameters differ from those for transport parameters. Moreover, values of a REV of the same property for different soils are not identical. For example, soil aggregation and the presence or absence of peds influence the value of the REV for the saturated hydraulic conductivity.

If the sampling volume is substantially increased above the REV, variations of the soil property are again evident. After briefly providing data on the variability of soil properties, we shall classify this kind of heterogeneity in relation to scale of observation.

Replicated measurements are frequently not statistically independent. In other words, soil properties measured at nearby locations vary less than those measured further apart. Because this spatial correlation impacts the results of classical statistics, the spatial variability of soils is often analyzed using regionalized variable analysis, geostatistics and applied time series analysis to identify different spatial and temporal structures within and between data sets.

Scaling techniques based upon similarity theory provide an additional opportunity to unify spatially variable field soil properties. Below we discuss statistical and scaling methods and demonstrate their practical use.

8.1 VARIABILITY OF SOIL PHYSICAL PROPERTIES

A table by Wilding (1985) summarizing the variability of selected soil properties expressed as a coefficient of variation is reproduced here as Table 8.1. Wilding differentiates between static soil properties (e.g. organic matter, texture, mineralogy, solum depth and soil color) and dynamic soil properties (e.g.

Table 8.1. Relative ranking of variability of soil properties that occur within landscape units of a few ha or less (Wilding, 1985).

variability of property	property
least (coefficient of variation < 15%)	soil color (hue and value) soil pH thickness of A-horizon total silt content plasticity limit
moderate (coefficient of variation 15 to 35%)	total sand content total clay content cation exchange capacity base saturation soil structure (grade and class) liquid limit depth in minimum pH calcium carbonate equivalent
most (coefficient of variation > 15%)	B2-horizon soil color (chroma) depth to mottling depth to leaching (carbonates) exchangeable H, Ca, Mg and K fine clay content organic matter content plasticity index soluble salt content hydraulic conductivity soil water content

hydraulic conductivity, soil water content, salt content, microorganisms, exchangeable cations and redox conditions). Moreover, he concludes that properties closely calibrated against a standard are less variable than those which are more qualitative (texture, color and pH versus structure, consistence and root abundance).

Based upon a large number of studies conducted during the last decade regarding spatial variability, it appears more appropriate to group soil physical properties into two classes (i) capacity parameters and (ii) transport parameters. Capacity parameters include the content of sand, silt or clay particles, organic matter content, porosity and soil water content after a sufficient period of time to remove any local perturbation caused by an event at the upper boundary of the soil profile. Such parameters usually denote static soil properties. Transport or dynamic parameters comprise hydraulic conductivity, soil water diffusivity and fluxes of water and solutes. Jury (1989) reviewing data of various authors found that soil bulk density has the smallest variability of all soil physical characteristics with a CV generally smaller than 10%. While most values of porosity have a CV of about 10%, the variability of textural categories and the 15-bar water content is characterized by larger CV values ranging between 15 to 50%. Saturated and unsaturated hydraulic conductivity values as well as transport parameters characteristic of solute behavior are in the majority of instances well above 100%.

The variation of soil physical properties is better understood and more completely described if the probability density function (*pdf*) is estimated or ascertained. This *pdf* for an idealized normally distributed set of observations x is defined as

$$pdf = \frac{1}{\sigma\sqrt{2\pi}} \exp\left[\frac{-(x-\mu)^2}{2\sigma^2}\right] \tag{8.1}$$

where μ is the mean and σ the standard deviation of the mean. For a set of n observations, the value of the $pdf = N/(n\Delta x)$ where N is the number of observations of x within a class size $[x \pm (\Delta x)/2]$. This function is graphically demonstrated in Fig. 8.1a for 120 observations of soil water content θ. The solid line is described by

$$N/n = \frac{\Delta\theta}{s\sqrt{2\pi}} \exp\left[\frac{-(\theta-m)^2}{2s^2}\right] \tag{8.2}$$

where m (= 0.433 $cm^3 \cdot cm^{-3}$) is the estimate of μ, s (= 0.0455 $cm^3 \cdot cm^{-3}$) the estimate of σ and the class $\Delta\theta$ = 0.01 $cm^3 \cdot cm^{-3}$. We see that the soil water content, typical of commonly found capacity parameters, is normally distributed and manifests the well-known Gaussian or bell-shaped curve. We note that the mean as well as the median and the mode are identical and occur at the center and top of the bell-shaped curve. One half of the observations are smaller and half are larger than the median value. The mode represents the value which occurs most frequently.

When transport coefficients (e.g. K_S, D and K for given θ or h) or fluxes for a given boundary condition (e.g. the infiltration rate at a particular time t) are plotted as a *pdf*, their curves do not resemble the Gaussian shape. Inasmuch as their values vary by orders of magnitude, a Gaussian shape is sometimes manifested when they are plotted on the horizontal axis having a logarithmic scale or when their logarithmic values are plotted on the axis. Such is the case for 120 observations of hydraulic conductivity K given in Fig. 8.1b. The solid line is that of (8.2) except that ln$K$ has been substituted for θ with a class of $\Delta(\ln K)$ = 1 and the other parameters calculated using the logarithmic values of K. Hence, we speak of a log-normal distribution (Nielsen et al., 1973).

A more objective test is usually used to determine if a distribution is log-normal. A better procedure consists of cumulative values of probability based upon relative number of observations plotted against the data arranged in monotonically increasing values. The cumulative probability function is

$$P\{u\} = \frac{1}{\sqrt{2\pi}} \int_{-\infty}^{u} \exp\left(-t^2/2\right) dt \tag{8.3}$$

where $u = \{[g(x) - \mu]/\sigma\}$ and $g(x)$ is the function that transforms the set of data x

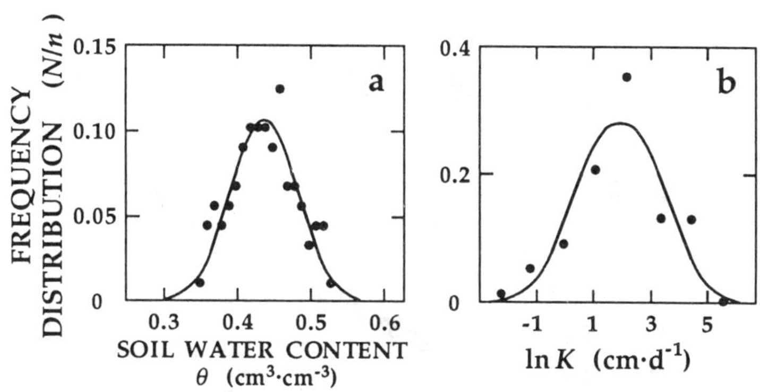

Figure 8.1. a. Probability density function of soil water content during steady infiltration manifests a normal distribution (Nielsen et al., 1973). b. Probability density function of lnK manifests a normal distribution.

into a normal distribution. For a set of n observations, the value of $P\{u\}$ is approximated by $(i - 0.5)/n$ with corresponding values of u obtained from tables of $P\{u\}$ for each observation $i = 1, 2, 3, \cdots, n$. If we choose $g(x) = x$ and the data fit a straight line, the values x are said to be normally distributed. If we choose $g(x) = \ln x$ and their logarithmic values fit a straight line, the original data x are said to be log-normally distributed. In Fig. 8.2a we see that a plot of our K values is strongly curvilinear while that of $\ln K$ is nearly a straight line. Apparently, the function $g(x) = \ln x$ transforms our 120 values of K into a normally distributed set of observations seen in Fig. 8.2b and described by

$$N/n = \frac{\Delta K}{K s_g \sqrt{2\pi}} \exp\left\{\frac{-[g(K) - m_g]^2}{2 s_g^2}\right\} \tag{8.4}$$

where $g(K) = \ln K$, $m_g = 2.28$, $s_g = 1.40$ and the class $\Delta K = 10$ cm·d^{-1}. Unlike a normal distribution, the mode, median and mean have different values. The mode, equal to $\exp(m_g - s_g^2)$ is 1.39 cm·d^{-1}, the median, equal to $\exp(m_g)$ is 9.74 cm·d^{-1} and the mean, equal to $\exp(m_g + 0.5 s_g^2)$ is 25.79 cm·d^{-1}. For a log-normal distribution, we note that only a few very large values yield a mean value substantially larger than that of the majority of the observations. Not recognizing that the observed values of K were not log-normally distributed would cause the mean to be underestimated by about 20% [(25.79 - 20.63)100/25.79 where the arithmetic mean is 20.63 cm·d^{-1}].

A still better procedure for ascertaining the type of *pdf* involves the Kolgomorov- Smirnov test or the χ^2 test. However, no test is considered an absolute proof regarding the type of *pdf*. In addition to normal and log-normal distributions, many other types of *pdf* such as the Weibull, Gamma or Erlang functions can be advantageously used to fit sets of sampled data.

When a property A is studied, an observational sample of A is the estimate $(A+\varepsilon)$ where ε is the error of the measuring method. The real *pdf* of A

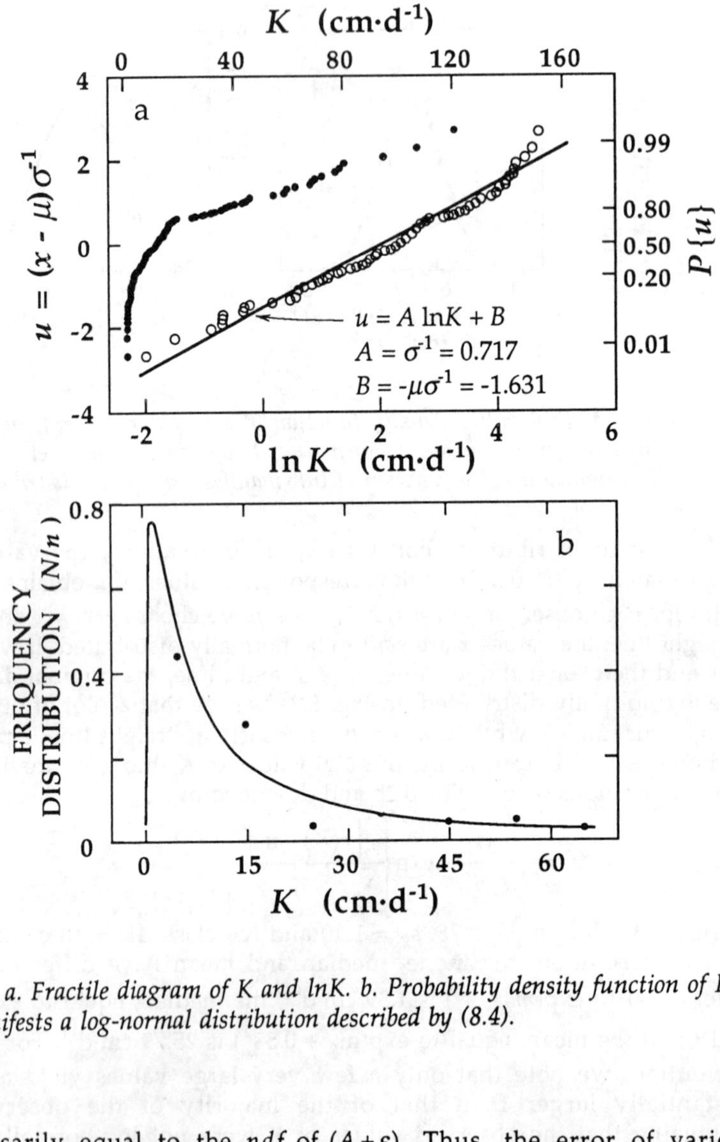

Figure 8.2. a. Fractile diagram of K and lnK. b. Probability density function of K manifests a log-normal distribution described by (8.4).

is not necessarily equal to the *pdf* of $(A+\varepsilon)$. Thus, the error of various experimental methods deforms our observation of A. For example, if K_S is determined by three different methods and if by each method we obtained the subset of K_S of the same population, we can compare the *pdf* of the subsets. When *pdf* are of the same type only for two different methods, only those estimates of K_S are mutually compatible.

8.2 CONCEPT OF SOIL HETEROGENEITY

Soil heterogeneity will be discussed with the aim of a realistic description of soil hydrological processes. Soil heterogeneity can have either stochastic or deterministic character. Stochastic heterogeneity concerns soils with randomly variable properties defined as a realization of a random function studied through observation and data sampling. The realization is unique and deterministic but unknown to us.

Deterministic heterogeneity is related to our knowledge regarding a distinct change of soil properties, as e.g. the existence of layers within a profile or boundaries between soil districts. With a distinct change of a soil taxon we expect a change of soil physical properties. Two neighboring taxons, each characterized in space by randomly varying hydraulic properties, are distinguishable if the first and second moments for each taxon are spatially distinct. One of the aims of statistics is to confirm this distance and to objectively define the boundary between the two taxons if such a boundary even exists. Hence, we use statistics to quantify soil hydraulic characteristics which are deterministically related to pedological characteristics. These hydraulic properties characterizing districts of pedotaxons are frequently given on soil maps. In each of the districts, random heterogeneity exists. Hence, a mosaic of deterministic districts each manifesting random heterogeneity comprises soil maps. We assume the links between hydraulic and pedological properties may be complemented by geomorphological and geobotanical features.

For field soil hydrology we start with the *macroscopic* or *Darcian scale* which for field conditions is related to the pedon, and therefore we speak of a pedon scale. Its size is determined by the REV related to transport characteristics, mainly K_S. Such observations for the solution field problems are frequently called "point" data. In relation to soil survey, pedon is considered a homogeneous unit having a size of about 10^0 m^2.

On the farmer's field we frequently deal with a set of pedons belonging to the district of the lowest taxonomic unit - pedotop described as a family or soil series. The stochastic character of soil heterogeneity of such an areal unit having a size of about 10^4 m^2 and possibly much more is included in the *soil series* (or *pedotop*) *scale*. Examples of variability at this scale were shown in section 8.1.

On the *watershed or regional scale*, the lowest pedotaxons are clustered to soil mapping units of higher pedotops. At this realistic megascale, objective, comprehensive studies of water and solute fluxes have yet to be experimentally measured in the field. A surrogate analysis is presently provided by introducing pedotransfer functions which relate soil hydraulic functions to basic and simply measured soil data such as the content of clay, silt, sand, organic matter and values of bulk density or porosity. Although those functions obtained by regression analyses can not replace direct measurements and estimates of soil physical properties [e.g. K_S or $h(\theta)$], they can improve interpolation between field experimental points, i.e. to extend pedon-size data to entire pedological districts. In the remainder of this chapter, we deal primarily with the soil series scale (pedotop scale).

8.3 SPATIAL VARIABILITY AND GEOSTATISTICS

Soils vary continuously in space, especially when we consider domains at the soil series scale. However, our measurements of soil physical properties yield sets of discrete values for particular "point" locations within our soil sampling area. For a more complete interpretation of those discrete measurements, the theory of regionalized variables transforms our point discrete data to the soil continuum.

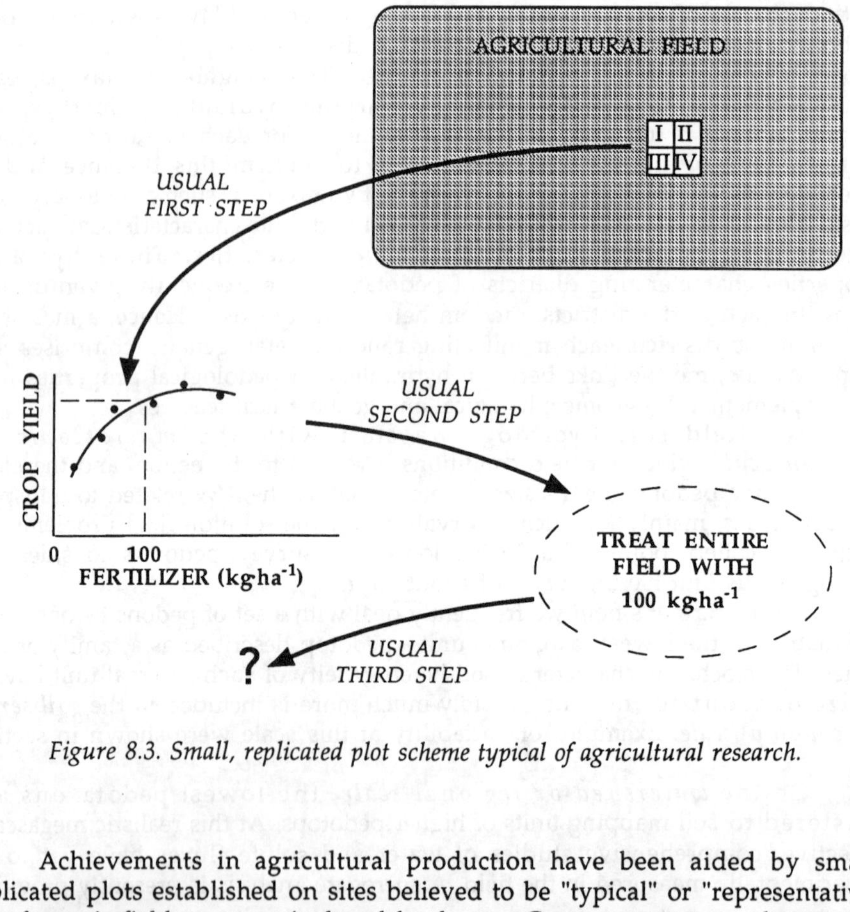

Figure 8.3. Small, replicated plot scheme typical of agricultural research.

Achievements in agricultural production have been aided by small, replicated plots established on sites believed to be "typical" or "representative" of a farmer's field or an agricultural landscape. Crop responses on these small plots treated with fertilizers, pesticides, irrigations, etc. were interpreted and formed the basis for uniformly treating an entire field. Figure 8.3 illustrates the all-too-common method today of conducting agronomic field research. In an agricultural field an experiment is located at a site which is assumed to be "representative" of the rest of the field as well as the soil of the agricultural region. With replicated treatments established in four statistical blocks, the first step is to establish the regression between crop yield and treatment. In this example, a maximum yield is measured at about 150 $kg{\cdot}ha^{-1}$ fertilizer. Being mindful of the desire not to waste fertilizer or pollute the subsoil environment with excess fertilizer and also noting that 100 $kg{\cdot}ha^{-1}$ fertilizer produced a yield close to the maximum, the second step is a recommendation to apply 100 $kg{\cdot}ha^{-1}$ fertilizer over the entire field. With an increased yield being obtained when the recommendation is followed, the farmer is happy. What is the usual third step taken in the above experimentation? It varies from repeating the same experiment or modifying its treatment levels, going to another field or soil condition or doing nothing more. Doing nothing more does not benefit the farmer nor does it benefit agriculture or the environment. Another sampling across the entire field could detect specific locations within the field where, for example, crop production could be increased, excessive levels of the fertilizer nutrient and deficiencies of other plant nutrients prevail or changes in soil quality indicative of achieving or denying sustainable agriculture are observed. Yes, another sampling across the entire field would ascertain if the farmer could make still further improvements in crop and soil management without necessarily imposing different treatments in still another replicated small plot experiment. Similar inefficient procedures can be found in research and in practical applications of soil hydrology.

Assuming steady-state conditions, deterministic concepts and mass balance equations have usually been applied for relatively short time periods – minutes, days, weeks or for times no longer than a growing season. Attempts to assess the impacts of agricultural methods and weather events between and during several growing seasons have been made through long-term experimental plots managed for decades, and in a few cases for more than a century. These kinds of experiments remain effective today when one wishes to ascertain the effect of a particular treatment relative to crop or animal production. On the other hand, they do little for improving our understanding of how agricultural practices impact on the quality of water leaving a cultivated field or rangeland. Moreover, they provide no direct information regarding the all-too-often subtle changes in soil quality occurring on a farm or within an agricultural region.

During the century, we have asked the question, "Does the treatment cause a significant perturbation from the expected mean?" We have used analysis of variance and regression techniques designed to minimize the impact of spatial or temporal heterogeneity in field soils and until recently, have not even taken the time to record where we take an observation within an experimental plot or field.

Regionalized variable analysis considers the distance between pairs of measured values as the main criterion in dealing with the variance. Here lies one of the main differences between classical "Fisher" statistics and geostatistics. Whereas the coordinate system is ignored in classical "Fisher" statistics, it is used in geostatistics to better answer the question, "How far apart should we take our samples?"

8.3.1 Autocorrelograms and Semivariograms

A geostatistical evaluation of data is performed using two tools; (i) the autocorrelogram and (ii) the semivariogram. Both tools can be used for data obtained in 1, 2 and 3 dimensions as well as in time. For the sake of simplicity here, we limit our description and their construction for sampling along a transect at regular intervals.

For an explanation of an autocorrelogram, we consider a transect of equidistant sampling and measurements of soil property A. We obtain a value of A at location x_1 designated as $A_1(x_1)$, and similarly $A_2(x_2)$, $A_3(x_3) \cdots A_n(x_n)$, see Fig. 8.4. We compute correlation coefficients for pairs separated by a specified distance h using the relation

$$r(h) = \frac{\text{cov}\left[A(x), A(x+h)\right]}{\sqrt{Var\left[A(x)\right]}\sqrt{Var\left[A(x+h)\right]}}. \tag{8.5}$$

For example, if we specify a distance $h = 0$, we find for pairs $[A(x_i), A(x_i)]$ a correlation coefficient $r_o = 1$. For pairs $[A(x_i), A(x_{i+1})]$ at a distance $h = 1$(nearest neighbors), we obtain a value of r_1. Next, we increase the distance between the neighbors to $h = 2$ for pairs $[A(x_i), A(x_{i+2})]$ and obtain a value for r_2 and so on. Details of the calculation are e.g. Webster, 1985; Webster and Oliver, 1990). Finally, we plot r_i versus h or $r(h)$ and we obtain the function illustrated in Fig. 8.4. The distance between neighbors h is called the lag. The distance over which

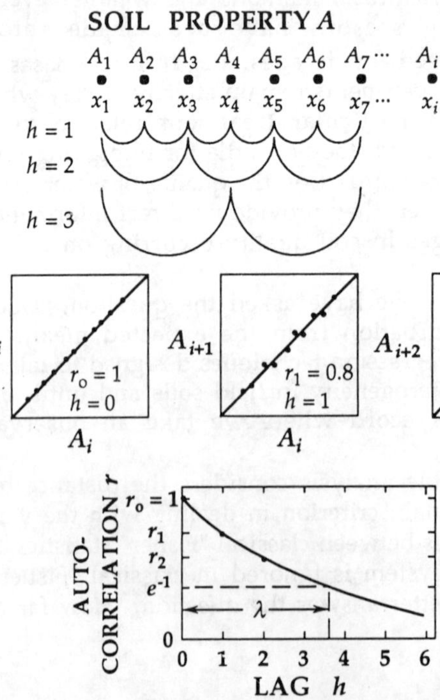

Figure 8.4. Derivation of the autocorrelogram with equidistant sampling along a transect.

a significant correlation exists is called the correlation length, scale, or range. Commonly, the correlation length λ is defined for a 1-dimensional transect by the relation

$$r = r_o \exp(-x/\lambda) \tag{8.6}$$

where the value of $r_0 = 1$ and that of r is diminished to e^{-1} at a distance of λ. Figure 8.5a presents soil water content θ measured in a field at 1-m intervals with a neutron moisture meter at the 50-cm soil depth within a 160-m long transect. Neglecting the locations of the observations ($n = 160$), the mean m and standard deviation s of the 160 observations are 0.136 and 0.0162 $cm^3 \cdot cm^{-3}$, respectively. Figure 8.5b, the autocorrelogram of the 160 observations, shows that the autocorrelation length λ is about 6 m. Sampling at intervals less than 6 m is somewhat unnecessary because the observations are related to each other. Sampling at intervals greater than 6 m does not allow meaningful interpolation between neighboring observations. It should be obvious that the functional relation between r and h depends upon the size of the sample, and that in general, the greater the sample size, the greater the value of the autocorrelation length λ.

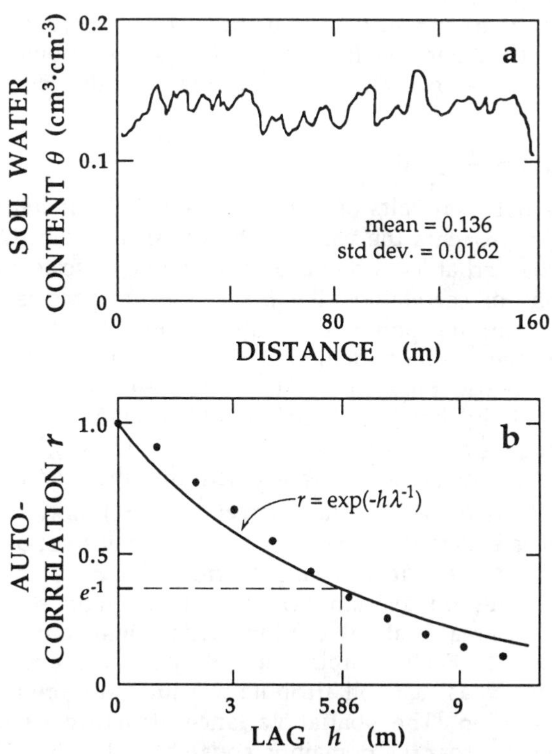

Figure 8.5. a. Values of soil water content θ measured with a neutron probe along a 160-m transect at 1-m intervals. b. Autocorrelogram of θ illustrating a correlation length λ of 5.86 m.

When we calibrate a soil moisture meter, we should try to sample at distances smaller than the value of λ. Or, when K_S is measured on samples of various sizes, autocorrelograms for each size provide estimates of λ. As the size of the sample increases, the value of λ or the scale of the autocorrelogram approaches an upper limit which defines the value of the REV. The same technique can be used to ascertain the magnitude of the REV of θ for a particular value of h. Or by using different sampling volumes, we can establish equal autocorrelation scales for both θ and solute content when we formulate transport equations for leaching.

The concept of autocorrelation is applicable only if second order stationarity exists, i.e. if the variance is constant across the field. Spatial trends across fields or within soil profiles can be found from autocorrelation functions. If there are shoulders within $r(h)$ or if $r(h)$ remains large as the separation distance increases, we can assume that trends exist. Because nonlinear trends yield different correlation scales for different sampling grid densities, trends along a transect can be examined by a regular dropping out of data to change the density of sampling. The correlation scale is also probably interrelated with transect spacing (Jury, 1989).

The concept of variance known from classical statistics is extended in geostatistics to consider the location of the observations $[A(x_i), A(x_i + h)]$ separated by a distance h. The equation for the construction of the semivariogram is

$$\gamma(h) = \frac{1}{2n} \sum_{i=1}^{n} \left[A(x_i + h) - A(x_i)\right]^2 . \tag{8.7}$$

As the distance between pairs of observations or lag h increases, $\gamma(h)$ rises and asymptotically approaches the value of the variance called the sill, see Fig. 8.6. The sill is approached at $h = \lambda$ denoted as the range or scale of the variogram as well as the zone of correlation. For $h < \lambda$, the variance is deformed by the position of the sampling points, or in other words, by the spatial dependence otherwise called the spatial structure. Methods for calculating λ are reviewed by Peck (1984). Semivariograms for spatially independent and dependent data as well as for a spatially changing domain are illustrated in Fig. 8.6. If the domain is spatially changing and not statistically homogeneous, $\gamma(h)$ increases and does not approach a sill. Methods for dealing with this situation are described in the literature (e.g. Webster, 1985). The intercept at $h = 0$ is called the nugget and usually appears as a consequence of fine scale estimates not being available. The nugget also includes the measurement error. Uehara et al. (1985) collected surface soil samples on a 1-km grid within a single soil series in Sudan extending over an area of about 400 km^2. Additional samples only 1 m apart were also collected. Each sample was analyzed for exchangeable sodium percentage inasmuch as that soil attribute was judged to be most likely to limit sugarcane production. The spatial variance structure of the exchangeable sodium percentage across the domain is apparent in Fig. 8.7. The variogram has a nugget of 6.15, a range of 3 km and a sill and sample variance of 21.6. Kriged contours of exchangeable sodium percentage having values between 5 and 21% had a mean estimation variance of 10.5. (Kriging and co-kriging are discussed in the next section.)

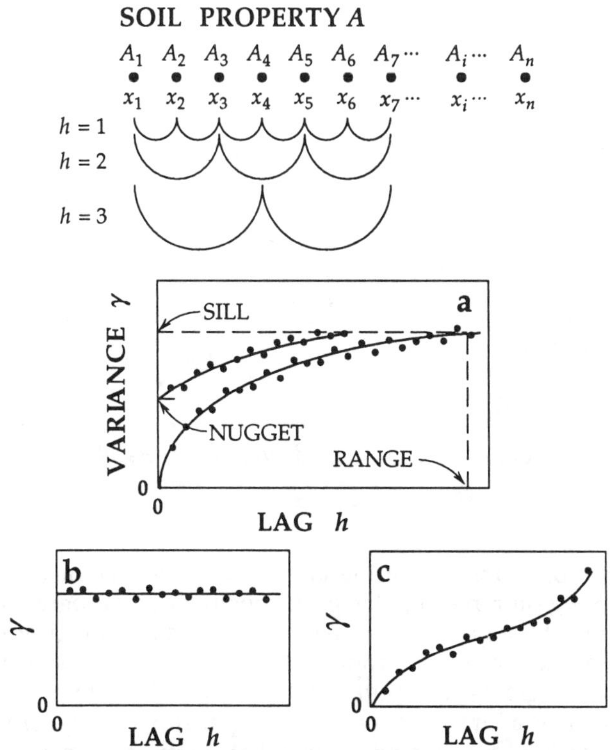

Figure 8.6. Derivation of the semivariogram with equidistant sampling along a transect: a. spatially dependent with and without a nugget, b. spatially independent and c. spatially changing domain.

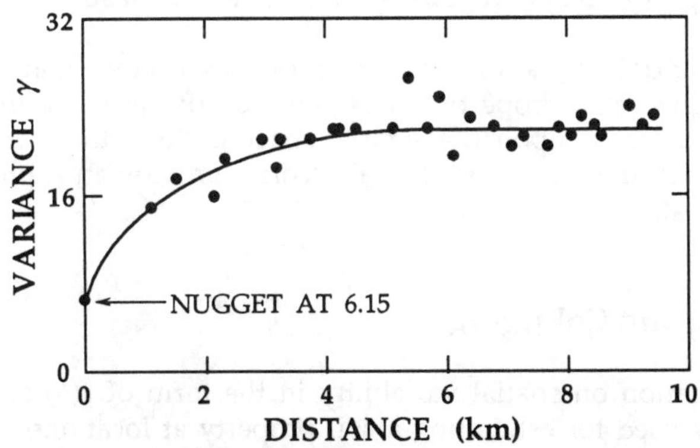

Figure 8.7. Semivariogram of exchangeable sodium percentage from samples taken on a 1-km grid over a 400-km² area of a mapping unit in Sudan (Uehara et al., 1985).

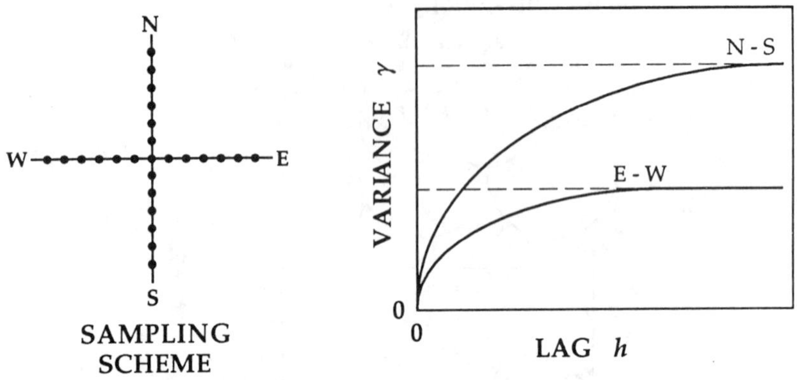

Figure 8.8. Semivariograms of two perpendicular transects of a non-isotropic domain.

In a non-isotropic domain variograms differ for different directions, see Fig. 8.8. For sampling on a rectangular grid, constructing variograms along the main two directions of the grid and on the two diagonals is a logical first choice to identify the presence of non-isotropic behavior. For the construction of the appropriate semivariograms, computed data $\gamma(h)$ are fitted to a simple curve, usually the segment of a circle, or that of an exponential or hyperbolic curve. Interpreting semivariograms is made somewhat more reliable if a couple of "rules of thumbs" are followed. First, the minimum number of samples along a transect should be in the range of 50 to 100 (Gutjahr, 1985; and Webster, 1985). And second, the estimation of $\gamma(h)$ is considered reliable for lags not exceeding 20% of the total transect length.

The term support refers to the size, shape and orientation of samples. An increase of support, called regularization, generally leads to a decrease of variation.

Reviewing data of various authors, Jury (1989) shows that the correlation length λ of a given soil property varies significantly. In some instances when the value of λ depends upon the sample spacing along the transect, we deal with a scale effect. In such cases, several theories provide an explanation (e.g. a violation of the stationarity hypothesis).

8.3.2 Kriging and Cokriging

Derived information on spatial variability in the form of $r(h)$ and $\gamma(h)$ can be advantageously used for estimating a soil property at locations where it is not measured. Kriging is a weighted interpolation named after D. G. Krige (1966) who devised it for estimating the gold content of ore in South Africa. Although its theoretical foundation was apparently recognized much earlier (Kolmogorov, 1941), its present day development is largely attributed to

Matheron (1965) and his associates. If $A(x_1)$ and $A(x_2)$ are measured values of A at locations x_1 and x_2, respectively, we seek an unbiased estimate of A in between x_1 and x_2. We interpolate with weights μ and v for each of the positions x_1 and x_2. Values of μ and v depend upon the covariance function or the semivariogram as well as upon the location of the interpolated value. Note that the weights do not depend upon the actual values of A. The kriging variance or the minimum square error σ_K^2 is a measure of the precision of the interpolated value. Many kriging formulations are available (e.g. Journel and Huijbregts, 1978). Here, we briefly introduce punctual kriging.

For n sampling points x_i (i = 1, 2, ··· n) in a field, A is estimated (or kriged) at x_o by

$$A^*(x_o)=\sum_{i=1}^{n}\lambda_i A(x_i) \tag{8.8}$$

where λ_i are the weights assigned to the sampling points and have a sum of unity. The kriging variance is

$$\sigma_K^2(x_o)=2\sum_{i=1}^{n}\lambda_i\bar{\gamma}(x_i,x_o)-\sum_{i=1}^{n}\sum_{j=1}^{n}\lambda_i\lambda_j\gamma(x_i,x_j)-\bar{\gamma}(x_o,x_o) \tag{8.9}$$

where $\gamma(x_i,x_j)$ is the semivariance of A between the ith and jth sampling points, $\bar{\gamma}(x_i,x_o)$ is the average semivariance between the ith sampling point and the field and $\bar{\gamma}(x_o,x_o)$ is the average variance within the field. The kriging variance is least when

$$\sum_{i=1}^{n}\lambda_i\gamma(x_i,x_j)+\mu_L=\bar{\gamma}(x_j,x_o) \tag{8.10}$$

for all j where μ_L is the Lagrange multiplier. The solution of (8.10) subject to the sum of the weights λ_i being unity provides values of the weights λ_i used in (8.8) as well as that of μ_L. These values also allow estimation of $\sigma_K^2(x_o)$ with

$$\sigma_K^2(x_o)=\sum_{i=1}^{n}\lambda_i\bar{\gamma}(x_i,x_o)-\bar{\gamma}(x_o,x_o)+\mu_L. \tag{8.11}$$

An example of kriging along a transect is given in Fig. 8.9. With kriging, additional optimal locations of sampling can be gained inasmuch as the kriged isolines depict more objectively the district of soils than an interpolation done by eye, or by linear interpolation between the measured data. The results of kriging depend upon the fitted semivariogram and can be easily validated as follows. An estimate of A is made for each location x_j for which an observation was obtained but is purposely left out of the kriging process. The procedure ("jack-knifing") is repeated for each of the measured j locations. The differences between the kriged and observed values are related to σ^2, and if equal to 1, the kriging procedure has been properly executed. The computational procedure is readily available in the literature (e.g. Journel and Huijbregts, 1978; and Webster, 1985).

In the derivation of the autocorrelogram we considered the spatial correlation of only one soil property with itself – hence, the term autocorrelation. The same principle can be applied to two properties A and B to determine to what extent property A at location x_i depends upon B at x_{i+1}. Here,

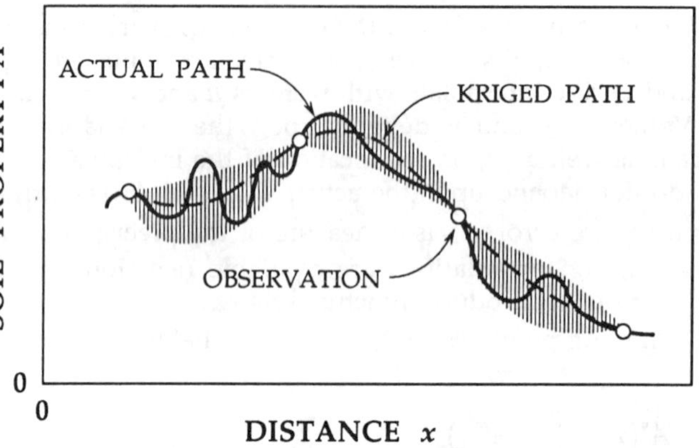

Figure 8.9. Demonstration of Kriged estimates yielding a smoother path than the actual path (Gutjahr, 1985). The shaded area represents $\pm\sigma_K$.

we deal with random fields of two properties, and when we transfer spatial variability information from one field to the other, their variables are co-regionalized or cross-correlated. The procedure is particularly useful when sampling locations for variable *A* are not identical to those for variable *B*, e.g. variable *B* is more densely sampled than variable *A*. Practical considerations of cost of analysis, reliability of particular measurement procedures, ease and access of sampling locations and time all contribute to the numbers of observations selected. For example, the available water content AWC in a soil profile and soil water content at pF = 2.5 are more difficult and expensive than the contents of sand, silt and clay. The semivariogram of these five properties and their cross-semivariograms are plotted in Fig. 8.10 and used in co-kriging procedures to obtain the contour isarithmic maps given in Fig. 8.11. The detailed computational procedure is described by Vauclin et al. (1983).

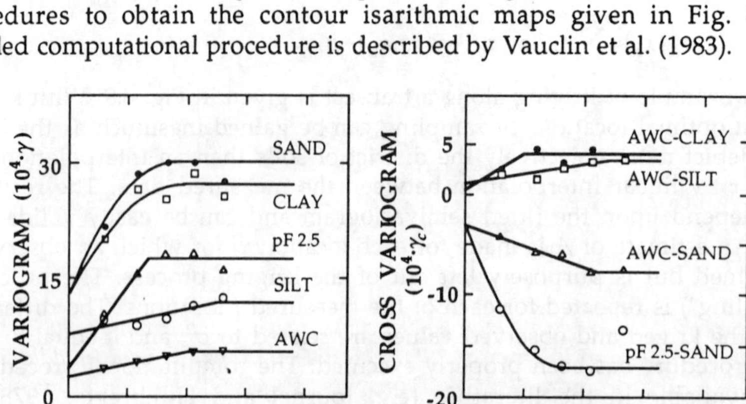

Figure 8.10. a. Semivariograms of five soil physical properties and b. Cross-variograms. AWC is available water content and pF 2.5 represents the soil water content at pF = 2.5.

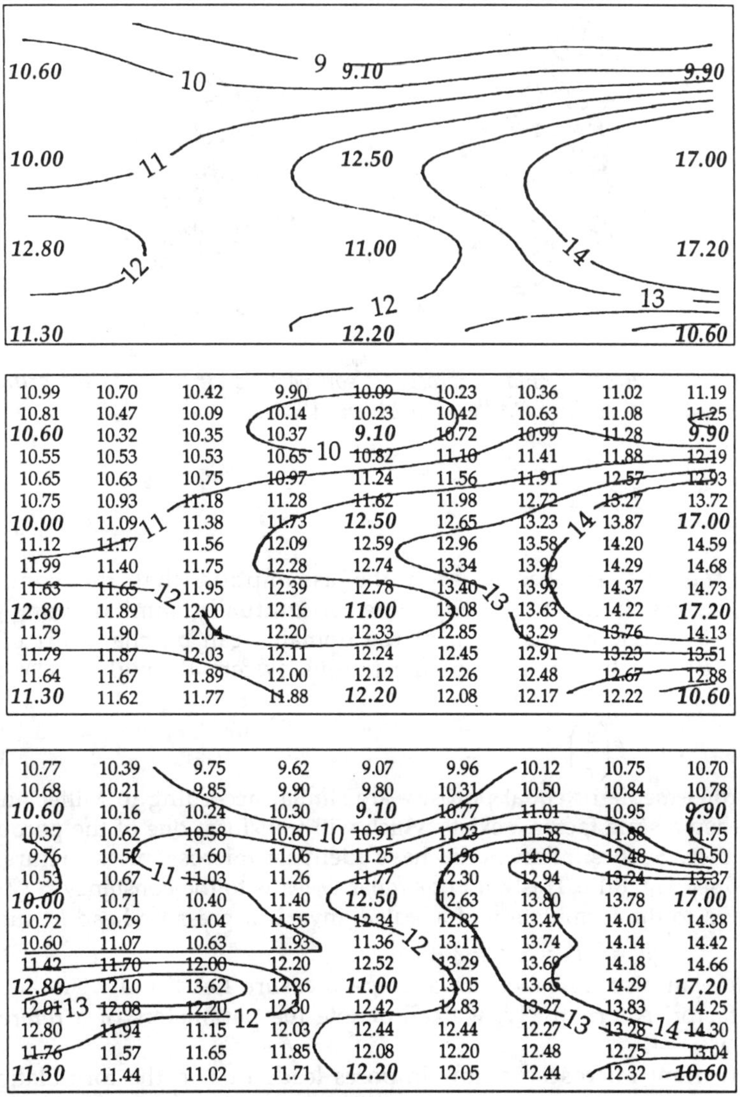

Figure 8.11. Isolines of available water content using data from Fig. 8.10 (top fig.).The accuracy of the map increases when kriging is used (middle fig.) and co-kriging is used (bottom fig.). The bold, horizontal numbers correspond to measured values (Vauclin et al., 1983).

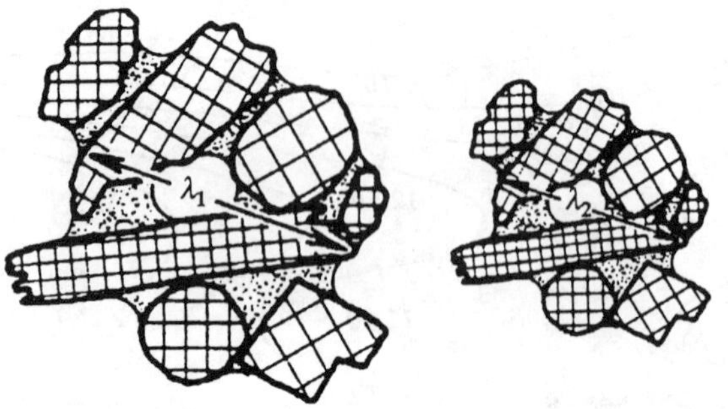

Figure 8.12. A self similar microscopic soil particle structure is the principle of Millers' scaling (Miller and Miller, 1956).

8.4 SCALING

Scaling theories are based upon the assumption that the continuously heterogeneous field is an ensemble of mutually similar homogeneous domains. We assume that each of the domains can be characterized by the SWRC $h(\theta)$ which is related to the porous system through (4.11), or more generally by

$$h = f\left(\frac{1}{r}\right). \tag{8.12}$$

Two porous media of equal porosity are similar according to Miller and Miller (1956) when a scale factor λ exists which will transform one of the porous media to the other. Such similar media have identical microscopic structures except for scale, see Fig. 8.12. This kind of similarity leads to the constancy $r_1/\lambda_1 = r_2/\lambda_2 = r_i/\lambda_i$ and to the formulation of a scaled, invariant pressure head h^* such that

$$\lambda_1 h_1 = \lambda_2 h_2 = \lambda_i h_i = \lambda^* h^* \tag{8.13}$$

where h^* can also be called an average pressure head and λ^* is an average scaling factor. Alternatively, we can denote the parameters of a reference soil with an asterisk.

Invariant microscopic flow in pores leads first to the formulation of an equation for the mean pore water velocity, see Hagen-Poiseuille (5.10) for laminar flow. Hence, the saturated hydraulic conductivity is

$$K_S = f(r^2) \tag{8.14}$$

and with the dimensionless term r/λ, we obtain

$$K_S^*/(\lambda^*)^2 = K_{Si}/\lambda_i^2 \tag{8.15}$$

and analogously for $K(\theta)$

$$K^*(\theta)/(\lambda^*)^2 = K_i(\theta)/\lambda_i^2 \tag{8.16}$$

where K_S^* and $K^*(\theta)$ are either average values or values of a reference soil. The original scaling of Miller and Miller (1956) is restrictive in two aspects. First, a microscopic length is physically interpreted, and second, the requirement of a constant porosity is rarely applicable.

Warrick et al. (1977) extended the use of the Millers' single scaling factor by introducing the degree of saturation (equivalent to the earlier relative soil water content θ_R) and eliminating the assumption of identical porosities. Thus, they scaled θ with the scaling factor θ_S. Additionally, their derivation of λ does not require a search for a microscopic physical length. For the derivation of λ, the sum of squares

$$\sum_{r,i}\left(h^* - \lambda_{ri} h_i\right)^2$$

was minimized for r locations. Using this scaling procedure, a large dispersion of experimental data of $h(\theta/\theta_S)$ and $K(\theta/\theta_S)$ was nicely coalesced into unique functions, see Fig. 8.13. Hence, soil heterogeneity is approximated by the stochastic character of λ which retains its universal meaning with relations

$$\lambda = \frac{h^*}{h} \qquad \lambda = \left(\frac{K}{K^*}\right)^{1/2} \qquad \theta_R = \frac{\theta}{\theta_S} \tag{8.17}$$

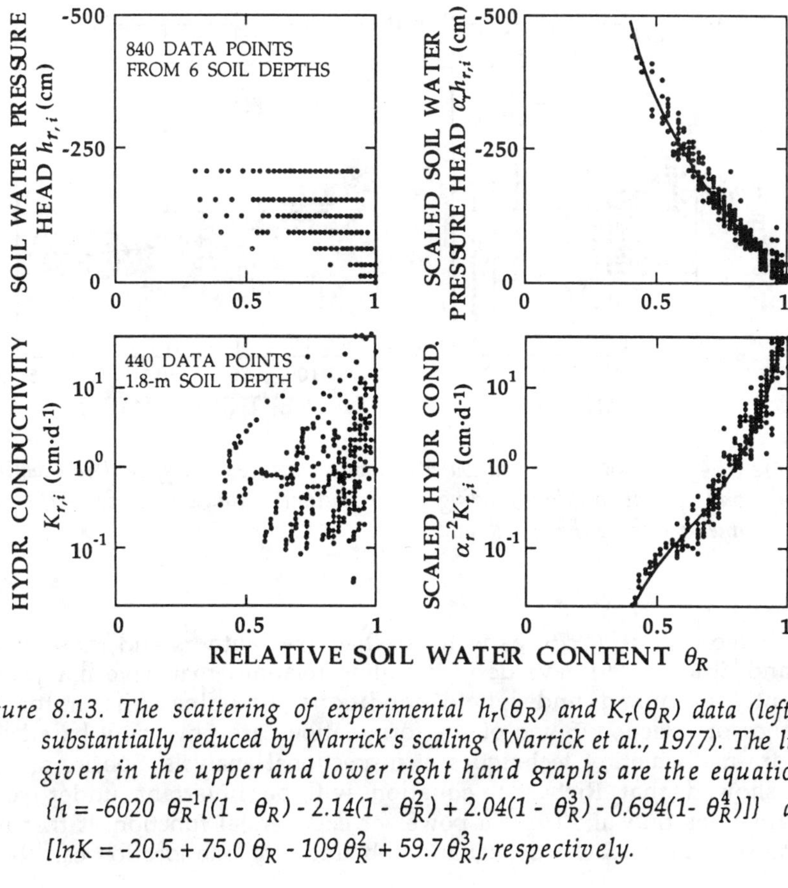

Figure 8.13. The scattering of experimental $h_r(\theta_R)$ and $K_r(\theta_R)$ data (left) is substantially reduced by Warrick's scaling (Warrick et al., 1977). The lines given in the upper and lower right hand graphs are the equations $\{h = -6020\ \theta_R^{-1}[(1-\theta_R) - 2.14(1-\theta_R^2) + 2.04(1-\theta_R^3) - 0.694(1-\theta_R^4)]\}$ and $[\ln K = -20.5 + 75.0\,\theta_R - 109\,\theta_R^2 + 59.7\,\theta_R^3]$, respectively.

The modification for estimating λ indicated in the Warrick et al. (1977) procedure was fully developed by Simmons et al. (1979). They rejected the assumption of microscopic geometrical similarity and based their method on the similarity between soil hydraulic functions.

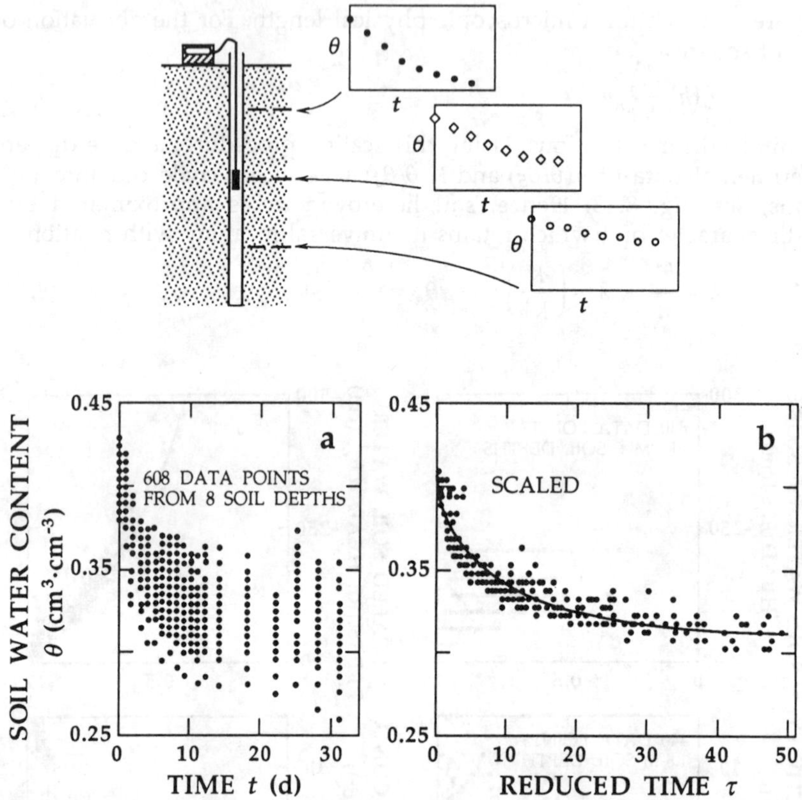

Figure 8.14. Neutron probe soil water contents measured at different soil depths and spatial locations during redistribution. Equation (8.18) is the solid line describing the scaled data.

Simmons et al. (1979) as well as others (see authors and those cited in Hillel and Elrick, 1990) have derived scaling relations that have not yet been sufficiently examined under field conditions to define criteria for their acceptance or rejection. Some of the formulations are physically based while others are mathematical techniques of inspectional analysis. Sposito and Jury (1990) showed that Richards' equation will be invariant under scaling transformations only if $K(\theta)$ is a power or exponential function. If that is the case, the solution of Richards' equation obtained for one location can then be scaled to other locations in the same field or domain. Assuming $K(\theta)$ is an exponential function and a unit hydraulic gradient exists during redistribution in the absence of evapotranspiration, Simmons et al. (1979) recommended that (7.28) be scaled with a common value of β in (7.27) for all locations within a field. Fig. 8.14a shows soil water content versus redistribution time for a total of 608 measurements (19 times at 32 locations) within 4 plots covered by plastic sheeting to prevent evaporation after steady state infiltration had ceased. Fig. 8.14b shows the data from Fig. 8.14a. scaled with reduced time τ and a common initial value of $\overline{\theta}_o$. The solid line in Fig. 8.14b is

$$\theta=\overline{\theta}_o-\frac{1}{\beta}\ln\left[1+\frac{\beta K^*\tau}{z^*}\right] \tag{8.18}$$

where $\overline{\theta}_o$ = 0.408 $cm^3 \cdot cm^{-3}$, β = 50, K^* = 5.29 $cm \cdot d^{-1}$, z^* = 120 cm and $\tau=\omega^2 z^*(az)^{-1}$ [a is defined in (7.?6) and ω is the scale factor for each location defined by $K_o=\omega^2 K^*$ where K is the scale mean of all K_o.]. The measured θ deviate from the solid line with a pooled standard deviation is 0.008 $cm^3 \cdot cm^{-3}$ – a value comparable to the neutron probe measurement error.

Methods based upon regression analysis are described as functional normalization techniques (Tillotson and Nielsen, 1984). With this approach being only macroscopic, geometrical similarity of the porous system is therefore not the condition for scaling. The idea of a universal λ for all hydraulic functions was abandoned as it was found that λ for scaling $h(\theta_R)$ is not necessarily identical with that of $K(\theta_R)$. Even if the two scaling factors are different, they are generally correlated. In order to differentiate from the previous universal single set of scaling factors λ_i, we shall now use symbols α_{1i} and α_{2i} for the two sets of scaling factors, the first one denoting the scaling of h and the second denoting the scaling of K. Hence, we have

$$\alpha_1=\frac{h^*}{h} \qquad \alpha_2=\left(\frac{K}{K^*}\right)^{1/2} \qquad \theta_R=\frac{\theta}{\theta_S} \tag{8.19}$$

A set of sampling locations is similar if the soil hydraulic functions can be scaled. Warrick (1990) reviewed the application of this scaling in three different regions, while Clausnitzer et al. (1992) demonstrated that simultaneous scaling with α_1 and α_2 is not always as successful as independent scaling using a single scaling factor λ (our notation here).

Still yet another scaling proposal of Vogel et al. (1991) is based upon the assumption that the spatial variability of soil hydraulic functions has two components, one being linear and the other being non-linear. Supposing that the linear component is dominant, he proposed linear scaling with

$$\alpha_h=\frac{h}{h^*} \qquad \alpha_K=\frac{K(h)}{K^*(h^*)} \qquad \alpha_\theta=\frac{\theta(h)-\theta_r}{\theta^*(h^*)-\theta_r^*} \tag{8.20}$$

Any of the above types of scaling should be tested for the measured set of functions. For the selected type of scaling the invariant form of Richards' equation is applied together with the scaled boundary conditions. We can denote the soils as Warrick, or Simmons, or Vogel similar. Once successfully scaled, computed fluxes can be "descaled" for any given sampling point. Of greater importance is the study of the variability of the scaling parameters by stochastic and geostatistical methods. Scaling yields higher quality data or more useful information when soil samples within one soil type are scaled independently from other soil types (Clausnitzer et al., 1992). The variability of soil hydraulic functions is appropriately expressed by scaling factors within each soil type where their *pdf* and correlation structure can be easily determined. Additional details on numerical procedures of scaling are described by Clausnitzer et al. (1992). Appropriate developments of dimensionless variables and scaled basic equations for various types of boundary conditions and soil hydrological problems have been assembled by Hillel and Elrick (1990).

Up to now we have demonstrated the scaling of soil hydraulic functions of field soils in order to ascertain reference soil parameters and the statistical character of the scaling factors. However, scaling techniques offer still a greater opportunity to formulate Richards' equation in an invariant form for the solution of elementary hydrological processes. Once a solution is known for a defined soil or boundary condition and is expressed in scaled variables, it is valid for all soils or boundary conditions within the given class of flow problem. Two different procedures are available. (i) Variables and soil hydraulic functions are scaled by the boundary condition with these scaled variables usually not being dimensionless. (ii) Variables are scaled to dimensionless forms using soil hydraulic functions. These solutions are similar to traditional expressions of solutions of flow problems in dimensionless variables.

The first procedure was used in the study of two scaling classes. (i) Infiltration with a constant flux at the soil surface (Neuman's boundary condition) and infiltration into a crust-topped soil (Kutílek et al., 1991). Variables z, t and θ scaled by the boundary condition, i.e. either using the flux density q_o (left-hand column) or using the resistance R (right-hand column) are

$$t = q_o^{\alpha} T^* \qquad t = R^a T^* \tag{8.21}$$

$$z = q_o^{\beta} Z^* \qquad z = R^b Z^* \tag{8.22}$$

$$\theta - \theta_r = q_o^{\gamma} \theta^* \qquad \theta - \theta_r = R^c \theta^* \tag{8.23}$$

The soil hydraulic functions expressed in a power form $D(\theta) = D_o(\theta - \theta_r)^n$, $h(\theta) = -p(\theta - \theta_r)^{-m}$ and $K(\theta) = K_o(\theta - \theta_r)^{\upsilon}$ similarly scaled are

$$D(\theta) = q_o^{n\gamma} D^*(\theta^*) \qquad D(\theta) = R^{cn} D^*(\theta^*) \tag{8.24}$$

$$h(\theta) = q_o^{-m\gamma} h^*(\theta^*) \qquad h(\theta) = R^{-cm} h^*(\theta^*) \tag{8.25}$$

$$K(\theta) = q_o^{\upsilon\gamma} K^*(\theta^*) \qquad K(\theta) = R^{c\upsilon} K^*(\theta^*) \tag{8.26}$$

And, Richards' equation in the diffusive form (5.68) transcribed into scaled variables invariant to q_o is

$$q_o^{(\gamma-\alpha)} \frac{\partial \theta^*}{\partial T^*} = q_o^{(n\gamma+\gamma-2\beta)} \frac{\partial}{\partial Z^*}\left[D^*(\theta^*)\frac{\partial \theta^*}{\partial Z^*}\right] - q_o^{(\gamma\upsilon-\beta)} \frac{\partial K^*(\theta^*)}{\partial Z^*}. \tag{8.27}$$

Similarly, (5.68) transcribed into scaled variables invariant to R is

$$R^{(c-a)} \frac{\partial \theta^*}{\partial T^*} = R^{(cn+c-2b)} \frac{\partial}{\partial Z^*}\left[D^*(\theta^*)\frac{\partial \theta^*}{\partial Z^*}\right] - R^{(c\upsilon-b)} \frac{\partial K^*(\theta^*)}{\partial Z^*}. \tag{8.28}$$

Boundary conditions are expressed in a similar manner. The exponents for the above conditions are

$$\alpha = \frac{-2m-n}{m+n+1} \qquad a = \frac{2m+n}{2m+n+1} \tag{8.29}$$

$$\beta = \frac{-m}{m+n+1} \qquad b = \frac{m}{2m+n+1} \tag{8.30}$$

$$\gamma = \frac{1}{m+n+1} \qquad c = \frac{-1}{2m+n+1} \tag{8.31}$$

The solution of the infiltration problem plotted in scaled variables $\theta^*(Z^*, T^*)$ is valid for the whole class, i.e. either for all variations of constant flux at the soil surface, see Fig. 8.15, or for all variations of resistance (except for $R = 0$), see Fig. 6.25.

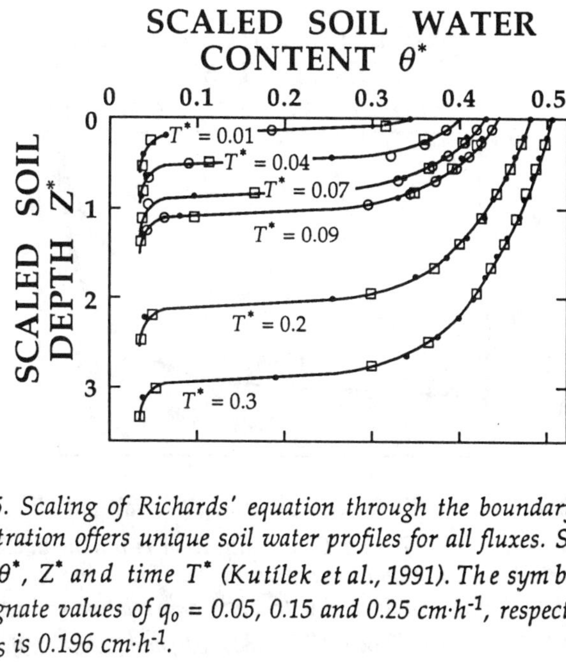

Figure 8.15. Scaling of Richards' equation through the boundary flux for NBC infiltration offers unique soil water profiles for all fluxes. Scaled variables are θ^*, Z^* and time T^* (Kutílek et al., 1991). The symbols □, • and o designate values of q_0 = 0.05, 0.15 and 0.25 $cm{\cdot}h^{-1}$, respectively.The value of K_S is 0.196 $cm{\cdot}h^{-1}$.

The second scaling procedure elaborated by Warrick and Hussen (1993) is applicable to a broader family of flow classes, e.g. redistribution and upward flow. They defined the dimensionless variables

$$Z^* = z/z_o \qquad T^* = t/t_o \tag{8.32}$$

and functions

$$\theta^* = \frac{\theta-\theta_r}{\theta_o-\theta_r}, \quad K^* = K/K_o, \quad h^* = h/z_o \tag{8.33}$$

where $\theta_o \le \theta_S$ and $K_o = K(\theta_o)$. When they considered the soil hydraulic characteristics

$$\frac{\theta-\theta_r}{\theta_S-\theta_r}=\left(\frac{h_A}{h}\right)^{\lambda} \qquad \frac{K}{K_r}=\left(\frac{\theta-\theta_r}{\theta_S-\theta_r}\right)^{n} \tag{8.34}$$

they defined

$$z_o=|h_A|\left(\frac{\theta_o-\theta_r}{\theta_S-\theta_r}\right)^{-1/\lambda} \tag{8.35}$$

and

$$t_o=\frac{(\theta_o-\theta_r)z_o}{K_o}. \tag{8.36}$$

Hence, in Richards' equation all variables and soil hydraulic functions are scaled with each equation being invariant and dimensionless. An example of a scaled soil water profile $\theta^*(Z^*, T^*)$ for infiltration with DBC is given in Fig. 8.16.

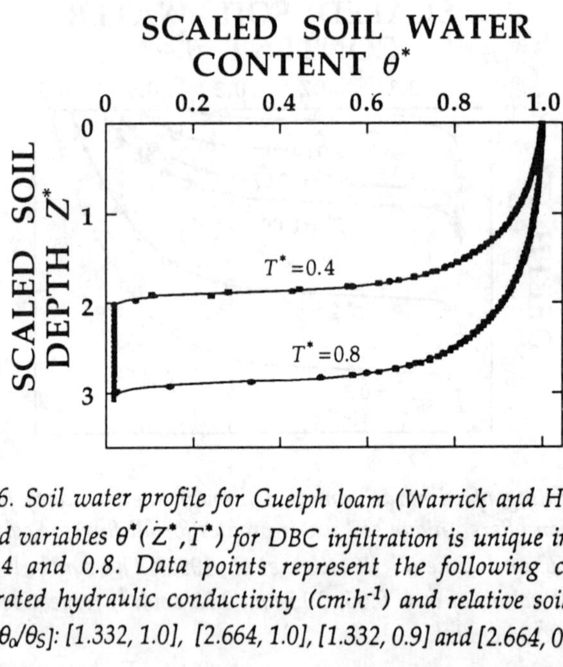

Figure 8. 16. Soil water profile for Guelph loam (Warrick and Hussen, 1993) in scaled variables $\theta^*(Z^*,T^*)$ for DBC infiltration is unique in $\theta^*(Z^*)$ for T^* = 0.4 and 0.8. Data points represent the following combinations of saturated hydraulic conductivity (cm·h^{-1}) and relative soil water content [K_S, θ_o/θ_S]: [1.332, 1.0], [2.664, 1.0], [1.332, 0.9] and [2.664, 0.9].

A scaling procedure is also advantageous for the solution of the inverse problem of infiltration (Warrick, 1993). With the soil hydraulic functions being scaled together with variables, the procedure is conveniently reduced to simple algebraic computations instead of repetitive numerical simulations.

8.5 STATE-SPACE EQUATIONS FOR MULTIPLE LOCATIONS

A state-space solution was introduced (section 7.3.2.2) in order to ascertain $K(\theta)$ within prescribed fiducial limits for a pedon. That state-space formulation described soil water redistribution with an approximate ordinary first-order differential equation, assumed $K(\theta)$ to be of exponential form and quantitively included the uncertainty of the neutron probe and its calibration. Here we present a similar analysis to compute evaporation between irrigation events as well as to determine a soil water diffusivity function from soil water profiles measured at multiple locations within a field. The uncertainties associated with soil heterogeneity as well as those of instrument calibration are separately identified.

Five neutron access tubes positioned every 18 m along a transect allowed soil water content to be monitored with a neutron probe at 15-cm depth intervals within the soil profile at each location. The level site was free of vegetation and equipped with a sprinkler irrigation system that was used to apply 15 small irrigations (each <20 mm) during a 3-month period. In addition to the neutron probe measurements of soil water to estimate evaporation, 20-min weighings of a 50-t capacity lysimeter in the same field were integrated to obtain daily values of evaporation.

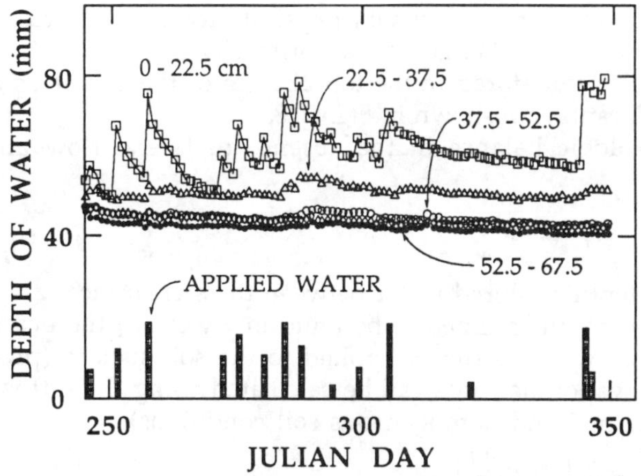

Figure 8.17. Depths of water stored within five depth intervals of a pedon measured with a neutron probe together with depths of water applied to the soil surface during the experiment. The depth interval 67.5 - 82.5 cm is represented with ◊ symbols.

The spatially averaged amounts of water stored within the five depth intervals of the soil profile measured with the neutron probe as a function of time together with the amounts of water applied with the sprinklers are shown in Fig. 8.17. Notice that the water stored in the lower depth intervals does not

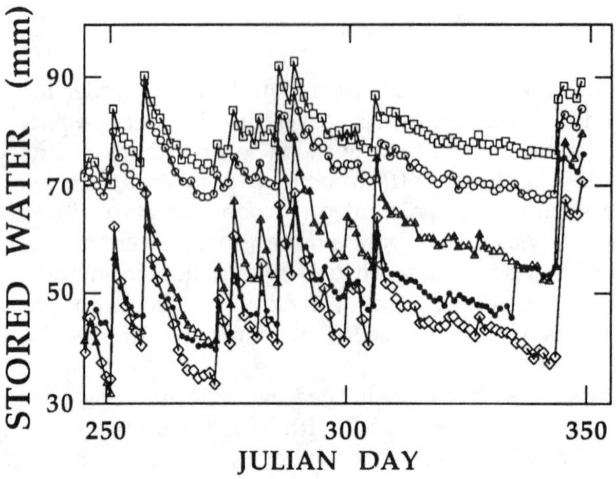

Figure 8.18. Depths of water stored within the 0 - 22.5 cm topsoil at each of five field locations measured with a neutron probe .

appreciably change during the entire experiment. Evaporation and infiltration from the applied water are the primary physical processes that create changes in stored water and those changes occur primarily in the 0 - 22.5 cm depth. Variations in soil water stored in the top 22.5 cm of the profile as a function of time as well as location are shown in Fig. 8.18.

The hydrologic balance in the absence of lateral flow and negligible drainage is

$$\frac{dS}{dt} = P_t - E_t \tag{8.37}$$

where S is the depth of stored water between the soil surface ($z = 0$) and some depth b (b is the depth assumed to be uniformly wet), E_t the evaporation rate and P_t the rate at which water is applied to the soil surface (precipitation or irrigation). The evaporation rate can be calculated using the diffusion equation (assuming isothermal and homogeneous soil conditions)

$$\frac{\partial\theta}{\partial t} = \frac{\partial}{\partial z}\left[D(\theta)\frac{\partial\theta}{\partial z}\right] \tag{8.38}$$

where $D(\theta)$ is the soil water diffusivity function. Gardner (1962) obtained the following approximate solution of the depth-integrated form of (8.38)

$$E_t = \frac{\pi^2 S D(S/b)}{4b^2} \qquad \frac{Dt}{b^2} > 0.3 \tag{8.39}$$

where the diffusivity is a function of stored water S. Combining (8.37) and (8.39) the hydrologic balance becomes the ordinary differential equation

$$\frac{dS}{dt} = P_t - \left(\frac{\pi}{2b}\right)^2 S D(S). \tag{8.40}$$

For an exponential diffusivity function $D(S/b) = A\exp(BS/b)$, the state-space equation for (8.40) is

$$dX_t = \left[P_t - \left(\frac{\pi}{2b} \right)^2 X_t A \exp(B X_t / b) \right] dt + v_S dt \qquad (8.41)$$

where X_t is the state variable representing the stored water S and $v_S dt$ is a stochastic noise owing to the uncertainties in the proposed equation. The noise $v_S dt$ arises owing to the various simplifying assumptions of the physical model such as neglecting thermal and salinity effects, swelling and shrinkage, hysteresis, etc. The corresponding observation equation at time t_k ($k = 1, 2, 3, \cdots$) is

$$Z_m(t_k) = X_{t_k} + v_m(t_k) \qquad (8.42)$$

where Z_m is the observed amount of stored water at time t_k and v_m is the observation noise.

The diffusivity parameters A and B were calculated assuming that the observation variance R_t can be estimated from spatially averaging the neutron probe readings from the five access tubes located 18 m apart. It can be seen in Fig. 8.19 that the observation variance R_t is much larger than the neutron probe calibration variance. Values of A and B were estimated to be 0.0292 $mm^2 \cdot d^{-1}$ and 32.59, respectively, with the state variance per unit time Q being 18.87 $mm^2 \cdot d^{-1}$. The value of Q is of the same order of magnitude as the neutron probe calibration variance, while the model uncertainty for stored water prediction on a daily basis is within the neutron probe noise. The agreement found between the state space estimation of cumulative evaporation and that measured with the lysimeter ($r^2 = 0.98$) indicates that the estimated diffusivity function provides an adequate description of evaporation and the soil water transport process.

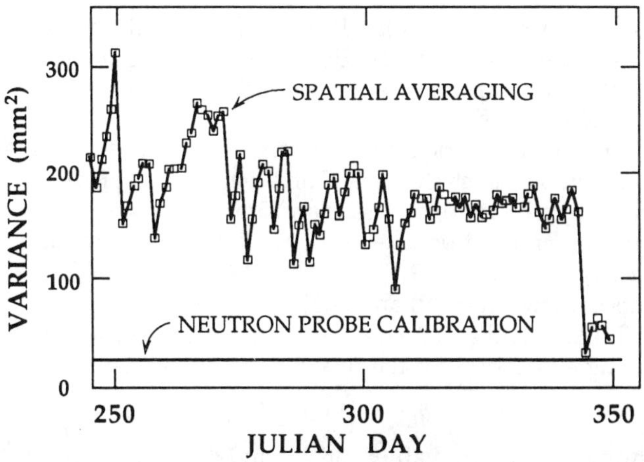

Figure 8.19. Observation variance owing to spatial averaging of measured soil water storage for the 0 - 22.5 cm topsoil as well as the variance of the neutron probe calibration.

State-space formulations that describe location-averaged phenomena could improve our understanding of the development of soil horizons, growth and water extraction of plant roots and chemical and microbiologically-induced reactions within soil water. We also expect that progress could be made using time-averaged equations to examine critical periods during which soil processes occur; these soil processes may be related to particular soil locations, mapping units or regions.

PROBLEMS

1. Derive the scaled cumulative infiltration term I^* for infiltration into the crust-topped profile. Use scaled variables and functions of Kutílek et al. (1991).
2. Values of soil property $A(x)$ have been measured along a transect at three locations: $A(1) = 7$, $A(2) = 5$ and $A(4) = 3$. a.Using a simple desk calculator, estimate $A(3)$ by kriging. Hint: After calculating the only value possible of $\gamma(h)$ with (8.7) for each of the lags (1, 2 and 3), estimate a smooth variogram $\gamma(h) = mh^b$ by regression. Using values of $\gamma(h)$ from the regression at $h = 1$, 2 and 3, calculate the weights λ_i with (8.10). b. What is the value of the kriging variance for $A(3)$?
3. You have measured infiltration with double ring infiltrometers every 5 m along a 500-m transect. Two different inner ring diameters were used. The diameter d was a. 50..5 cm and b. 25.2 cm. The diameter of the outer ring $D = 2d$. The evaluated semivariograms are in Fig. 8.6a with nugget for $d = 50.5$ cm and in Fig. 8.6b for $d = 25.2$ cm. What are your conclusions regarding an appropriate size of the ring and the range of the variograms?
4. You are calibrating a neutron probe. Recognizing that the autocorrelogram in Fig. 8.5 had been previously determined in your field, sketch the locations of 4 soil sampling sites for measuring the soil water content around the access tube.
5. Estimate the range of the variograms for sand, pF = 2.5, silt and AWC in Fig. 8.10. Would you expect the range of a variogram of quasi-steady state infilration rate to be more or less than 25m?
6. How will you scale infiltration rate and cumulative evaporation if the scaling factors are expressed by (8.19)?
7. Show the relation between (8.39) and (7.3).
8. a. Starting with (7.25) and (7.27) and assuming $a = 1$ and $b = 0$ in (7.26), derive the following expression for the flux density at soil depth L :

$$q_L = K_o\left(1+\beta K_o t L^{-1}\right)^{-1}.$$

   b. Assuming the mean values of K_o and β are 31.9 cm·d^{-1} and 61.1, respectively, calculate q_L at a soil depth of 1.8 m for $t = 1$ and 10 d.

9. The probability distribution function of K_o in problem 7 is divided into five 20%-fractile groups having median values of $K_o = 1.98$, 5.73, 11.9, 24.9 and

72.0 cm·d^{-1}. Similarly, the five 20%-fractile groups of β have median values of β = 22.1, 35.8, 49.9, 69.6 and 112.8. Assuming that the two probability distribution functions are independent, substitute into the equation of problem 7 each of the five median values of K_o with each of the five median values of β. to obtain 25 estimates of q_L (each of 4% probability) for t = 1 d. From the 25 values, calculate the mean value $\bar{q}_L$. Repeat the calculations for t = 10 d. Considering the 25 estimates of q_L and the mean $\bar{q}_L$ and comparing these values with those of problem 7, discuss the implications for sampling a field to estimate $\bar{q}_L$.

9 TRANSPORT OF SOLUTES IN SOILS

In Chapter 3 we acknowledged that the liquid phase of a soil is never just pure water, and in Chapter 4 and subsequent chapters we rigorously defined soil water potential to include consideration of the quantity and diversity of solute species in the soil solution. Nevertheless, up to now, we did not consider the physical, chemical and biological processes within a soil profile that distribute, dilute or concentrate solute species within the liquid phase of a soil. Inasmuch as the relative concentration of solutes in the liquid phase governs not only the retention and transport of water within soils but also contributes to our understanding of managing the quality of water within soils as well as that moving below the recall of plant roots deeper into the vadose zone, we present here both microscopic and macroscopic considerations of solute behavior in soils. We limit our discussion to the soil solution and consider transport in the soil air only when it plays an important part in transport of the soil solution.

Although it has been known for at least a century that water and solutes do not travel uniformly within field soils (Lawes et al., 1881), relatively few soil hydrologists individually or collectively made a concerted effort to study and understand the topic until recent times. A few studies now considered classic kept the topic alive until the 1950s. In 1900, Means and Holmes (1901) provided a lucid description of the chemical and physical processes occurring within soils during and after rainfall and irrigation events. They understood the complexities of molecular diffusion and convection in a reactive, structured soil manifesting a heterogeneous pore size distribution. Later, Slichter (1905) noted of a water soluble chemical or "tracer" added to ground water, "its appearance ··· is gradual" when measured at a downstream well. He explained that the gradual appearance of the tracer was caused by the fact that the central thread of water in each capillary pore of the soil moves faster than the water along the walls of the pore. Thirty years later, Kitagawa (1934) studying the dispersion of sodium chloride from a point source in a water-saturated sand expressed the mixing process as a function of the average pore water velocity. Approximately 10 and 20 years later, Bosworth (1948) and Taylor (1953) examined the contributions of molecular diffusion in cylindrical capillary tubes. During the succeeding forty years, investigations have accelerated owing to the growing importance of water quality. In this chapter we limit our discussion to miscible displacement - that is, when two miscible fluids are brought into contact, the initial abrupt interface between the two becomes blurred. The abrupt interface disappears owing to a mixing process dependent upon the properties of the two fluids, the properties of the porous medium, the nature of the microscopic velocity distributions of the fluids and the reactions occurring within each fluid as well as between each fluid and the porous medium.

9.1 SOLUTE INTERACTIONS

We consider here those physical interactions of solutes that play a primary role in the distribution of solutes within the soil profile.

9.1.1 Molecular Diffusion

Thermal energy provides a continual, never ending movement of gas and liquid phases of a soil system. In 1855, Fick provided a theoretical basis for this movement by showing that molecular diffusion obeyed the same mathematical laws that Newton had derived for momentum flux and Fourier for heat flux. The solid matrix of the soil complicates matters by altering both the diffusion path length and the cross sectional area available for diffusion as well as providing an electric field and reactive surfaces that further alter molecular movement.

Fick's first law of diffusion states that a gaseous or solute species moves or diffuses relative to a mixture or solution in the direction of decreasing concentration of that species just as heat flows by conduction in the direction of decreasing temperature. Hence,

$$q = -DA\frac{dC}{dx} \tag{9.1}$$

where q is the diffusive flux [MT^{-1}], A the cross sectional area [L^2], C the concentration [ML^{-3}], x the space coordinate [L] and D the molecular coefficient [L^2T^{-1}]. Because concentration in a porous medium can be expressed in a number of ways, C should be of the same quantity reference as q, and the volume should be of the same length reference as x and A. Jackson et al. (1963) provide details for expressing the frame of reference as the entire bulk soil system or either one of the two fluid phases. Temperature gradients, pressure gradients and external forces also contribute to the diffusive flux. Gaseous diffusion coefficients in soil air are almost composition independent, increase with temperature and vary inversely with pressure. Coefficients in soil water depend upon concentration, solute species and usually increase with temperature. Comprehensive treatments of molecular diffusion are available (e. g. Bird et al., 1960; Crank, 1956; and Currie, 1960).

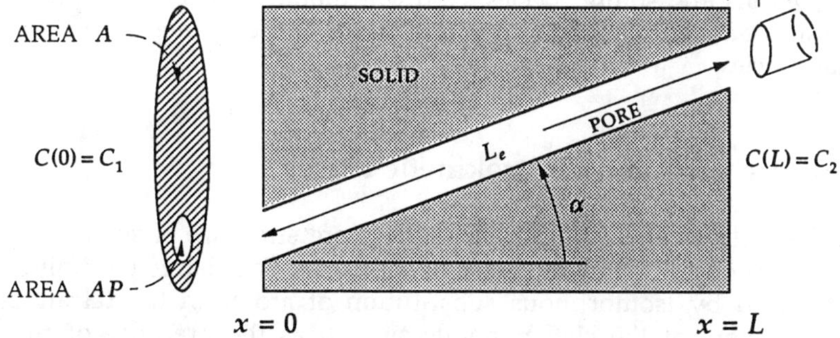

Figure 9.1 Idealized cylindrical pore within a solid matrix through which diffusion occurs.

The complicated geometry of a soil relative to diffusion path length and cross sectional area is usually described in terms of two parameters - tortuosity and porosity. We assume here that isothermal conditions prevail, the bulk volume of the soil remains constant, the soil solids are inert and for convenience, we consider gaseous diffusion in one dimension within a completely dry soil schematically shown in Fig. 9.1. For steady state conditions with $C(0) = C_1$ and $C(L) = C_2$, the diffusion flux for the cylindrical pore of length L_e from (9.1) becomes

$$q = -D_o A_e \left(\frac{C_2 - C_1}{L_e} \right) \tag{9.2}$$

where D_o is the gaseous diffusion coefficient, $A_e = APL/L_e$ the cross sectional area of the cylindrical pore, P the porosity and A the cross sectional area of the bulk soil. Similarly, for the bulk soil of length L (9.1) becomes

$$q = -DA\left(\frac{C_2 - C_1}{L} \right). \tag{9.3}$$

Equating (9.2) and (9.3) we have

$$D = D_o P (L / L_e)^2 \tag{9.4}$$

which is of a form suggested by Buckingham (1904) where $(L/L_e)^2$ is the called tortuosity and equals $cos^2\alpha$. Penman's (1940) commonly used value of $(L/L_e)^2$ = 0.66 yields an angle of 0.61 or 35° between the pore and the solid particle of soil. Marshall (1958) and Millington and Quirk (1959) empirically raised the power of P to 3/2 and 4/3, respectively, and deleted $(L/L_e)^2$ in (9.4) to account for the tortuosity of the average diffusion path.

Equations similar to (9.4) are easily derived for diffusion coefficients in partially water saturated soil as well as for solute diffusion in saturated and unsaturated soils. For example, Sallam et al. (1984) experimentally showed that the product $P(L/L_e)^2$ in (9.4) for gaseous diffusion in unsaturated soils is more nearly equal to $a^{3.10}P^{-2}$ rather than $a^{10/3}P^{-2}$ (Millington and Quirk, 1959) where a is the air-filled porosity. Even for isothermal conditions, in addition to the concentration gradient, we have oversimplified our discussion here by neglecting pressure gradients and external forces acting unequally on the various gaseous and solute species. And, we should remember that the value of a diffusion coefficient depends upon the nature of the counter-diffusing gaseous or solute species.

9.1.2 Electrostatic and Electrokinetic Forces

Electric force fields always exist within the pore structure of soils owing to the electric charge possessed by the walls of soil pores. The charge per unit pore wall area is caused by isomorphous substitution of atoms in the tetrahedral and octahedral layers of the clay minerals as well as the presence of the Si-O-H (silanol) group on quartz, kaolin minerals and other surfaces like organic matter (-OH and -COOH). The magnitude of the former is fixed while that of the latter depends upon pH and concentration of the soil solution. In general, small highly charged ions cause the viscosity of the soil solution to increase while large monovalent ions cause the viscosity to decrease. The electrostatic fields of the ions cause polarization and a binding of surrounding water molecules which alter the kinetic properties of soil water. The hydrophilic nature of most soil particles is attributed to the attraction of hydrated cations by the electrostatic field of soil particles and to the hydrogen bonding of water to the clays (Low, 1961). The mobility of both water and ions in the region of the pore walls is reduced below that in bulk solutions (Kemper, 1960; and Dutt and Low, 1962). The impact of the electric field on ions and water is more pronounced in clayey soils and depends upon the ionic concentration and distance from the pore wall.

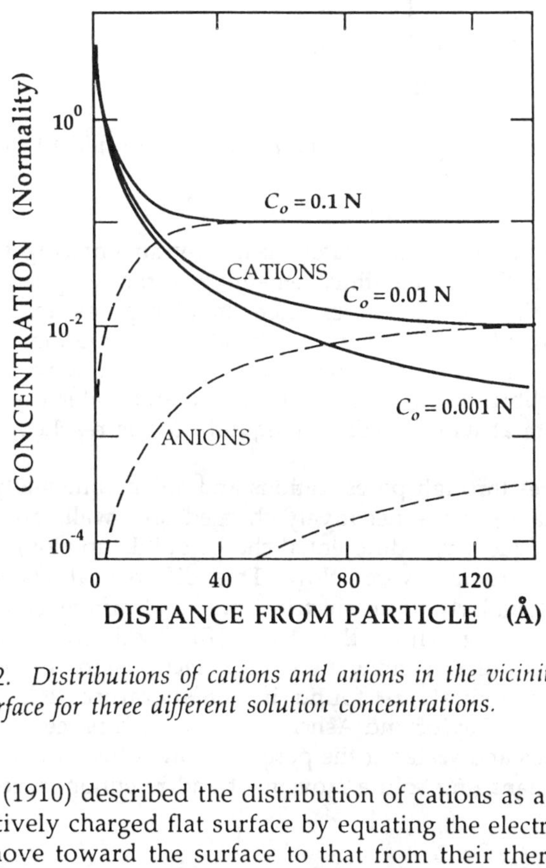

Figure 9.2. Distributions of cations and anions in the vicinity of a clay particle surface for three different solution concentrations.

Gouy (1910) described the distribution of cations as a function of distance from a negatively charged flat surface by equating the electrostatic force causing cations to move toward the surface to that from their thermal motion causing them to diffuse away from the surface. We see in Fig. 9.2 that the extent of the unequal distribution of cations and anions away from the surface depends inversely upon the total concentration C_o of the solution. And, we note from Fig 9.3 for cylindrical pores with a wall having a net negative charge and filled with a solution of concentration C_o that the concentration distribution across the pores depends upon the magnitude of the pore radius. In the center of large pores the concentrations of cations and anions are identical while in the center of small pores owing to the electric field, the cationic concentration exceeds that of anions.

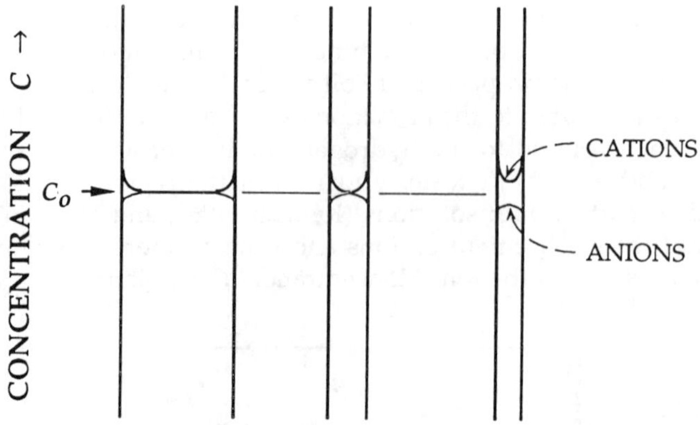

Figure 9.3. Distributions of cations and anions within capillary tubes having three different radii.

When redistribution of soil water occurs by air entering larger water-filled pores, solutes within water films remaining in the "emptied" large pores diffuse and mix with those in water-filled smaller pores achieving a new equilibrium which manifests different concentration distributions in all pores - large and small. Recognizing that the rates of water and solute redistribution are interdependent, equilibrium distributions of solute and water (hysteresis) both depend upon the rate at which hydraulic equilibrium is reached (Davidson et al., 1966).

As water moves through pores, cations and anions unequally distributed across the pores because of the negatively charged pore walls are swept along with the water. Consequently, a differential charge builds up along the length of flow which tends to retard water flow. This differential charge is called streaming potential. Similarly, if an electrical potential difference is established across a soil, ions moving within the electric field will create a water flux. Analytical descriptions of streaming potential, electroosmosis, electrophoresis and sedimentation potential are found in the literature (e.g. Nerpin and Tchudnovskii, 1967, and Taylor and Ashcroft, 1972). Each process contributes to the behavior of solutes and water at the pore scale and offers an opportunity for understanding and managing solute movement and retention in soil profiles.

9.1.3 Other Reactions

Constituents in the gas, liquid and solid phases of soil continually reacting with each other through a variety of chemical and biological pathways contribute to the presence and behavior of particular solutes in soil profiles. Applicable equilibrium and nonequilibrium chemical concepts such as oxidation-reduction, solubility-precipitation, association-dissociation, acid-base and exchange-adsorption are described by Freeze and Cherry (1979) and Luckner and

Schestakow (1991). Descriptions of microbiological reactions and pathways are available (e. g. Bazin et al., 1976) as well as those involving root systems of higher plants (e. g. Nye and Tinker, 1977). A full understanding of solute transport requires a knowledge of the information contained in these and other references.

9.2 MISCIBLE DISPLACEMENT IN A CAPILLARY

The oldest approach to analytic descriptions of miscible displacement in soils and other porous media is that of the displacement of a fluid by a second miscible fluid within a capillary tube. Consideration was focused upon the average displacement velocity and geometric boundaries with little concern for chemical or microbial processes.

9.2.1 Displacement without Molecular Diffusion

From the definition of viscosity, the force per unit area τ required to shear a fluid of viscosity η is

$$\tau = -\eta \frac{dv}{dr} \tag{9.5}$$

where v is the velocity of the fluid and r the coordinate normal to the force. The velocity distribution within a horizontal capillary tube of radius a during steady, uniform flow caused by a pressure difference Δp across its length L is the result of an equilibrium between the pressure and shearing forces acting on the fluid. Hence, for the capillary tube

$$\pi r^2 \Delta p + 2\pi r L \eta \frac{dv}{dr} = 0 \tag{9.6}$$

Integrating (9.6) between the limits $r = r$ and $r = a$ (where $v = 0$), we have the well known parabolic velocity distribution

$$v(r) = 2v_o\left(1 - \frac{r^2}{a^2}\right) \tag{9.7}$$

where $v(0) = v_o = a^2 \Delta P / 8\eta L$. The volumetric flow rate Q ($cm^3 \cdot s^{-1}$) through the capillary is easily obtained by integrating $v(r)$ with the areal cross-section of the capillary

$$Q = \int_0^a 2\pi r v(r) dr = 4\pi v_o \int_0^a \left(r - r^3 / a^2\right) dr \tag{9.8}$$

or

$$Q = a^2 \pi v_o = \frac{a^4 \pi \Delta P}{8\eta L}. \tag{9.9}$$

And the average velocity $\overline{v}$ of fluid flowing within a capillary tube is simply

$$\overline{v} = \frac{1}{\pi a^2} \int_0^a 2\pi r v(r) dr = \frac{4v_o}{a^2} \int_0^a \left(r - r^3 / a^2\right) dr \tag{9.10}$$

which leads to $\overline{v} = v_o$.

If we assume no molecular diffusion and rely solely on (9.7) to describe the fluid velocity, what will be the distribution of a second fluid of concentration C_o as it displaces a fluid of zero concentration initially within a capillary? Consider the solution C_o enters the tube at $x = 0$ at time $t = 0$. The concentration C averaged over the cross section of the capillary at distance x is

$$C_a = \frac{C_o}{\pi a^2}\int_0^a 2\pi r C(r)dr$$

or as a function of distance and time is

$$C_a(x,t) = C_o(1 - x/2v_o t). \tag{9.11}$$

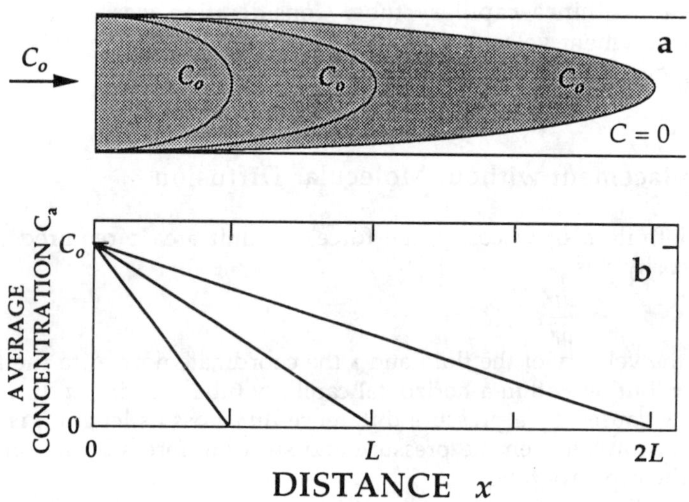

Figure 9.4. Parabolic velocity distributions of an invading solution C_o within a capillary (upper) give rise to linear average concentration distributions along the tube (lower).

In Fig. 9.4, the paraboloid of the displacing fluid C_o within the capillary gives rise to a linear concentration distribution. When the invading front of C_o has reached a distance $2L$, the average concentration across the plane normal to the capillary at a distance of L is $C_o/2$. Interestingly, the average concentration of the fluid moving across the plane L at that instant is not $C_o/2$ but $3C_o/4$. The average concentration of fluid moving past $x = L$ (see Fig. 9.5a) is

$$C_a = \frac{\textit{mass of solute moving past } x = L}{\textit{volume of fluid moving past } x = L}$$

$$C_a = \frac{C_o\int_0^{r'}\left[2v_o\left(1 - r^2/a^2\right)\right]2\pi r\,dr}{\int_0^a\left[2v_o\left(1 - r^2/a^2\right)\right]2\pi r\,dr} \tag{9.12}$$

where $r' = a(1 - L/2v_o^2)^{1/2}$, the radial boundary between the displacing fluid ($C = C_o$ for $r < r'$) and the original fluid ($C = 0$ for $r > r'$). Integrating (9.12) leads to

$$C_a(t)/C_o = 0 \qquad v_o t/L \le 0.5$$
$$= 1 - L^2/4v_o^2 t^2 \qquad v_o t/L > 0.5 \tag{9.13}$$

or

$$C(p)/C_o = 0 \qquad p \le 0.5$$
$$= 1 - 1/4p^2 \qquad p > 0.5 \tag{9.14}$$

where $p = v_0 t/L$ and is the ratio of the volume of fluid passing $x = L$ to the volume of the capillary between $0 \le x \le L$. Pore volume of effluent or simply pore volume is the name commonly used for p. The value of $C = 3C_0/4$ for $p = 1$ and approaches unity as $p \to \infty$, see Fig. 9.5b. Even for such a simple geometry as a capillary tube, the concentration distribution within the tube (Fig. 9.4b) is not easily reconciled with the shape of the concentration elution curve (Fig. 9.5b).

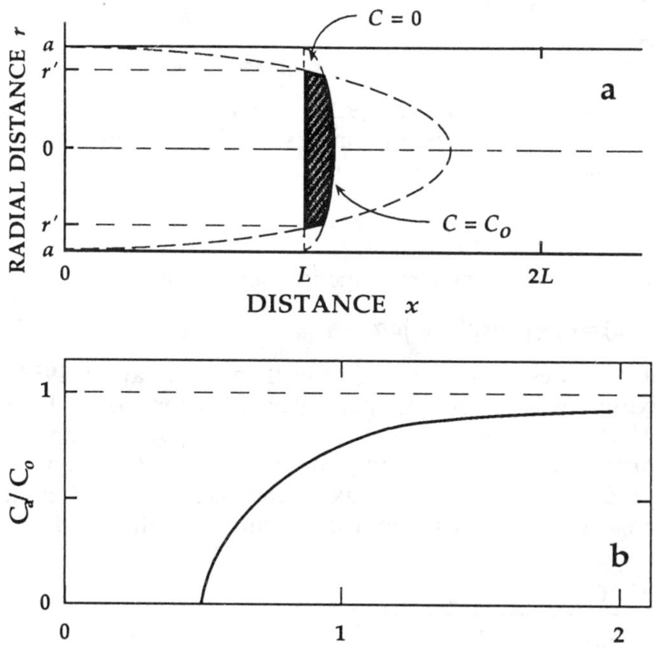

Figure 9.5. a. Cross sectional diagram of an elementary volume of fluid moving within a capillary. b. Average relative concentration of fluid leaving a capillary of length L as a function of pore volume of effluent.

9.2.2 Displacement with Molecular Diffusion

Taylor (1953) provided insights to separate dispersion of a solute within a capillary caused by molecular diffusion from that by the velocity distribution of the fluid. Assuming that the solute concentration within a capillary is symmetric about its central axis and the velocity of the fluid is described by (9.7), the mixing of a solute by both diffusion and convection are described by

$$\frac{\partial C}{\partial t} = D_m\left(\frac{\partial^2 C}{\partial r^2}+\frac{1}{r}\frac{\partial C}{\partial r}+\frac{\partial^2 C}{\partial x^2}\right) - 2v_o\left(1-\frac{r^2}{a^2}\right)\left(\frac{\partial C}{\partial x}\right) \tag{9.15}$$

where we assume the molecular diffusion coefficient D_m is independent of concentration. The following initial and boundary conditions approximate our assumption that the displacing solution C_o enters a long capillary at a uniform rate at $x = 0$ starting at time $t = 0$:

$$\begin{array}{llll} C=0 & t=0 & x>0 & 0<r<a \\ C=C_o & t\geq 0 & x=0 & 0<r<a \\ C=0 & t\geq 0 & x\rightarrow\infty & 0<r<a \\ \dfrac{\partial C}{\partial r}=0 & t\geq 0 & 0<x<\infty & r=0 \text{ and } r=a \end{array} \tag{9.16}$$

Details of a numerical solution of (9.15) subject to (9.16) are given by Amanthakrishnan et al. (1965). The limiting case for a very small fluid velocity when $v \rightarrow 0$ is the analytic solution

$$(C/C_o)_a = erfc\left[\frac{x}{\sqrt{4Dt}}\right] \tag{9.17}$$

where erfc is the complementary error function defined by

$$erfc(z) = \frac{2}{\sqrt{\pi}}\int_z^\infty \exp(-\alpha^2)d\alpha. \tag{9.18}$$

For small velocities considerably greater than zero, Taylor (1953) assumed that solute mixing by longitudinal molecular diffusion could be neglected compared with that owing to convection and suggested an alternative experimental procedure for measuring the molecular diffusion coefficient. Assuming that $\partial^2C/\partial x^2 = 0$ in (9.15), making the Galilean transformation [$x_1 = (x - v_o t)$] to change the coordinate from the cylinder to the fluid and further assuming that

$$\frac{\partial(C/C_o)_a}{\partial x_1} = \frac{\partial(C/C_o)_{x_1}}{\partial x_1},$$

he found that the flux density of solute at x_1 is

$$q = -\frac{a^2 v_o^2}{48D_m}\frac{\partial C}{\partial x_1}. \tag{9.19}$$

This equation shows that a solute is dispersed relative to a plane which moves with a velocity v_o exactly as though it were being mixed by molecular diffusion [compare with (9.1)] but with an apparent diffusion coefficient D_a defined by

$$D_a = \frac{a^2 v_o^2}{48D_m}. \tag{9.20}$$

Aris (1956) extended the above analysis for irregularly shaped capillaries which also included the mixing contribution from longitudinal molecular diffusion. His analysis, allowing a non parabolic local velocity distribution within the irregular capillaries and a concentration-dependent molecular diffusion coefficient, gives the apparent diffusion coefficient

$$D_a = D_m + \frac{\alpha \bar{a}^2 \bar{v}_o^2}{D_m} \tag{9.21}$$

where α is a dimensionless number which depends upon the cross section, $\bar{a}$ the effective radius and $\bar{v}_o$ the mean flow velocity.

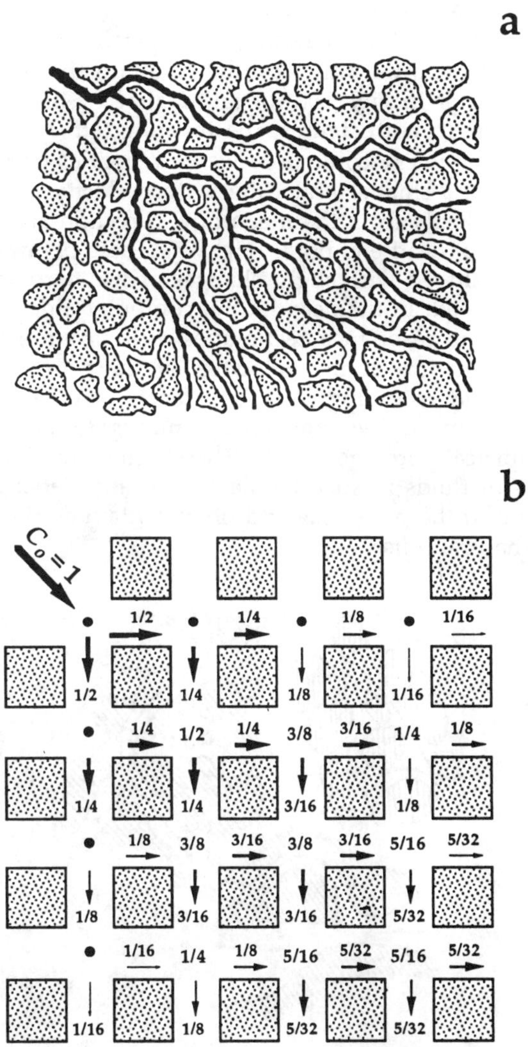

Figure 9.6. Spreading or dispersion of a solute caused by convective transport in a simplified soil a. and a 2-dimensional network of square soil particles b.

9.3 MISCIBLE DISPLACEMENT IN SURROGATE POROUS MEDIA

The mixing and interactive processes described in section 9.1 for simple, well-defined geometries and materials provide a basis for understanding transport in soils. Unfortunately, the rigor of solutions exemplified by those of Taylor (1953) gives way to that of empirical or statistical formulations owing to our inability to mathematically define the geometry of the soil pore system or to measure parameters descriptive of the displacement processes that can be translated from the pore scale through intermediate scales (including that of a REV) to that of a pedon or field. Miscible flow in a porous medium differs from that in a single capillary owing to meandering paths of water and solutes within and between irregularly shaped pores. The spreading or dispersion of the solute caused by convective transport with the water can be qualitatively visualized in Fig. 9.6a for a simplified soil, or in Fig. 9.6b for an idealized 2-dimensional network of square soil particles. In both figures, the invading stream of solute partitions itself according to the microscopic pore water velocities occurring between the soil particles. At still a smaller scale, the water velocity is zero at the particle surface, departs markedly from the mean flow direction and approaches a zero value in the vicinity of dead-end pores (Fig. 9.7). These pathways and pore water velocities, severely altered with slight changes of water content, have yet to be quantitatively evaluated. In the near future, computer-aided micro tomography and nuclear magnetic resonance techniques will provide an opportunity to ascertain the exact nature of the velocities at the pore scale. Without such observations, our understanding of certain facets of miscible displacement in soils has been enhanced by considering surrogate porous media having simplified or empirical pore geometries. These equations developed to describe the displacement of fluids in such media are usually deficient of physical and mathematical rigor at the pore scale and often contain empirical coefficients not easily related to natural soils.

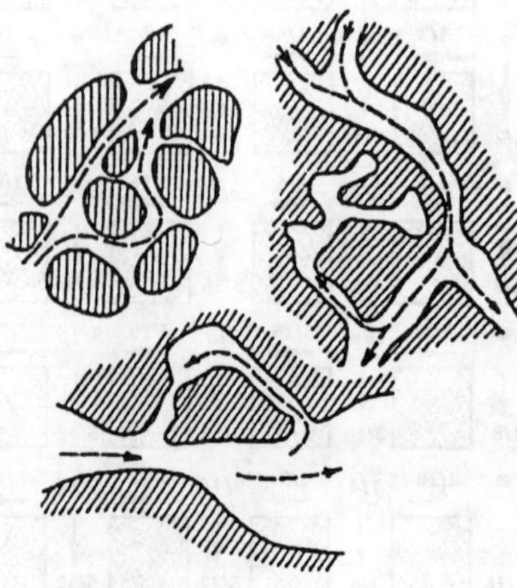

Figure 9.7. Examples of microscopic pathways and dead-end pores.

9.3.1 Displacement in Capillary Networks

Descriptions of idealized soil pores having capillary shapes include some mechanism for transport between parallel capillaries or allow one or more capillary tubes of differing radii to be joined at their ends at common junctions (e. g. Marle and Defrenne, 1960). The concept of random networks of capillary tubes provides insights to the meandering paths of displacing fluids.

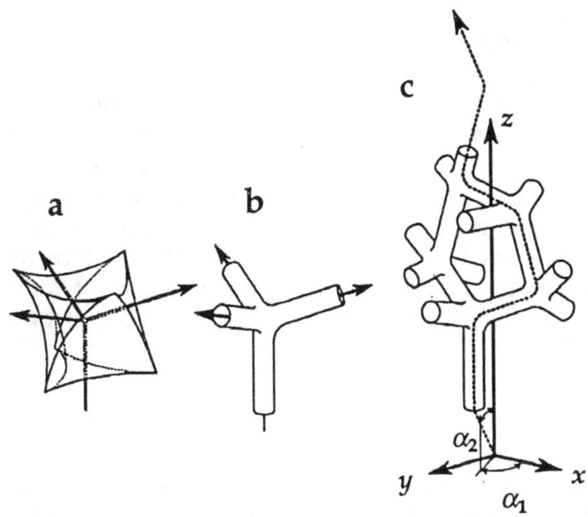

Figure 9.8. Diagram of pore space. a. Tetrahedral pore between four spheres, b. pore schematized by a canal bifurcation and c. random path chosen by a solute particle through the canal system.

The capillary tube network (de Josselin de Jong, 1958) illustrated in Fig. 9.8 stems from the tetrahedral pore between four closely packed spheres being represented by a junction of four capillaries. The randomness of the capillaries originates from the assumption that their positions are dictated by a random arrangement of soil particles. With z being the direction of principal flow, the direction of each segment of the capillary network is described by the angles α_1 and α_2. de Josselin de Jong neglected molecular diffusion and assumed that the velocity of a fluid particle was that of the mean velocity across the capillary diameter. Assuming that every capillary segment is of length l and the fluid velocity within each segment is proportional to cos α_2, the residence time t_i for a fluid particle in segment i is $t_i = t_R/\cos\alpha_2$ where t_R is the minimum residence time of a fluid particle traveling in a segment oriented in the direction of principal flow. Hence, a fluid particle traveling in a segment oriented in the direction (α_1, α_2) for a time Δt with a velocity $(l/t_R)\cos\alpha_2$ will deviate from the mean flow path by

$$\Delta z = \frac{l \Delta t \cos^2 \alpha_2}{t_R} \tag{9.22}$$

and

$$\Delta r = \frac{l \Delta t \sin \alpha_2 \cos \alpha_2}{t_R}. \tag{9.23}$$

These deviations lie on a sphere centered at $l\,\Delta t/2t_R$ having a radius $l\,\Delta t/2t_R$.

The dispersion of many fluid particles having passed through a large number of capillary segments resembles that of Brownian movement, but translated in the direction of the principal flow. Hence, the maximum concentration of an injected solute travels at the mean velocity of the fluid $l/(3t_R)$. With this simple capillary network composed of segments of invariant length, de Josselin de Jong was the first to show that the transverse apparent diffusion coefficient is smaller than the longitudinal apparent diffusion coefficient. He also showed that the magnitude of the longitudinal apparent diffusion coefficient depends upon the distance traveled. Subsequently, Saffman (1959, 1960) derived a similar, but more general model where the path of the fluid particle was regarded as a random walk with the direction, length and duration of each step being random variables. He accounted for pure molecular diffusion and the interaction between molecular diffusion and fluid convection.

9.3.2 Miscible Displacement as a Random Walk Process

Statistical concepts have been applied to solute and water transport through porous media at the pore scale primarily because of the difficulty of integrating differential equations of motion with poorly or undefined complex boundary conditions. Danckwerts (1953), Scheidegger (1954) and others have assumed that a simple random walk stochastic process can be used to describe transport in a fluid-saturated homogeneous, isotropic porous medium generally considered chemically inert. Here the exact nature of the path followed by fluid particles theoretically obeying Navier-Stokes equation in the porous system is not known. The velocity or position of a water or solute particle is the random variable and as the particle passes through the porous system, it eventually encounters all situations that are possible at any one given time. The probability distribution function w for any water particle at various points along the random path (Scheidegger, 1954) is given by

$$w(x,y,z,t) = \frac{1}{(4\pi Dt)^{3/2}} \exp\left\{\frac{-\left[(x-v_x t)^2+(y-v_y t)^2+(z-v_z t)^2\right]}{4Dt}\right\} \tag{9.24}$$

where v_x, v_y and v_z are the Darcy velocity components [components of q in (5.32) each divided by θ] in directions x, y and z, and $D = \sigma^2/2t$ where σ^2 is the variance of $w(x, y, z, t)$. Day (1956) described in detail the connection between $w(x, y, z, t)$ and the macroscopic concentration of a solute being displaced in a saturated sand.

Recognizing that (9.24) is proportional to the solute concentration and knowing that (9.24) satisfies classical diffusion equations, Danckwerts (1953) used solutions of

$$\frac{\partial C}{\partial t} = D\frac{\partial^2 C}{\partial X^2} \tag{9.25}$$

subject to appropriate initial and boundary conditions to describe the displacement of solutes through fluid-saturated porous media where $X = (x - v_x t)$. He noted that the value of D must be determined empirically and would presumably depend upon the viscosity, density and velocity of the fluid, and on the size and shape of the solid particles. He called D the "diffusivity" while Scheidegger named it the "factor of dispersion".

9.3.3 Displacement in a Representative Elementary Volume

The random capillary models described above were made somewhat more physically realistic [Bear and Bachmat, 1967] by deriving the idea of a representative elementary volume at the macroscopic scale from microscopic quantities at the pore scale averaged over many pores. Molecular diffusion and convection of solutes and water flowing within individual pores are described at the pore scale while the spreading or dispersion of solutes with water as it curves around and between soil particles through sequences of pores occurs at the macroscopic scale.

Bear and Bachmat envisioned the porous medium as a network of randomly interconnected narrow channels of varying length, cross section and orientation. The chemically inert, non compressible liquid of variable viscosity and density saturating the pores obeys Poiseulle's law and has two components - a solvent and a solute. No surface phenomena between the solid particles and the liquid take place. After deriving and averaging mass conservation and movement equations for the liquid in and across a channel, these local equations were averaged in a REV to obtain macroscopic equations. Details of all assumptions and the various averaging processes that lead to macroscopic equations containing average non random variables and parameters assigned to the centroid of the REV are given by Bear (1969). We repeat here their final equation of mass conservation in one direction avoiding the second-rank tensorial notation necessary for a three dimensional analysis

$$\frac{\partial C}{\partial t} = \frac{\partial}{\partial x}\left[\left(D_c + D_m\right)\frac{\partial C}{\partial x}\right] - \frac{\partial(vC)}{\partial x} \tag{9.26}$$

where v and C are average values within the REV and D_c and D_m are the coefficient of convective (or mechanical) dispersion [L T^{-2}] and coefficient of molecular diffusion [L T^{-2}], respectively. Combining the latter coefficients into a single term D_a (commonly called the hydrodynamic dispersion coefficient or the apparent diffusion coefficient), we have

$$\frac{\partial C}{\partial t} = \frac{\partial}{\partial x}\left[D_a\frac{\partial C}{\partial x}\right] - \frac{\partial(vC)}{\partial x} \tag{9.27}$$

for which many investigators have sought theoretical or experimental relationships between the value of D_a (which embraces solute mixing at both the pore scale and the pore-sequence scale within an REV) and the value of v (the average pore water velocity usually estimated by the ratio of the Darcian flux density and water content θ).

The results of several studies summarized by Pfannkuch (1962) related the value of D_a to the Peclet number of molecular diffusion Pe equal to vd/D_m where d is the mean solid particle size or other characteristic length of the porous medium. Recognizing that the effects of both molecular diffusion and convection on solute mixing in typical one-dimensional experiments were difficult separate, Fried and Combarnous (1971) suggested five ranges of Peclet numbers or zones to delineate the relative magnitudes of each process. For laboratory or field soils, we suggest that the following four zones will generally suffice except under some field conditions when turbulent flow down large fissures and cracks prevails during periods of rapid infiltration:

Zone 1	Pe < 0.3	$D_a = D_m$	$D_c \ll D_m$
Zone 2	0.3 < Pe < 5	$D_a = (D_m + D_c)$	$D_c \approx D_m$
Zone 3	5 < Pe < 20	$D_c < D_a < (D_c + D_m)$	$D_c > D_m$
Zone 4	Pe < 20	$D_a = D_c$	$D_c \gg D_m$

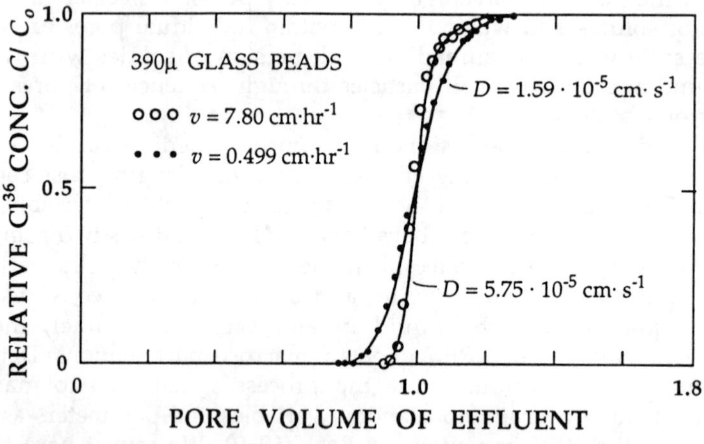

Figure 9.9. Breakthrough curves of ^{36}Cl for the displacement of 0.10 N NaCl at two flow rates through a uniform column of glass beads.

Zones 1 and 2 are illustrated in Fig. 9.9 for a miscible displacement experiment conducted with a column uniformly packed with 390μ glass beads initially saturated with 0.10 N NaCl. This initial solution was displaced at two different pore water velocities (0.499 and 7.80 cm·h^{-1}) with a chemically identical solution containing Na^{36}Cl to observe the mixing process and to match the concentration distribution measured at x = 30 cm with the solution of (9.27) subject to conditions

$$C = 0 \qquad x > 0 \qquad t = 0$$
$$C = C_0 \qquad x = 0 \qquad t > 0 \tag{9.28}$$
$$C = 0 \qquad x \to \infty \qquad t \geq 0$$

where C_0 is the concentration of ^{36}Cl. For each of the two pore water velocities, Fig. 9.9 shows measured values of the relative concentration C/C_0 and smooth curves satisfying (9.27) and (9.28) fitted to the data through the selection of the only unmeasured entity D_a in

$$\frac{C(x,t)}{C_0} = erfc\left(\frac{x - vt}{\sqrt{4D_a t}}\right) + \exp\left(\frac{vx}{D_a}\right) erfc\left(\frac{x + vt}{\sqrt{4D_a t}}\right) \tag{9.29}$$

where erfc the complementary error function has been defined earlier by (9.18). Values of D_a for each value of v satisfy the equation (Fried and Combarnous, 1971)

$$\frac{D_a}{D_m} = 0.67 + \alpha\left(\frac{vd}{D_m}\right)^n \tag{9.30}$$

where $D_m = 1.98 \cdot 10^{-5}$ cm$^2\cdot$s^{-1} (Wang, 1952) and $d = 3.9 \cdot 10^{-2}$ cm. Experimentally determined values of $\alpha = 0.51$ and $n = 1.02$ agree with those suggested by Fried and Combarnous (1971). For the slower pore water velocity (0.499 cm$\cdot$h^{-1}) with Pe = 0.27 (corresponding to the upper end of Zone 1), 83% of the mixing is caused by molecular diffusion. For the faster velocity (7.80 cm$\cdot$h^{-1}) with Pe = 0.43 (corresponding to the lower end of Zone 2), only 23 % of the mixing is caused by molecular diffusion.

Equation (9.30) is frequently reduced to

$$D_a = 0.67 D_m + \beta v^n \tag{9.31}$$

for Zones 3 and 4 with the molecular diffusion term neglected and the value of n taken as unity. In such instances, β is called the dispersivity.

Various investigators have used (9.27) or many similar diffusion type equations such as (9.25) with particular assumptions regarding the role of longitudinal and transverse molecular diffusion to theoretically or experimentally relate D_a to some function of v such as (9.30) or (9.31). Twenty five years ago Simpson (1969), writing a review article on the relationship between pore water velocity and the value of the longitudinal dispersion coefficient, stated, "The critical experiments remain to be performed: a systematic investigation of the effect on dispersion of changes in the molecular diffusivity". Somewhat later Sposito et al. (1979) presented a critical review of different theories used to describe solute transport through porous media. His remarks relative to theories based upon fluid mechanics and kinematic approaches employing various statistical techniques reveal major opportunities to improve our understanding of miscible displacement in surrogate porous materials. Earlier reviews by Fried and Combarnous (1971) and Bear (1969) also provide additional insights for future research.

9.4 ONE-DIMENSIONAL LABORATORY OBSERVATIONS

The displacement of one fluid by another miscible fluid in a soil studied either in the laboratory or the field offers theoretical and experimental challenges that have only partially been explored despite 30 years of recent investigation. For water infiltrating into a deep, homogeneous, water-saturated soil, we see in Fig. 9.10 that a solute of concentration C_o maintained at a point on the soil surface is dispersed vertically and horizontally. The velocity of the soil solution varies in both magnitude and direction owing to the distribution of irregularly shaped pores within the soil. Along transect A-A', the initial concentration C_o at $z = 0$ gradually diminishes to zero. Similarly, the concentration distribution normal to the average flow along transect B-B' gradually broadens with soil depth.

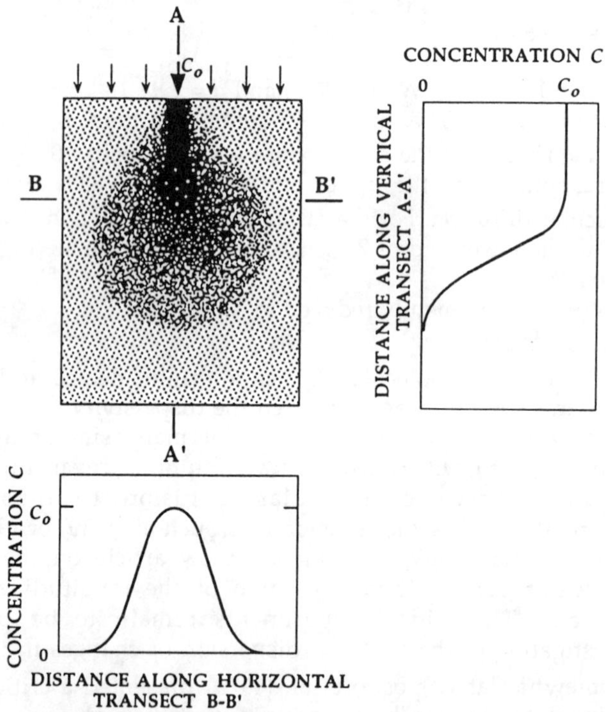

Figure 9.10. A solution of concentration C_0 being introduced at one point on the surface of a uniform, water-saturated soil during steady state infiltration. Right: Concentration distribution along vertical transect A-A'. Bottom: Concentration distribution along horizontal transect B-B'.

One dimensional soil columns studied in the laboratory provide a simple means of quantifying the mixing, spreading or attenuation of the solute schematically presented in Fig. 9.10. An apparatus is required to maintain steady state flow and invariant soil water content conditions when the initial soil

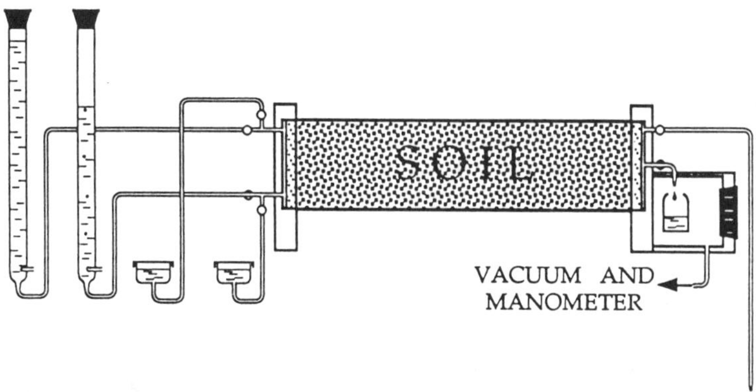

Figure 9.11. Laboratory apparatus for conducting miscible displacement experiments.

solution is invaded and eventually displaced by a second miscible solution. No mixing of the two solutions should occur at their boundary before entering the soil column, and samples of effluent to be analyzed for solute concentration has to be collected without disturbing the steady state flow conditions. A cross sectional sketch of a typical apparatus is given in Fig. 9.11. Details of its construction and operation are available (e. g. Nielsen and Biggar, 1961).

9.4.1 Breakthrough Curves

Without first considering specific physical, chemical or biological mechanisms involved in miscible displacement, much can be learned from a general description of continuous flow systems (Danckwerts, 1953). Let the volume of the soil column occupied by soil solution be V_o [L^3] and the rate of inflow and outflow of the soil solution be Q [L^3 T^{-1}]. If the initial soil solution identified by a solute concentration C_i is suddenly displaced by an incoming solution C_o, the fraction of this incoming solute in the effluent at time t [T] will be $(C - C_i)/(C_o - C_i)$, or for an initial concentration of zero, simply C/C_o. Plots of C/C_o versus pore volume of effluent (Qt/V_o), commonly called breakthrough curves, describe the relative times taken for the incoming solution to flow through the soil column. Note that the definition of pore volume of effluent is not restricted to water-saturated conditions but is applicable to all soil water contents. Any experimentally measured breakthrough curve may be considered one or a combination of any of the five curves shown in Fig. 9.12.

For Fig. 9.12a-c, the solute spreads only as a result of molecular diffusion and microscopic variations of the velocity of the soil solution, i. e. there is no interaction between the solute, water and soil particle surfaces. In these cases

$$\frac{Q}{V_o}\int_0^{\infty}(1-C/C_o)dt = 1 \tag{9.32}$$

regardless of the shape of the curve. This equation expresses the fact that the original soil solution occupied exactly one pore volume or that the quantity of solute within the soil column that will eventually reach a chemical equilibrium with that in the influent and effluent is $C_o V_o$. Note also that the area under the breakthrough curve up to one pore volume (area A in Fig. 9.12b and c) equals that above the curve for all values greater than one pore volume (area B), regardless of the shape of the curve. This latter statement is a direct result of (9.32), i. e.

$$\frac{Q}{V_o}\int_0^{V_o/Q}(C/C_o)dt = \frac{Q}{V_o}\int_{V_o/Q}^{\infty}(1 - C/C_o)dt. \tag{9.33}$$

Danckwerts (1953) defined holdback H_b as the left hand term of (9.33) having a range $0 \le H_b < 1$ for non reacting solutes. The concept of holdback is a useful qualitative description whenever interactions between solute, water and soil solids are minimal. It indicates the amount of the soil water or solutes not easily displaced. Values of H_b for unsaturated soils have been evaluated to be 3 to 4 times greater than those for saturated soils.

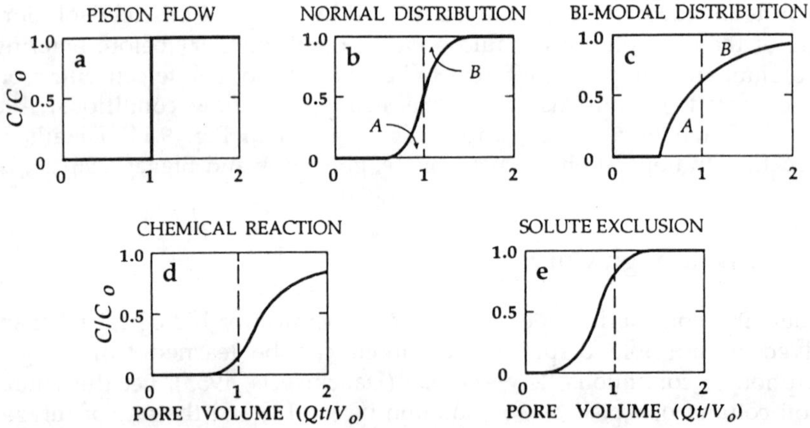

Figure 9.12. Types of breakthrough curves for miscible displacement. C/Co is the relative concentration of the invading fluid measured in the effluent and pore volume is the ratio of the volume of effluent to the volume of fluid in the sample.

Piston flow (Fig. 9.12a) never occurs owing to solute mixing that takes place by molecular diffusion and variations in water velocity at the microscopic level within soil pores. A breakthrough curve obtained from a water-saturated Oakley sand (Fig. 9.13a) is characteristic of the longitudinal dispersion shown in Fig. 9.12b. Evidence for lack of solute-solid interaction is the fact that the areas described by (9.33) are nearly identical. A water-saturated soil composed of equal-sized aggregates manifesting a bimodal pore water velocity distribution yields the breakthrough curve given in Fig. 9.13b that illustrates the curve of Fig. 9.12c. With the areas of (9.33) for this curve being comparable, any interaction between the solute and the soil particles is negligible.

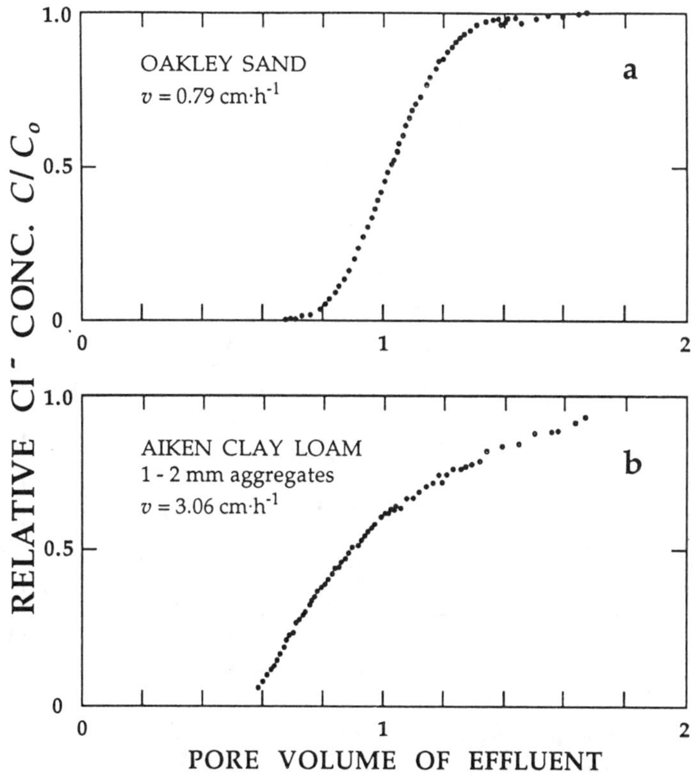

Figure 9.13. Chloride breakthrough curves for water-saturated a. Oakley sand and b. Aiken clay loam.

Figure 9.14 presents breakthrough curves for Cl^- and tritium (as tritiated water) from Columbia silt loam at two water contents and an average pore water velocity of approximately 0.4 $cm \cdot h^{-1}$. At a saturated water content of 0.482 $cm^3 \cdot cm^{-3}$ (Fig. 9.14a), the Cl^- curve was measured to the left of one pore volume (given by the vertical broken line) and the tritium curve to the right of one pore volume. The translation of the Cl^- curve to the left of one pore volume is characteristic of the curve in Fig. 9.12d and results from the repulsion of the negatively charged Cl ions from the negatively charged soil particle surfaces. The translation of the tritium curve to the right of one pore volume is characteristic of the curve in Fig. 9.12e and results from adsorption and exchange of tritium in the soil. Despite these different interactions, the close proximity of the curves at $C/C_o = 0$ and their relative shapes near $C/C_o = 1$ indicate the more complete mixing of the tritium having the greater molecular diffusion coefficient. A solute having the greater diffusion coefficient mixes more completely with the water in stagnant and slowly conducting zones, thus delaying its appearance in the effluent. In this case, the initial breakthrough of the solute having the greater molecular diffusion coefficient is translated to the

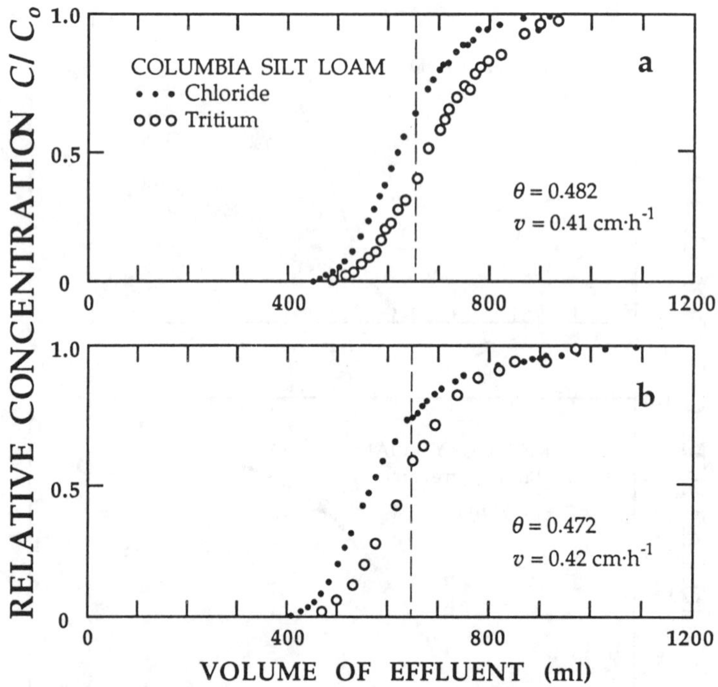

Figure 9.14. Chloride and tritium breakthrough curves for Columbia silt loam at soil water contents a. 0.482 cm³·cm⁻³ and b. 0.472 cm³·cm⁻³. The vertical broken lines designate 1 pore volume.

right of that of the other solute. It should not be expected that the relative behavior of solutes described above would be the same for all velocities and different soils. For example, if one solute has a diffusion coefficient much greater than the other, it would be possible for it to not only invade the nearly stagnant zones but also diffuse downstream ahead of the other solute. In this case the faster diffusing solute will appear in the effluent earlier that the more slowly diffusing solute.

In Fig. 9.14b, decreasing the water content by only 0.01 $cm^3 \cdot cm^{-3}$ translated both Cl^- and tritium curves to the left of one pore volume. Unsaturating the soil alters the pore water velocity distribution, allows some of the solute to arrive downstream earlier and increases the magnitude of holdback manifested by area A in Fig. 9.12. Desaturation eliminates larger flow channels and increases the volume of water within the soil which does not readily move. These almost stagnant water zones act as sinks to molecular diffusion. Later we shall discuss the opportunity afforded by controlling the water content and pore water velocity to change the leaching efficiency of field soils.

A more illustrative example of curves b, c and d in Fig. 9.12 are given in Fig. 9.15 from two columns of Oakley sand initially fully saturated with Ca^{2+}

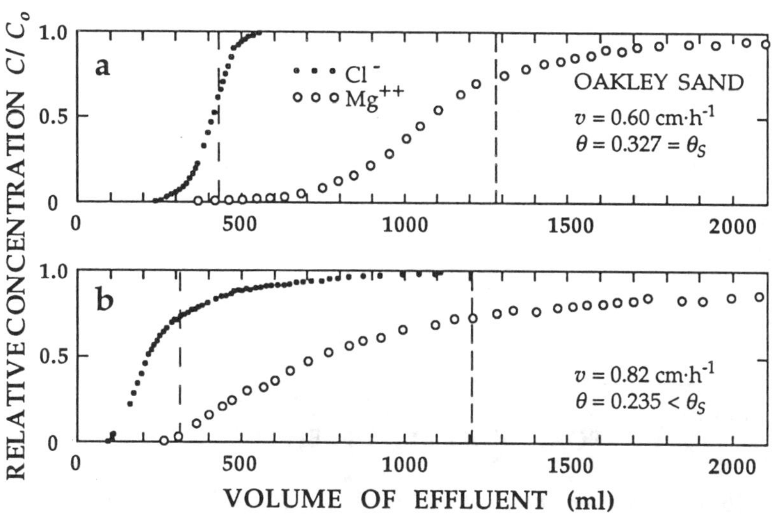

Figure 9.15. Chloride and magnesium breakthrough curves for initially Ca-saturated Oakley sand at soil water contents a. 0.327 cm³·cm⁻³ and b. 0.235 cm³·cm⁻³. The two vertical, broken lines designate 1 pore volume for the chloride and the cation exchange capacity, respectively.

[cation exchange capacity of 3.75 me·(100g)$^{-1}$] and 0.1 N calcium acetate flowing through them at a constant velocity while maintaining a constant water content. Breakthrough curves of Cl^- and Mg^{2+} for 0.10 N $MgCl_2$ displacing the calcium solution at water contents of 0.327 and 0.235 cm^3·cm^{-3} are presented in Fig. 9.15a and 9.15b, respectively. The positions of the Cl^- curves appear in the vicinity of the vertical lines representing the volume of solution in each soil column, while those of Mg^{2+} appear further to the right in the vicinity of another vertical line accounting for the cation exchange capacity. For the greater soil water content (Fig. 9.15a) the Mg^{2+} curve has the same characteristics as the Cl^- curve (similar to Fig. 9.12b) even though it is flatter and is displaced to the right (similar to Fig. 9.12d). Desaturating the soil water content (Fig. 9.15b) produces a Cl^- curve similar to Fig. 9.12c and a Mg^{2+} curve having the shape of Fig. 9.12c but a position illustrated in Fig. 9.12d. Depending upon the concentration and velocity of the displacing solution and the prevailing soil water content, the shapes and positions of breakthrough curves are governed by the characteristics of the cation exchange process as well as the mixing which occurs by the pore water velocity distribution, molecular or ionic diffusion and the spatial and temporal interaction of these processes.

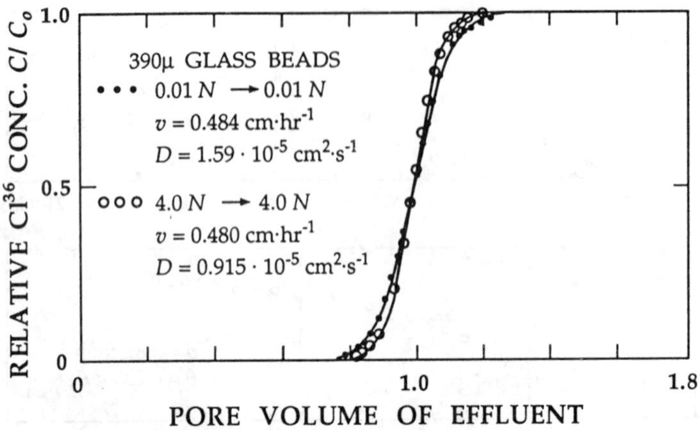

Figure 9.16. Breakthrough curves of ^{36}Cl for 0.01 N NaCl displacing 0.01 N NaCl and for 4.0 N NaCl displacing 4.0 N NaCl in horizontal columns of water-saturated glass beads.

9.4.2 Magnitude of the Diffusion Coefficient

The majority of inorganic cations, anions and solutes in soil solutions have diffusion coefficients in the order of 10^{-5} $cm^2 \cdot s^{-1}$ while organic cations, anions and solutes usually manifest much smaller values. These coefficients are moderately temperature dependent and slightly concentration dependent. The importance of their different magnitudes is apparent only at relatively small pore water velocities and often difficult to measure except under rigorously controlled laboratory conditions.

For a 30-cm long column of 390 µ glass beads, the results of two experiments observing the displacement of 0.01 N NaCl by an identical solution identified with a tracer of $^{36}Cl^-$ and the displacement of 4.00 N NaCl by a similarly observed identical solution containing $^{36}Cl^-$ are given in Fig. 9.16. With the pore water velocity for each experiment conducted at 25 ±1°C being about 0.48 $cm \cdot h^{-1}$ (Zone 1 described in section 9.3.3) and the densities and viscosities of the displacing and displaced solutions being virtually identical, distinctly separate curves for each experiment are obtained owing primarily to the different self diffusion coefficients of $^{36}Cl^-$ ($1.98 \cdot 10^{-5}$ and $1.24 \cdot 10^{-5}$ $cm^2 \cdot s^{-1}$ for the concentrations of 0.01 and 4.00 N NaCl, respectively). At greater pore water velocities (Zones 2 through 4) where mixing by convection progressively dominates the displacement process, the importance of diffusion lessens. Accordingly, for most infiltration and redistribution events following rainfall or irrigation in field soils, a nominal value of the diffusion coefficient is assigned to either an inorganic or organic solute without discriminating between solute species or their concentrations. The impact of temperature on diffusional mixing remains obscure from investigation.

9.4.3 The Impact of Density and Viscosity

The existence of concentration gradients of inorganic salts in the soil solution responsible for solute transfer by diffusion or as a result of convection guarantees that the displacing and displaced solutions do not generally have identical values of density or viscosity no matter how close their values. In soils, it is not uncommon to experience solutions of unequal density and viscosity. During the extraction of water from soil profiles by plants or by evaporation at the soil surface, the density and viscosity of the soil solution increase continually. Conversely the infiltration of rain or many irrigation waters causes the soil solution to be diluted. Fertilizers and other agrochemicals also alter these properties of the soil solution. The density and viscosity of the soil solution also differ from those of the bulk solution owing to the interaction of water and the soil particle surfaces especially in unsaturated soils or those soils having large clay contents (Dutt and Low, 1962).

When two superposed solutions of unequal density are accelerated in a direction perpendicular to their interface, the surface may be stable or unstable. Differences in density provide unbalanced forces while differences in viscosities account for unequal drag forces. For example, unstable flow occurs for particular velocities vertically downward when a dense, more viscous fluid displaces a less dense, less viscous fluid. Here, the unbalanced forces tend to accelerate the denser fluid into the less dense fluid below with the viscous drag of the lower fluid unable to counter-balance this acceleration. With this action "fingers" of the more viscous fluid invade those pore sequences occupied by the less dense fluid. For a 30 cm long vertical column of 390 μ glass beads, the breakthrough curve in Fig. 9.17 obtained for 0.01 N NaCl flowing upward at a velocity of 6.07

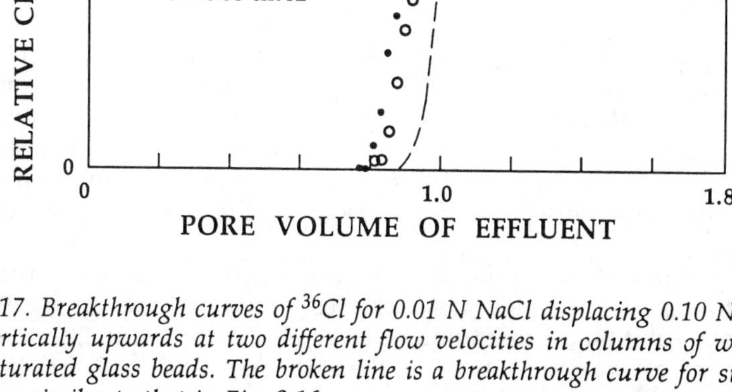

Figure 9.17. Breakthrough curves of ^{36}Cl for 0.01 N NaCl displacing 0.10 N NaCl vertically upwards at two different flow velocities in columns of water-saturated glass beads. The broken line is a breakthrough curve for stable flow similar to that in Fig. 9.16.

$cm \cdot h^{-1}$ and displacing the more dense and viscous 0.10 N NaCl exhibits unstable flow compared to that for stable flow (a curve from Fig. 9.16 shown as a broken line).

The stability of the displacement depends upon the viscosities and densities of the fluids, the permeability of the soil, and the direction and velocity of the displacement. If Darcy's equation is obeyed for steady movement with average pore water velocity v vertically upwards through a soil of permeability K_p, the interface between the two solutions will be unstable for

$$(\mu_2 - \mu_1)v + K_p(\rho_2 - \rho_1)g < 0 \tag{9.34}$$

where ρ is the density, μ the viscosity and g the gravitational acceleration (Taylor, 1950). The subscripts 1 and 2 refer to the displaced and displacing fluids, respectively. Except for signs owing to the direction of the velocity, (9.34) also applies to movement vertically downward. Inequality (9.34) has been used extensively for immiscible fluids (e.g. Oatmans, 1962) and for miscible fluids (e. g. Wooding, 1959, and Brigham et al., 1961). For the immiscible fluids water displacing air during ordinary infiltration at the soil surface, the same principles apply for the potential development of fingers of water moving ahead of the average wetting front into the unsaturated soil profile. For miscible fluids, the thickness of the interface region is not constant and tends to increases owing to molecular diffusion. Hence, with the viscosity and density of the interfacial fluid being neither those of the displacing nor displaced fluid and with the pore geometry only implicitly considered through the value of K_p, (9.34) holds only approximately for soils.

9.4.4 Influences of Solution Concentration and pH

Recalling from section 9.1.2, the surface charge characteristics of soil particles and colloids are of two general types - one having a constant surface charge and a variable surface potential, and the other having a constant surface potential and a variable surface charge (Bolt, 1979).

The interplay of these chemical effects on transport during the displacement of pulses of solution containing $^{36}Cl^-$ through a water-saturated Oxisol (Nkedi-Kizza, 1979) is shown in Fig. 9.18. In Fig. 9.18a, as the concentration of the soil solution decreases from 0.1 to 0.001 N $CaCl_2$, the $^{36}Cl^-$ elution curves shift to the right with their maxima decreasing. At a pH of 4, Cl^- is adsorbed. Owing to the fact that an equal number of negative and positive exchange sites exists at pH 3.6 for this soil, we expect $^{36}Cl^-$ to be exchanged for their non radioactive isotopes on the clay surfaces. Differences in shapes and positions of the curves in Fig. 9.18a are a result of the concentration of the soil solution rather than caused by hydrodynamic and geometric aspects of the flow regime.

As the pH of the soil increases above 3.6, the relative proportion of negative to positive exchange sites increases. Thus, as shown in Fig. 9.18b for a constant soil solution concentration of 0.001 N, the $^{36}Cl^-$ elution curve shifts to the left as the pH increases. At pH 9, the early arrival of $^{36}Cl^-$ is indicative of a solute that is repelled from the predominantly negatively charged clay surfaces.

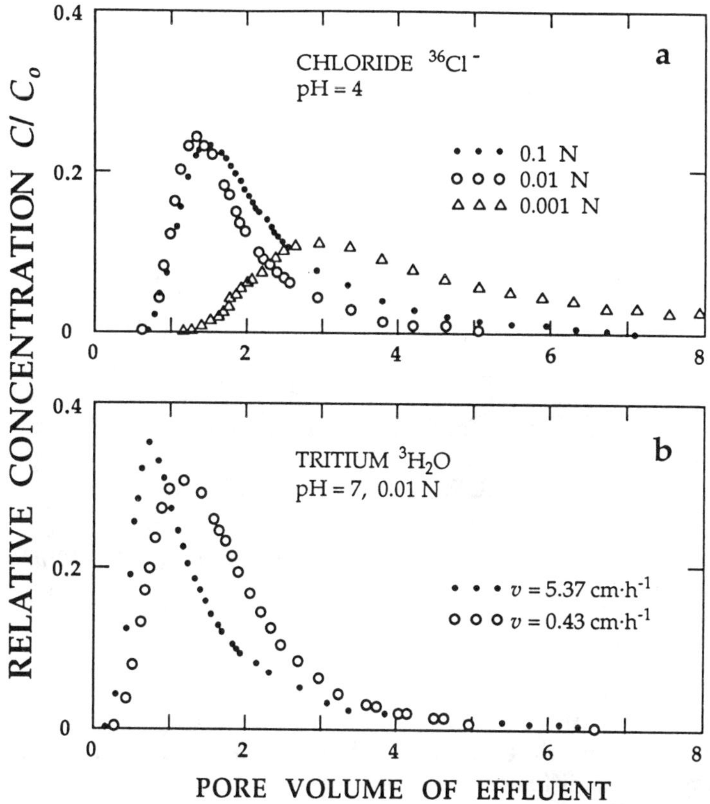

Figure 9.18. Breakthrough curves for a water-saturated, Ca-saturated Oxisol. a. ^{36}Cl in a $CaCl_2$ solution displacing a solution of an identical chemical composition for three different concentrations at a pH of 4. b. Tritium in a 0.01 N $CaCl_2$ solution displacing a solution of an identical chemical composition at pH = 7 for two different pore water velocities.

Although arid soils usually are dominated by constant charge colloids and tropical soils by those of constant potential, all soils are mixtures of both, and hence their behavior under conditions that induce shifts in pH cannot be ignored.

9.4.5 Influence of Displacement Length

The mixing and attenuation of a solute by convection depend upon the pore size distribution and the number of bifurcations experienced by the soil solution as water flows through its system of microscopic pores (recall Fig. 9.6). The greater the total macroscopic displacement length, the greater will be the

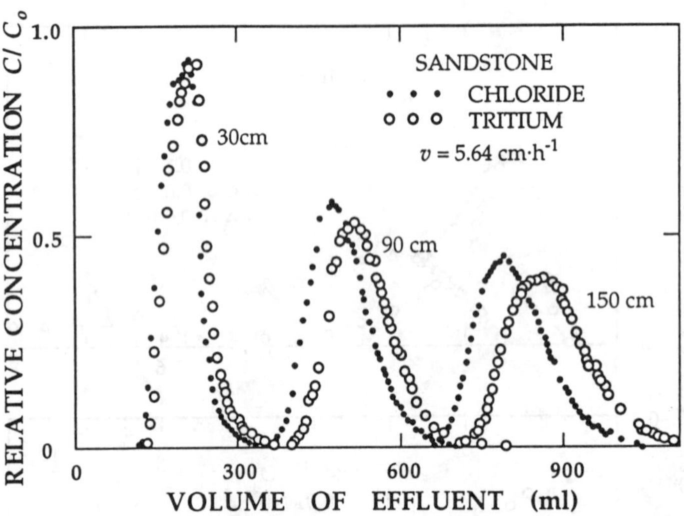

Figure 9.19. Chloride and tritium breakthrough curves for 30, 90 and 120 cm long , water-saturated sandstone columns.

opportunity for both convective and diffusive mixing. As the displacement length increases, both the number of bifurcations in the pore system and the time for molecular diffusion increase.

Corey et al. (1963) examined such a displacement as well as the nature of the pore structure of a uniform, consolidated sandstone by using five column lengths. Breakthrough curves for 75-ml pulses of tritiated $CaCl_2$ solution (Fig. 9.19) displaced through three of the columns illustrate the progressive attenuation of the initial concentration C_o as the solute is displaced through greater macroscopic lengths. Having traveled 150 cm, the relative concentration was reduced to about 0.4. A practical implication of such attenuation is the dilution of a solute within a soil profile without leaching any of its total mass beyond a desired depth.

Note in Fig. 9.19 that the chloride appears earlier in the effluent than the tritium and the separation between the data increases with column length. This behavior demonstrates the differences of the interactions between each of the solutes and the sandstone matrix. By conducting displacement experiments with homogeneous columns of different lengths, the relative amounts of convective and molecular diffusive mixing can also be ascertained when a series of different average pore water velocities are employed. Corey et al. avoided the difficulty of packing long soil columns with sufficent uniformity by using a naturally occurring homogeneous sandstone.

9.5 THEORETICAL DESCRIPTIONS

With soil physicists continuing their propensity to focus on geometric considerations, soil chemists focusing on chemical reactions and ignoring geometric effects and soil microbiologists and plant nutritionists limiting their investigations to biotic absorption, our understanding of the leaching process in soil profiles remains incomplete. This deficiency reflects our present-day inability to integrate simultaneously the most relevant physical, chemical and biological processes in a unified theoretical framework. Only recently have there been attempts to model transport in multicomponent systems with consideration of microbial activity and chemical reactions (e.g. Yeh and Tripathi, 1991, Richter et al., 1992, and Šimůnek and Suarez, 1994). Several theoretical opportunities provide a basis for future experiments and analyses.

9.5.1 The Convective-Diffusion Equation

Here we derive the cornerstone of most theoretical descriptions of solute transport in porous media recognizing its form is tentative in several aspects besides being fraught with uncertainties of applicable temporal (Skopp, 1986) and spatial scales (Dagan, 1986) that are not easily resolved.

We begin with the prism element having edges of length Δx, Δy and Δz given previously in Chapter 5 (Fig. 5.14). The difference between the mass of solute entering the prism and that leaving the prism equals the difference of the solute stored in the prism in time Δt providing that we account for any appearance (source) or disappearance (sink) of the solute within the prism by mechanisms other than transport. Hence, we obtain similar to (5.62) the equation of continuity of solutes S [M L^{-3}]

$$\frac{\partial S}{\partial t} = -\left(\frac{\partial J_x}{\partial x} + \frac{\partial J_y}{\partial y} + \frac{\partial J_z}{\partial z}\right) + \sum_i \phi_i \qquad (9.35)$$

where J_x, J_y and J_z are the flux densities of solute in directions x, y and z and ϕ_i [ML^{-3}T^{-1}] the i-th source or sink occurring within the prism usually considered irreversible during the time period over which the equation applies.

In general, soil solutes exist in both gaseous and aqueous phases as well as being associated with the solid organic and inorganic phases of the soil. Here, we neglect the fact that non aqueous polar and non polar liquids can also reside in soils and participate in the displacement process. We assume that at least some solutes in the soil solution are sufficiently volatile to consider their content or transport in the gaseous phase. Hence, the total solute concentration S in (9.35) is

$$S = \rho_T C_S + \theta C + (P - \theta) C_G \qquad (9.36)$$

where ρ_T is the soil bulk density [M L^{-3}], C_S the solute adsorbed or exchanged on the soil solids [M (M of dry soil)$^{-1}$], θ the volumetric soil water content, and C the solute in solution [M L^{-3}], P the porosity and C_G the solute in the soil air [M L^{-3}].

The solute flux density J in (9.35) relative to the prism $\Delta x \Delta y \Delta z$ is difficult to define unambiguously owing to the fact that the representative elementary volumes of each of the terms in J and S are not necessarily equal nor known, particularly for structured field soils. Each of the directional components of J is comprised of contributions of solute movement within the gaseous and liquid phases as well as along particle surfaces of the solid phase. We assume that solute movement along soil particle surfaces is nil or can be accounted for by functions relating the concentration of solutes in solution to that associated with the solid phase in (9.36). Hence, the solute flux density consists of three terms, one describing the bulk transport of the solute moving with the flowing soil solution, the second describing the solute moving by molecular diffusion and meandering convective paths within the soil solution and the third accounting for molecular diffusion in the gaseous phase. For the z-direction, we have

$$J_z = q_z C - \theta\left(D_{c_z} + D_{m_z}\right)\frac{\partial C}{\partial z} - (P-\theta)D_G\frac{\partial C_G}{\partial z} \tag{9.37}$$

where q_z is the Darcian flux, D_{c_z} and D_{m_z} are the coefficients of convective dispersion and molecular diffusion in the soil solution, respectively, D_G the molecular diffusion coefficient in the soil air and C_G the gaseous solute concentration in the soil air. Equations for J_x and J_y are identical to (9.37) when z has been replaced by x and y, respectively. We continue the analysis here for only the vertical soil profile direction z avoiding vectorial and tensorial notation.

Substituting (9.36) and (9.37) into (9.35), we obtain for a solute of the soil solution that does not volatilize into the soil air

$$\frac{\partial(\rho_T C_S)}{\partial t} + \frac{\partial(\theta C)}{\partial t} = \frac{\partial}{\partial z}\left[\theta\left(D_c + D_m\right)\frac{\partial C}{\partial z}\right] - \frac{\partial(qC)}{\partial z} + \sum_i \phi_i . \tag{9.38}$$

The first term of (9.38) describes the rate at which a solute reacts or exchanges with the soil solids. Its exact form (Helfferich, 1962) continues to be debated. We discuss both equilibrium and kinetic rate terms commonly used to describe this adsorption-exchange process in the next section.

With $D_a = (D_c + D_m)$, (9.38) reduces to

$$\frac{\partial(\rho_T C_S)}{\partial t} + \frac{\partial(\theta C)}{\partial t} = \frac{\partial}{\partial z}\left(\theta D_a \frac{\partial C}{\partial z}\right) - \frac{\partial(qC)}{\partial z} + \sum_i \phi_i . \tag{9.39}$$

Although the source-sink term ϕ_i in (9.38) or (9.39) has most often been considered in the absence of the rest of the equation in many disciplines, it is often approximated by zero- or first-order rate terms

$$\phi_i = \gamma\theta + \gamma_s\rho + \mu\theta C + \mu_s\rho_T C_s \tag{9.40}$$

where γ and γ_s are rate constants for zero-order decay or production in the soil solution and solid phases, respectively, and μ and μ_s are similar first-order rate constants for the two phases. For radioactive decay, physicists may safely assume that μ and μ_s are identical as well as assuming that both γ and γ_s are nil. Microbiologists, considering organic and inorganic transformations of soil solutes in relation to growth, maintenance and waste metabolism of soil microbes as a Michaelis-Menten process, often simplify their considerations to that of ϕ_i in (9.40). McLaren (1970) provided incentives to study such reactions as functions of both space and time in soil systems - a task not yet achieved by soil microbiologists, especially when the individual characteristics of each microbial species is quantified and not lumped together as a parameter of the entire microbial community. Agronomic or plant scientists consider ϕ_i as an irreversible sink and source of solutes taking place in the vicinity of the rhizosphere of cultivated or uncultivated plants as a function of soil depth and time as well as some empirical function defining the root distribution.

For a solute that does not appreciably react with the soil particles, does not exist in the soil air and does not appear or disappear in sources or sinks, respectively, (9.39) reduces to

$$\frac{\partial C}{\partial t} = \frac{\partial}{\partial z}\left(D_a \frac{\partial C}{\partial z}\right) - \frac{\partial (vC)}{\partial z} \tag{9.41}$$

which is identical to (9.27) of Bachmat and Bear. For steady state flow in a homogeneous soil at constant water content, (9.41) reduces still further to

$$\frac{\partial C}{\partial t} = D_a \frac{\partial^2 C}{\partial z^2} - v\frac{\partial C}{\partial z} \tag{9.42}$$

which has been extensively used to develop empirical relations between the apparent diffusion coefficient D_a and the average pore water velocity v.

An intensively measured, field scale miscible displacement experiment conducted to ascertain the distributions of D_a and v was reported by Biggar and Nielsen (1976). During steady state water flow conditions, they measured the leaching of water soluble salts at six soil depths to 1.8 m within 20 subplots of a 150-ha field. For times $0 < t \leq t_1$ the soil was leached steadily with water having a concentration C_o (chloride or nitrate). For times $t < 0$ and $t > t_1$ the soil was leached with water having a concentration C_i. In view of the large pore water velocities measured in their experiment, the appropriate initial and boundary conditions for the solution of (9.42) were

$$\begin{aligned} C&=C_i \qquad z>0 \qquad t=0 \\ C&=C_o \qquad z=0 \qquad 0<t\leq t_1 \\ C&=C_i \qquad z=0 \qquad t>t_1 \end{aligned} \tag{9.43}$$

The measured concentration $C(z, t)$ was fitted to the solution of (9.42) subject to (9.43)

$$\begin{aligned} \frac{C-C_i}{C_o-C_i} = \frac{1}{2}\left\{ erfc\left[\frac{z-vt}{(4D_a t)^{1/2}}\right] + \exp\left(\frac{vz}{D_a}\right) erfc\left[\frac{z+vt}{(4D_a t)^{1/2}}\right]\right\} - \\ -\frac{1}{2}\left\{ erfc\left[\frac{z-v(t-t_1)}{[4D_a(t-t_1)]^{1/2}}\right] + \exp\left(\frac{vz}{D_a}\right) erfc\left[\frac{z+v(t-t_1)}{[4D_a(t-t_1)]^{1/2}}\right]\right\} \end{aligned} \tag{9.44}$$

at different locations within the field to obtain 359 values of (D_a, v) which are plotted against each other in Fig. 9.20. Values of D_a were logarithmically normally distributed with a mode of 4.0, a median of 85.1 and a mean of 367.6 $cm^2 \cdot d^{-1}$. Values of v were also logarithmically normally distributed having a mode of 4.3, a median of 20.3 and a mean of 44.2 $cm \cdot d^{-1}$ (see Fig. 9.25). The solid

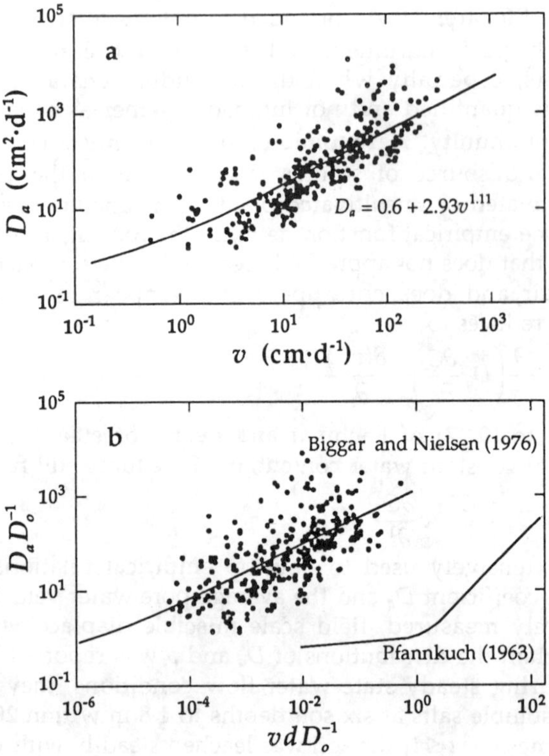

Figure 9.20. a. Field-measured values of the apparent diffusion coefficient D_a plotted against average pore water velocity v. b. Values of $D_aD_o^{-1}$ plotted against Peclet number vdD_o^{-1} where D_o is the molecular diffusion coefficient and d is the mean soil particle diameter.

line in Fig. 9.20a obtained by regressing $\ln(D_a - 0.6)$ versus $\ln v$ is of the same form as (9.31). The value of 0.6 represents the molecular diffusion coefficient for the unsaturated soil taking into account the average tortuosity and the average soil water content during displacement. The same data plotted as $D_aD_o^{-1}$ versus vdD_o^{-1} (Peclet number) in Fig. 9.20b where d is the mean soil particle diameter can be compared to those reported by Pfannkuch (1962) measured by several investigators in the laboratory using homogeneously packed columns. With the value of the coefficient $\beta = 2.93$ being two orders of magnitude greater that found in the majority of laboratory column studies, the impact of soil structure and the large range of particle diameters and pore sizes on the dispersion of solutes in natural field soils is abundantly evident.

9.5.1.1 Solutes in Continual Equilibrium with the Solid Phase

Instantaneous adsorption or exchange reactions included in the first term of (9.38) are described by equilibrium isotherms $C_s(C)$ of several different forms - mass action, linear, Freundlich, Langmuir or other functional forms (van Genuchten et al., 1974; Bolt, 1979; and Travis and Etnier, 1981).

The most common approach for modeling the first term of (9.38) has been to assume instantaneous adsorption or exchange as well as simple linearity between C_s and C [$C_s = kC$ where k is the slope of the isotherm $C_s(C)$ often referred to as the distribution coefficient K_d]. If we put this linear isotherm in (9.39) we have

$$\frac{\partial(\rho_T kC)}{\partial t} + \frac{\partial(\theta C)}{\partial t} = \frac{\partial}{\partial z}\left(\theta D_a \frac{\partial C}{\partial z}\right) - \frac{\partial(qC)}{\partial z} + \sum_i \phi_i \tag{9.45}$$

which reduces to

$$R\frac{\partial C}{\partial t} = D_a \frac{\partial^2 C}{\partial z^2} - v\frac{\partial C}{\partial z} + \sum_i \phi_i \theta^{-1} \tag{9.46}$$

for steady state flow in a homogeneous soil and where R is a retardation factor ($= 1 + \rho_T k\theta^{-1}$) . Equation (9.46) has been solved for many conditions for both finite and semi-infinite systems (van Genuchten and Alves, 1982; Javandel et al., 1984). While the simplicity of a linear isotherm is a convenient feature for mathematically modeling, its limitations are clearly apparent owing to adsorption and exchange processes usually being nonlinear and depending upon the competing species in the soil solution (see Reardon, 1981; Miller and Benson, 1983; and Valocchi, 1984).

9.5.1.2 Solutes Not in Equilibrium with the Solid Phase

Diffusion-controlled or chemically controlled kinetic rate reactions included in the first term of (9.38) have been described in a variety of equations. We consider here three of the more popular formulations.

The most simplest formulation is that of a first order linear kinetic reaction where (9.39) is replaced by two coupled equations. Here, for steady state flow through a homogeneous soil without sources or sinks, we have

$$\begin{aligned} \frac{\rho_T}{\theta}\frac{\partial C_s}{\partial t} + \frac{\partial C}{\partial t} &= D_a \frac{\partial^2 C}{\partial z^2} - v\frac{\partial C}{\partial z} \\ \frac{\partial C_s}{\partial t} &= \alpha(kC - C_s) \end{aligned} \tag{9.47}$$

where α is a first order rate constant. The success of this and similar rate models (e. g. Davidson and McDougal, 1973; van Genuchten et al., 1974) has been best when miscible displacement experiments have been carried out at relatively slow velocities when mixing is dominated by molecular diffusion. Nevertheless, under such conditions the values of α and k may indeed be biased owing to the use of an average value of v which does not embrace the spatial distribution of the solute influencing the rate reaction within soil pores at the microscopic scale.

The second formulation gives more consideration to the microscopic pore water velocity by defining a bimodal distribution which partitions the soil water into mobile and immobile phases. In the mobile phase where soil water flows, solute behavior is described by a convective-diffusion equation. Inasmuch as water is stagnant in the immobile phase, solutes move in and out of this phase only by molecular diffusion. Zones of stagnant water derive from thin liquid films around soil particles, dead-end pores, non moving intra-aggregate water or isolated regions associated with unsaturated conditions. Miscible displacement equations based on first order exchange of solute between mobile and immobile phases initially discussed by Coats and Smith (1964) were extended by van Genuchten and Wierenga (1976) to include Freundlich type equilibrium adsorption-desorption processes. Their equations are

$$\begin{gathered}\theta_m R_m \frac{\partial C_m}{\partial t} + \theta_{im} R_{im} \frac{\partial C_{im}}{\partial t} = \theta_m D_{a_m} \frac{\partial^2 C_m}{\partial z^2} - \theta_m v_m \frac{\partial C_m}{\partial z} \\ \theta_{im} R_{im} \frac{\partial C_{im}}{\partial t} = \alpha(C_m - C_{im})\end{gathered} \tag{9.48}$$

where the subscripts m and im refer to the mobile and immobile phases, respectively. The retardation factors account for equilibrium type adsorption processes similar to that in (9.46) while the mass transfer coefficient α embraces a diffusion coefficient and an average diffusional path length. Although (9.48) was used successfully by van Genuchten and Wierenga and by Gaudet et al. (1977) as well as many others more recently to describe laboratory column studies, its use in structured field soils has been limited owing to the difficulty of obtaining reliable values of α which depend upon the geometry of the soil pore structure (van Genuchten, 1985). For laboratory experiments, the value of α may well be confounded with nonlinear isotherm and chemically kinetic exchange effects. Moreover, the fraction of θ considered to be immobile is sensitive to hysteresis, the concentration of the soil solution, the soil water content and the soil water flux. From the experimental studies of Nkedi-Kizza mentioned earlier in section 9.4.1.4, Fig. 9.21 clearly shows the impact of the solution concentration and the average pore water velocity on the amount of soil water considered to be immobile.

The third formulation considers the first term of (9.38) having two components - one for exchange sites (type 1) on a fraction of the soil particle surfaces that involve instantaneous, equilibrium reactions and another for type 2 exchange sites involving first order kinetics or those assumed to be time-dependent (Selim et al., 1976; Cameron and Klute, 1977). Following Nkedi-Kizza et al. (1984) we have

$$\begin{gathered}\left(1 + F\rho_T k \theta^{-1}\right)\frac{\partial C}{\partial t} + \rho_T \theta^{-1} \frac{\partial C_{s_2}}{\partial t} = D_a \frac{\partial^2 C}{\partial z^2} - v\frac{\partial C}{\partial z} \\ \frac{\partial C_{s_2}}{\partial t} = \alpha\left[(1-F)kC - C_{s_2}\right]\end{gathered} \tag{9.49}$$

where F is the mass fraction of all sites being occupied with type 1 sites, and where subscript 2 refers to type 2 sites. With values of α and F usually being dependent upon the average displacement velocity v, values of F appropriate for (9.49) cannot be obtained from equilibrium batch studies. The use of an

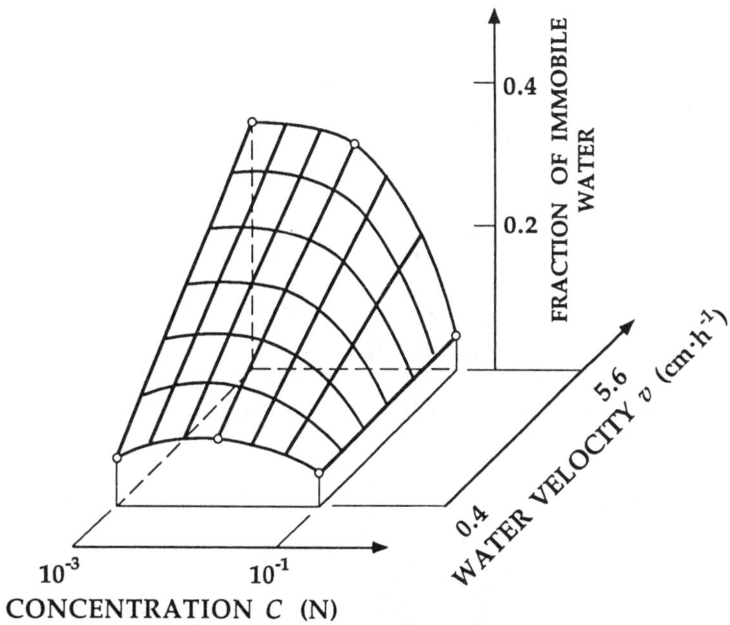

Figure 9.21. Fraction of immobile water versus solute concentration and average pore water velocity.

average value of v masks any effect of the microscopic pore water velocity distribution on the exchange process.

Comparison of (9.48) and (9.49) shows that they can be put in the same dimensionless form by means of equation-specific dimensionless parameters. With this information Nkedi-Kizza et al. (1984) proved that effluent curves from laboratory soil columns alone cannot be used to differentiate between the specific physical and chemical phenomena that cause an apparent non equilibrium situation during miscible displacement. The similarity of the two sets of equations allows an oftentimes satisfactory empirical description of the mixing of solutes at the macroscopic scale by either equation without ascertaining the exact nature of the particular chemical or physical process at the microscopic scale. The exact nature of the processes awaits further research using micro tomography or other techniques of observation at the microscopic scale.

9.5.1.3 Dual-Porosity Models for Structured Soils

Dual porosity models (also called bi-modal porosity models) assume that a soil can be separated into two distinct pore systems superpositioned over the same soil volume with each system being a homogeneous medium having its own water and solute transport properties (Dykhuizen, 1987). We assume that the same type of mathematical expressions can be used to describe both systems of pores inasmuch as similar capillary effects take place in the pores of both systems (Othmer et al., 1991). With the two systems exchanging water and solutes in response to hydraulic and concentration gradients, the soil is characterized by two water velocities, two hydraulic heads, two water contents and two solute concentrations. Gerke and van Genuchten (1993) provide a comprehensive review of various theoretical and experimental attempts to deal with water and solute movement in saturated and unsaturated structured soils during steady and transient water flow conditions. The basis of their numerical simulations demonstrating the complicated nature of solute leaching in structured, unsaturated soils during transient water flow are summarized here to provide future opportunities for field research in soil hydrology.

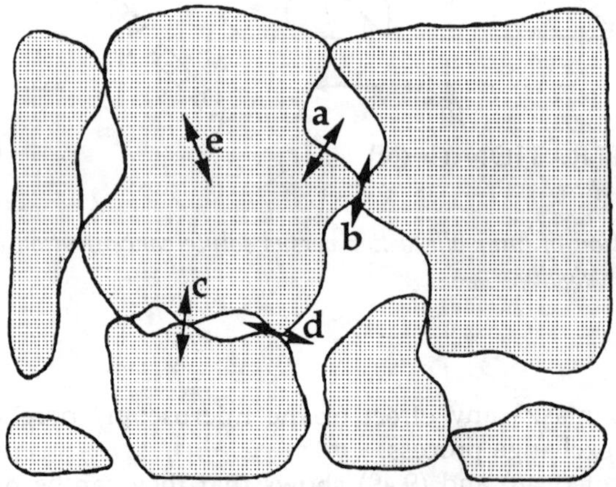

Figure 9.22. Schematic diagram of a structured soil at the microscopic level. Macropores, interaggregate pores and fracture pores appear between shaded areas representing soil aggregates. Arrows represent water and solute movement through a. the surface of an aggregate, through b. the fracture pore, between c. aggregates, between d. continuous and stagnant pores and inside e. an aggregate.

Microscopically, a structured soil (see Fig. 9.22 and our discussion in sections 2.4, 4.3.3 and 5.3.2) consists of soil aggregates (shaded irregular areas) surrounded by inter-aggregate pores (dotted areas) which form a more or less continuous network. The pore network is geometrically a combination of inter-aggregate and intra-aggregate pores further complicated by the presence of distinct mineral and organic particles discussed earlier. Here we use subscript M to denote the inter-aggregate system and subscript m to denote the intra-aggregate system. We refer the reader to the original article of Gerke and van Genuchten (1993) for a complete description of the hydraulic properties of the structured soil and the details of the solute transport equations only briefly described here.

Solute transport with linear adsorption and first order decay is described by

$$\begin{aligned}\frac{\partial(\theta_M R_M C_M)}{\partial t} &= \frac{\partial}{\partial z}\left(\theta_M D_{a_M}\frac{\partial C_M}{\partial z} - q_M C_M\right) - \theta_M \mu_M C_M - \frac{\Gamma_s}{w_M} \\ \frac{\partial(\theta_m R_m C_m)}{\partial t} &= \frac{\partial}{\partial z}\left(\theta_m D_{a_m}\frac{\partial C_m}{\partial z} - q_m C_m\right) - \theta_m \mu_m C_m + \frac{\Gamma_s}{1-w_M}\end{aligned} \tag{9.50}$$

where μ is a first order decay coefficient, Γ_s a solute mass transfer term to which both molecular diffusion and convective transport contributes and w_M is the ratio of the volume of the interaggregate pores to that of the total volume of all pores. With the exception of Γ_s which is defined as the mass of solutes per unit volume of bulk soil per unit time, all variables in (9.50) are defined relative to the partial volume of each pore system.

Although this model and its numerical solution can simulate transport related to specific chemical and physical properties of the soil, its large number of parameters not easily measured experimentally remain a topic of future laboratory and field research.

9.5.1.4 Consecutive Convective-Diffusion Equations

Equation (9.39) is oftentimes sequentially repeated when organic or inorganic products are of interest, especially when the products form a consecutive chain of reactions, e. g. for nitrogen (urea $\rightarrow NH_4^+ \rightarrow NO_3^- \rightarrow N_2$). In such cases a set of simultaneous equations stemming from (9.39) becomes

$$\begin{aligned}\frac{\partial(\rho_T C_{S1})}{\partial t} + \frac{\partial(\theta C_1)}{\partial t} &= \frac{\partial}{\partial z}\left(D_{a_1}\frac{\partial C_1}{\partial z}\right) - \frac{\partial(q C_1)}{\partial z} + \phi_1 \\ \frac{\partial(\rho_T C_{Sj})}{\partial t} + \frac{\partial(\theta C_j)}{\partial t} &= \frac{\partial}{\partial z}\left(D_{a_j}\frac{\partial C_j}{\partial z}\right) - \frac{\partial(q C_j)}{\partial z} - \phi_{j-1} + \phi_j\end{aligned} \tag{9.51}$$

where $j = 2, 3, \cdots n$ when n is the number of species considered in the reaction chain. The equations are linked to each other by their mutual ϕ_j terms. These equations have been applied to consecutive decay reactions of soil nitrogen species (e. g. Cho, 1971; Misra et al., 1974; Starr et al., 1974), organic phosphates (Castro and Rolston, 1977) and pesticides (Bromilow and Leistra, 1980).

We illustrate the application of (9.51) with an investigation conducted by Wagenet et al. (1977) who extended the mathematical analysis of Cho (1971) and the experimental techniques of Mansell et al. (1968) and Misra et al. (1974) to trace the fate of nitrogen applied as a pulse of 95% ^{15}N-enriched urea fertilizer to an unsaturated soil column during steady state leaching conditions. The concentration of oxygen within the partially air-filled pores was simultaneously controlled at a desired constant value (Wagenet and Starr, 1977). The transport

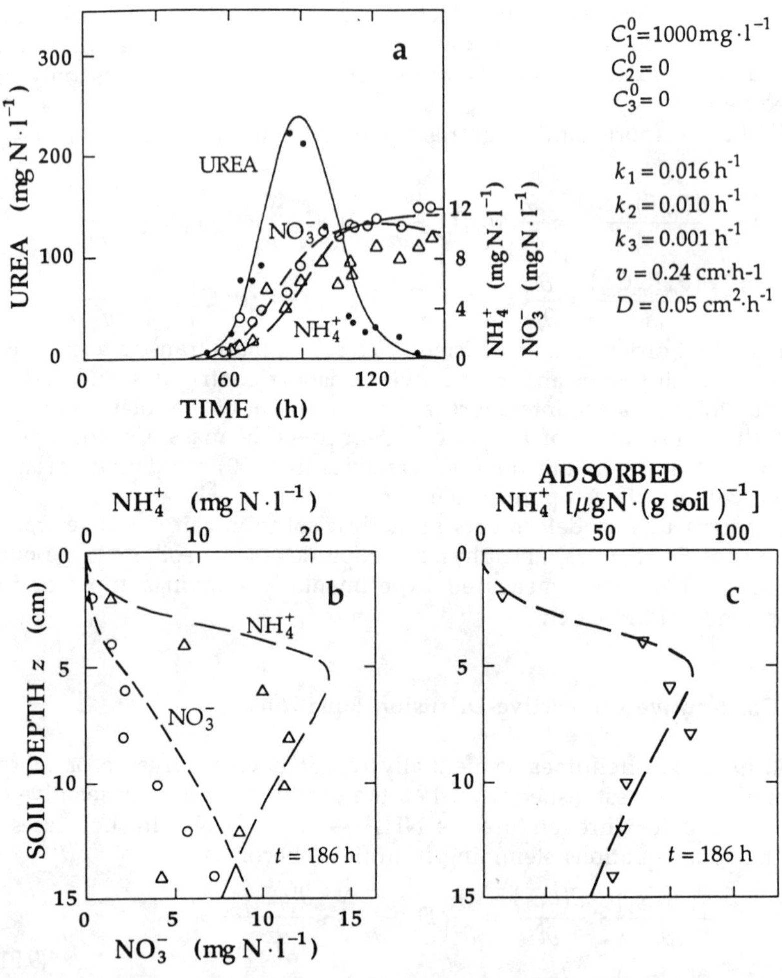

Figure 9.23. Concentration distributions of different nitrogen species identified in the effluent and soil from a pulse of ^{15}N-enriched urea solution displaced through a column under controlled laboratory conditions.

and transformation of urea, NH_4^+ and NO_3^- were identified by isotopic dilution techniques. Based upon complementary experiments, the enzymatic hydrolysis of urea C_1, the microbial oxidation of ammonium C_2 and the microbial reduction of nitrate C_3 were each considered first order rate reactions. For their study, (9.51) become

$$\left(1+R_1\right)\frac{\partial C_1}{\partial t} = D_a\frac{\partial^2 C_1}{\partial z^2} - v\frac{\partial C_1}{\partial z} - k_1 C_1$$

$$\left(1+R_2\right)\frac{\partial C_2}{\partial t} = D_a\frac{\partial^2 C_2}{\partial z^2} - v\frac{\partial C_2}{\partial z} + k_1 C_1 - k_2 C_2 \tag{9.52}$$

$$\frac{\partial C_3}{\partial t} = D_a\frac{\partial^2 C_3}{\partial z^2} - v\frac{\partial C_3}{\partial z} + k_2 C_2 - k_3 C_3$$

where R_1 and R_2 are the retardation factors for urea and ammonium [see (9.46)], respectively, D_a assumed to be identical for each of the chemical species and k_1, k_2 and k_3 are the rate constants describing the processes of urea hydrolysis, ammonium oxidation and nitrate reduction, respectively.

Elution curves for the three chemical species identified by the ^{15}N tracer introduced as a 100-ml pulse of urea solution (C_1^o = 1000 $mg{\cdot}l^{-1}$) into a soil column having 20% oxygen in its air-filled pores are shown in Fig. 9.23a. Theoretical and measured concentrations of NH_4^+-N and NO_3^--N in the soil solution and those of adsorbed NH_4^+-N in the soil column at the conclusion of the displacement process (t = 186 h) are given in Fig. 9.23b and c, respectively. Details of the methodology to ascertain values of v, D_a, R_1, R_2, k_1, k_2 and k_3 used in the solutions of (9.50) subject to appropriate initial and boundary conditions are given by Wagenet et al. The rate of nitrification in the presence of 20% oxygen was one order of magnitude greater than that of denitrification. Of the 100 mg ^{15}N applied to the soil, a mass balance of the ^{15}N in the chemical species measured in the effluent and in the soil at the conclusion of the experiment was within less than 3 mg.

With consecutive equations such as (9.52), field studies in the presence of higher plants provide opportunities to better understand agro-ecosystems. For example, Mishra and Misra (1993) learned how liming a cultivated field of corn modified the values of k_2 and k_3 as a function of soil depth and time in the presence and absence of crop roots. A better understanding of microbial-induced transformations of other chemical species for transient flow and nonisothermal conditions await investigation in both the laboratory and the field.

9.5.2 Chromatographic Formulations

Descriptions of the transport of fluids with their dissolved constituents through beds of reactive porous solids based upon chromatographic plate formulations stem primarily from those derived by chemical engineers nearly one-half century ago (e.g. Wilson, 1940; DeVault, 1943; Thomas, 1944, Glueckauf, 1949; Lapidus and Amundson, 1952; and Heister and Vermeulen, 1952). Subsequent studies in soil hydrology focused first on laboratory soil columns and later were applied in the field for reclaiming saline and sodic soils. For example, to describe cation exchange and ionic distributions within soil columns, Rible and Davis (1955) used DeVault's theory while Bower et al. (1957) used that of Heister and Vermeulen. Van der Molen (1956) predicted the reclamation of saline soils during winter rainfall periods using the theory of Glueckauf. Dutt and Tanji (1962) introduced computer-based chromatographic formulations that were followed by those of Kovda and Szabolcs (1979) and Oster and Frenkel (1980) which were applied to soil reclamation. Shaffer (1977) extended the model of Dutt to apply to cropped, tile-drained field soils.

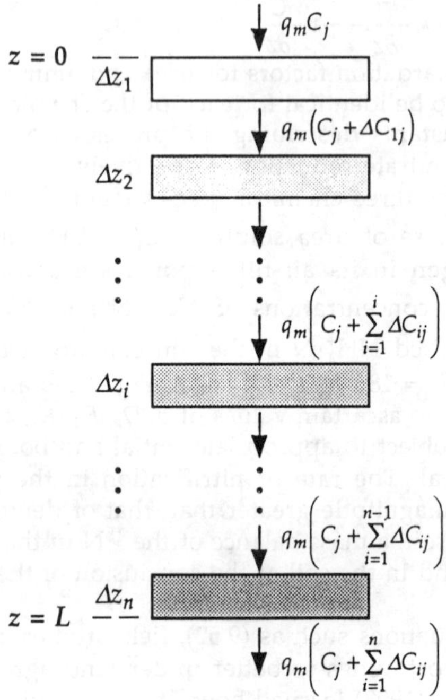

Figure 9.24. Illustration of a chromatographic formulation to describe the displacement and mixing of the soil solution during leaching.

The chromatographic formulation introduced by Dutt and Tanji (1962) is illustrated in Fig. 9.24 where a vertical, homogeneous soil column of length L of unit cross-sectional area is made up of n segments (plates) each of length Δz_i. The concentration C_j of a number of solute species j entering the column in each leaching aliquot ΔQ_m (where m is the number of the aliquot) changes as the solutes mix, react and pass through each segment. The first aliquot infiltrates into the first segment, and fills it to some prescribed soil water content. The second aliquot of infiltrating water displaces the soil water from the first segment into the second segment, and so forth. If the amount of solution ΔQ in each segment Δz_i is identical during infiltration, the final concentration C_j^* of solute species j in the first aliquot ΔQ_1 leached from the column will be

$$C_j^* = C_j + \sum_{i=1}^{n} \Delta C_{ij} \tag{9.53}$$

where ΔC_{ij} is the change in concentration of solute j when the aliquot ΔQ_1 is passed through segment Δz_i. As $n \rightarrow \infty$, the last term in (9.53) is the integral of the change in solute concentration from $z = 0$ to $z = L$. Assuming that the solution is in chemical equilibrium with the soil in each segment, the difference in concentration between the equilibrium solution and that entering each segment is calculated. If n in (9.53) is considered finite, the average concentration of the aliquots ΔQ_m is calculated by progressively equilibrating the solution of ΔQ_m with each of the n segments assuming that piston flow (Fig. 9.12a) takes place within each segment. The dispersion of the solutes associated with pore water velocity distributions and molecular diffusion are implicitly and empirically included by choosing the number of segments or plates n. Tanji et al. (1972) utilized the concept of holdback H_b [see (9.33)] to allow only a fraction of the soil water in each segment to be displaced into the next after chemical equilibrium. They also designated variable segment thicknesses Δz_i corresponding to soil sampling depth intervals or soil horizons as well as choosing the value of n based upon the dispersion of measured chloride breakthrough curves.

Because transient perturbations of the pore water velocity distribution and soil water content are ignored during infiltration and redistribution, chromatographic formulations of solute transport have more recently emphasized the need to study the kinetic aspects of chemical reactions, exchange processes and dissolution and precipitation. Even when these kinetic aspects have been articulated, the enigma of choosing the number and thickness of the segments to reconcile the omission of the pore water velocity distribution remains the objective of future research.

9.5.3 Stochastic Considerations

Because of the naturally occurring heterogeneity of field soils, deterministic formulations of solute transport processes presented above generally must be modified to describe pedon or field scale solute transport. Contemporary research efforts are based upon the consideration that transport phenomena are intrinsically erratic processes susceptible to quantitative characterization by stochastic models. Common to all stochastic models of pedon or field scale transport is the assumption that parameters observed in the field are functions with values distributed in space represented as random variables with discrete values assigned according to a probability distribution. The probability distribution functions at each point in space are usually unknown and cannot be evaluated from only one or a few observations within close proximity of the location. Reviews by Jury (1983) and Dagan (1986) provide pedagogic details. Many other statistical approaches are also described in the literature [e.g. continuous Markov processes (Knighton and Wagenet, 1987), random walk formulations (Kinzelbach, 1988), moment analyses (Cvetkovič, 1991) and hierarchical methods (Wheatcraft and Cushman, 1991). Here, we consider three approaches that are available to deal with spatial and temporal variability in addition to the state-space equations previously described in Chapters 7 and 8.

9.5.3.1 Monte Carlo Simulations

Monte Carlo simulations of a solution of a deterministic equation such as (9.42) allow coefficients to be random variables of the nature expected within a heterogeneous field soil. The variable may be independent, spatially or temporally correlated and perhaps manifest a variance structure. Based upon an initial sampling, parameters selected for the assumed probability density function (pdf) permit repeated solutions ($i = 1, 2, \cdots$) of the deterministic equation [e.g. $C_i(z, t)$] to be calculated. These solutions $C_i(z, t)$ are then used to calculate sample moments (mean, variance) which are assumed to represent the statistical properties of the underlying stochastic process.

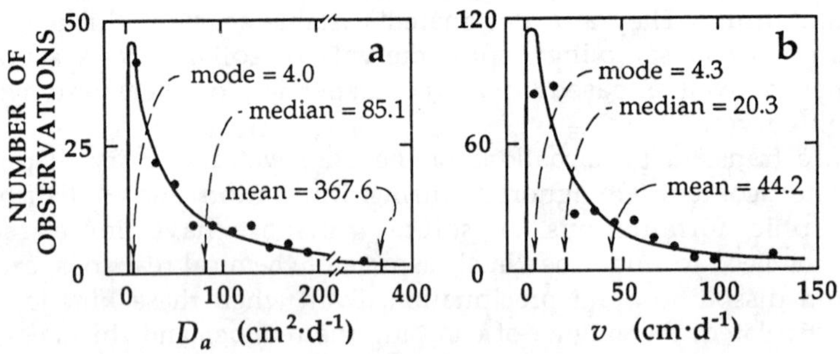

Figure 9.25. Probability distribution functions of D_a and v presented in Fig. 9.20.

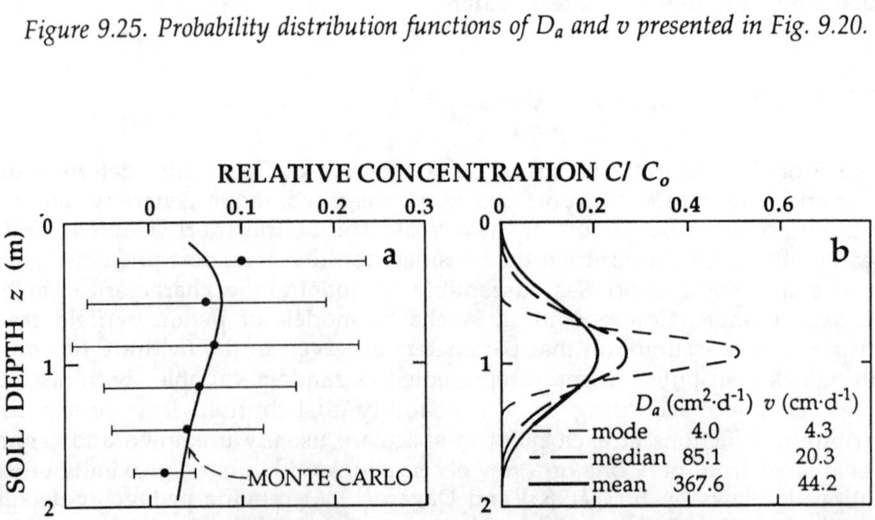

Figure 9.26. Solute concentration distributions within a field soil. a. Monte Carlo simulation and b. Deterministic equation (9.44). Measured values and their standard deviations are represented by solid circles and horizontal bars, respectively.

Amoozegar et al. (1982) relied on the initially measured observations of D_a and v shown in Fig. 9.20a and their pdfs shown in Fig. 9.25 together with the normal pdf of θ to estimate the solute concentration distribution $C(z, t)/C_o$ expected at any location within a field. The average solute concentration distribution $\bar{C}(z,t)/C_o$ expected for the entire field was derived from repeated calculations of (9.44) using the following Monte Carlo step-wise procedure:

1. Draw a random value from the normal distribution with mean zero and standard deviation 1 (Maisel and Gnugnoli, 1972).
2. Find random values of $\ln D$, $\ln v$ and/or θ from their respective statistical distributions represented by the equation $y = \beta\sigma + \mu$ where β is the value with mean zero and standard deviation 1, and y is the random value with mean μ and standard deviation σ.
3. Calculate the random value of C/C_o from (9.44) using the above values of D_a, v and θ for each z and t..
4. Repeat steps 1 through 3 above 2000 times and calculate the mean value of C/C_o for each z and t..

The average solute concentration distribution $\bar{C}(z,t)/C_o$ expected for the entire field at time t = 2 d for a pulse input of solute concentration C_o for t_1 = 0.4 d is shown in Fig. 9.26a together with the field values measured by Biggar and Nielsen (1976). Note that these distributions differ markedly from those in Fig. 9.26b calculated from (9.44) using constant values of D_a, v and θ. The spreading of the solute averaged over the entire field is dominated by variations in pore water velocity (Bresler and Dagan, 1979, and Dagan and Bresler, 1979).

Two kinds of solute distributions are of interest. The first kind is that of $C(z, t)/C_o$ realized at any location within a field, and the second is that of $\bar{C}(z,t)/C_o$ obtained by averaging a large number of $C_i(z, t)/C_o$ across a field. The former is important because it is associated with a particular soil pedon location, a single crop plant or a small community of plants. With site-specific crop and soil management practices, each location can indeed be treated and managed individually across the entire field in order to account for local variations of soil properties (Robert et al., 1993). The latter is important because it is the expectation of solute retention and emission of solutes from an entire field considered as a single domain. Although there is general appreciation of the latter, a farmer is also appreciative of the former owing to the desire to provide optimum growing conditions for each and every crop plant within the cultivated field.

Present-day research within soil mapping units focusing on auto- and cross-correlation lengths of soil properties and variables [such as those in (9.44)] will eventually allow the delineation of the extent of a field characterized by a single simulation. The opportunities afforded by Monte Carlo simulations hinge upon the development of methods to measure and ascertain pdfs of the transport coefficients within prescribed limits of vadose zone depths and times.

9.5.3.2 Stochastic Continuum Equations

Recognizing the paucity of solute concentration data usually available in soils and other subsurface environments coupled with their natural geometric complexities and heterogeneity, Gelhar et al. (1979) initiated stochastic continuum formulations to describe transport processes in water-saturated aquifers. Unlike the above Monte Carlo methods which assume that the random parameters or variables have no spatial correlation, stochastic continuum formulations assume that a random variable can be represented by the sum of its mean and a spatially correlated random fluctuation.

We illustrate the approach with Darcy's equation that describes the mean vertical water flux density $\bar{q}$ in a saturated soil (5.5) as

$$\bar{q} = -\bar{K}_S \frac{d\bar{H}}{dz}. \tag{9.54}$$

where , $\bar{q}$, $\bar{K}_S$ and $\bar{H}$ are expected mean values. Assuming that the values of z and its derivative are measurable and known deterministically within desired, prescribed limits of accuracy, (9.54.) can be written as

$$q = (\bar{q}+\tilde{q}) = -(\bar{K}_S+\tilde{K}_S)\frac{d(\bar{H}+\tilde{H})}{dz} \tag{9.55}$$

where each of the terms $\tilde{q}$, $\tilde{K}_S$ and $\tilde{H}$ are random, spatially autocorrelated functions and have a zero mean value. Subtracting (9.54) from (9.55) yields

$$\tilde{q} = -\bar{K}_S\frac{d\tilde{H}}{dz} - \tilde{K}_S\frac{d\bar{H}}{dz} - \tilde{K}_S\frac{d\tilde{H}}{dz}. \tag{9.56}$$

The terms in such an equation evaluated by deriving a first-order equation for the fluctuations are solved with Fourier transforms (Gelhar et al., 1979).

When a random parameter such as D_a in (9.42) is represented by the sum of its mean value and a random fluctuation, a mean transport model with additional terms is obtained. By solving the stochastic equation of the local-scale water and solute transport, the functional form of D_a for macroregions is related to the statistics describing the variability. A macro-scale value of D_a is reached asymptotically as distance and/or time increase. From such an analysis the long-term, large-scale solute transport can be described using the stochastically derived value of D_a in the deterministic equation (9.42).

A limited number of field aquifer experiments (e.g. Sudicky et al. 1985, and Freyberg, 1986) have shown that the value of D_a increases with solute travel time and travel distance and gradually approaches a constant asymptote consistent with the analysis proposed by Gelhar et al. The application of stochastic continuum research for unsaturated soils appears promising (e.g. Russo, 1993, and Yeh et al., 1985a, b and c) but not yet sufficiently developed to be a proven field technology.

9.5.3.3 Stochastic Convective Equations

The displacement and attenuation of a solute distribution within a vertical soil profile during infiltration can be considered the result of a stochastic convective flow process with its formulation based upon a solute travel time probability density function. The advantage of such formulations is that they do not require an explicit accounting of all of the various physical, chemical and biological processes occurring in the complex, heterogeneous soil environment. Although many stochastic convective models have been used in different scientific disciplines, those initiated by Simmons (1982) and Jury (1982) stimulated research in soil hydrology during the past decade. Here we introduce the stochastic convective concept as a transfer function (Jury, 1982) which can easily be obtained for a nonreactive solute by a single, simple field calibration to measure the travel time distribution. It is assumed that no dispersion of the solute takes place other than that which is represented by the travel time variations within the soil.

Assuming that the depth reached by a solute applied in water at the soil surface depends upon the net amount of water applied, the probability that the solute will reach depth L after a net amount of water I has been applied to the soil surface is

$$P_L(I)=\int_0^I f_L(I')dI' \tag{9.57}$$

where $f_L(I)$ is the probability density function. $f_L(I)$ is the average concentration at soil depth $z = L$ in response to a narrow pulse (Dirac δ-function) of solute $C_{IN} = C_o\delta(I)$ applied at $I = 0$ at the soil surface. A set of observations of $f_L(I)$ can be obtained by measuring the soil solution concentration at depth L at various locations within a field to determine the amount of uniformly applied water I required to move the solute pulse from the soil surface to depth L. The average concentration $C_L(I)$ at $z = L$ for arbitrary variations of solute C_{IN} applied at the soil surface is

$$C_L(I)=\int_0^\infty C_{IN}(I-I')F_L(I')dI'. \tag{9.58}$$

The integrand is the probability $f_L(I')$ of reaching $z = L$ between I' and $(I' + dI')$, and multiplied by the concentration $C_{IN}(I - I')$ of solution displacing at I'. For spatially variable water application rates Jury used a joint probability function in (9.58).

We assume that the distribution of physical processes contributing to the probability density $f_L(I)$ between $z = 0$ and $z = L$ is the same for all soil depths. Hence the probability that an applied solute will reach any depth z after an amount of water $I = I_1$ has infiltrated the soil surface is equal to the probability of reaching $z = L$ after $I = I_1Lz^{-1}$ has infiltrated. For example, the probability of reaching a depth of 50 cm with 10 cm of infiltrated water is equal to the probability of reaching 100 cm with 20 cm of infiltrated water. Hence, from (9.57)

$$P_z(I)=P_L\left(\frac{IL}{z}\right)=\int_0^{ILz^{-1}} f_L(I')dI'. \tag{9.59}$$

To predict the average solute concentration as a function of any depth $z \neq L$, we relate the probability density function $f_z(I)$ to the reference density function f_L by

$$f_z(I)=\frac{L}{z}f_L\left(\frac{IL}{z}\right) \tag{9.60}$$

and obtain

$$C(z,I)=\int_0^\infty C_{IN}(I-I')\frac{L}{z}f_L\left(\frac{I'L}{z}\right)dI'. \tag{9.61}$$

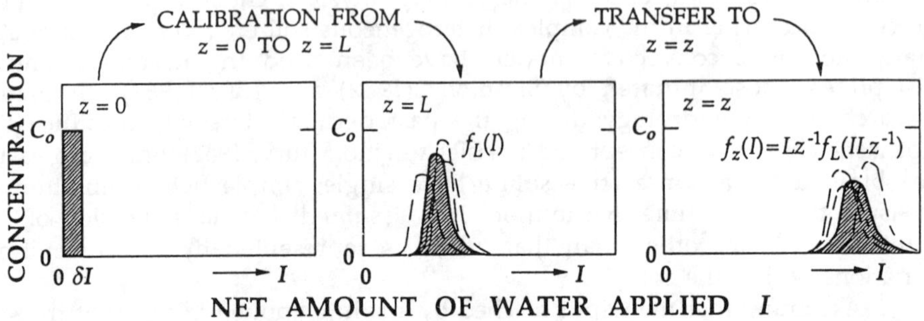

Figure 9.27. The concept of a transfer function calibrated at depth L to predict the arrival of a solute pulse from the soil surface at depth z as a function of the net amount of water applied at the soil surface. The broken curves each represent a distribution measured at a specific location within the field.

The concept is illustrated in Fig. 9.27 where the broken lines in the graphs for z = L and z = z represent concentration distributions measured at individual locations within the field.

If the transport properties are statistically similar for all depths, only one calibration f_L is needed. On the other hand, if they are dissimilar owing to strongly developed horizons or textural and structural differences, additional calibrations f are required or a more robust calibration f_L for all depths can be made at depth L below the strongly stratified soil.

Relatively few distributions of solute transport parameters have been measured in the field. Some of them appear to be lognormal. For such distributions the travel time density function $f_z(t)$ is

$$f_L(t)=\left[\frac{1}{(2\pi)^{1/2}\sigma t}\right]\exp\left[\frac{-(\ln t-\mu)^2}{2\sigma^2}\right] \tag{9.62}$$

where μ is the mean of the distribution of $\ln t$ and σ^2 the corresponding variance. On the other hand (9.44) for the same boundary and initial conditions yields the travel time density function $f_z(t)$

$$f_L(t)=\left[\frac{L}{2\left(\pi D_a t^3\right)^{1/2}}\right]\exp\left[\frac{-(L-vt)^2}{4D_a t}\right]. \tag{9.63}$$

Although the above functions are nearly identical when calibrated at the same depth (Simmons, 1982), they yield distinctly different predictions of solute movement at depths $z \neq L$. Jury and Sposito (1985) have shown that the variances of the travel time for (9.62) and (9.63) are

$$\mathrm{var}_{z_{TFM}}[t]=z^2L^{-2}\exp(2\mu)\left[\exp(2\sigma^2)-\exp(\sigma^2)\right] \tag{9.64}$$

and

$$\mathrm{var}_{z_{CDE}}[t]=2Dzv^{-3}, \tag{9.65}$$

respectively. Inasmuch as $\mathrm{var}_{z_{TFM}}[t]$ increases as the square of the depth z while that of (9.44) increases linearly with depth, the lognormal transfer function model predicts the greater amount of solute spreading for the same depth. Present-day field experimentation for different soils and local conditions is sufficiently limited to preclude a preferred choice of the two models (e.g. see Jury and Sposito, 1985).

Presently, stochastic-convective formulations are being extended to include the transport of adsorbing and decaying solutes, two component chemical nonequilibrium models, physical nonequilibrium models and other nonlinear processes. See a review by Sardin et al. (1991) as well as a more recent contribution by Roth and Jury (1993).

9.6 IMPLICATIONS FOR WATER AND SOLUTE MANAGEMENT

Although our understanding and theoretical description of solute transport in soils remain incomplete, we have nevertheless sufficient knowledge to derive a few principles or guidelines for managing solute retention or leaching in the field. Whether solutes accumulate or leach depends primarily upon the processes by which they enter, react and leave the soil profile relative to their association with water. Here, we make no attempt to discuss the well-known principles of managing salinity and drainage from irrigated agricultural lands used for crop production inasmuch as excellent texts are abundantly available. Disregarding horizontal surface and subsurface water flow, we focus our attention on transport owing to the water content and flux density conditions at the soil surface occurring naturally owing to local weather conditions and being deliberately modified by irrigation.

Summarizing the more important points of this chapter, we conclude the following regarding the relative movements of soil water and its dissolved constituents:

a. As water moves more slowly through a soil, there is a greater opportunity for more complete mixing and chemical reactions to take place within the entire microscopic pore structure owing to the relative importance of molecular diffusion compared with that of convection.

b. Microscopic pore water velocity distributions manifest their greatest divergence for water-saturated soil conditions. Hence, under water-saturated conditions, the greatest proportion of water moving through the soil matrix occurs within the largest pore sequences.

c. Under water-saturated soil conditions, when the average pore water velocity is large compared with transport by molecular diffusion, the relative amount of solute being displaced depends upon the solute concentration of the invading water.
d. The concept of preferential flow paths occurs at all degrees of water-unsaturation even though their existence is usually only demonstrated for macropores near water-saturation. At each progressively smaller water content, the larger pore sequences remaining full of water establish still another set of preferential flow paths.
e. Any attempt to measure the solute concentration based on extraction methods carried out either in the laboratory or the field will be dependent upon the rate of extraction and the soil water content during the extraction process.
f. Inasmuch as rainfall infiltration usually occurs at greater soil water contents and greater average pore water velocities than does evaporation at the soil surface, the amount of solutes transported near the soil surface per unit water moving through the soil surface is greater for evaporation than for infiltration.

Each of the above six points have been verified one way or another in numerous publications before and after the observations made in a field experiment conducted by Miller et al. (1965) which we describe below.

A level 0.4-ha site of Panoche clay loam was divided into 0.004-ha plots statistically replicated in five complete blocks. Potassium chloride uniformly applied to the soil surface was leached and redistributed within the profile with four different methods of water application: a. the soil surface continuously ponded with water, b. the soil surface intermittently ponded with repeated applications of 15-cm of water, c. the soil surface intermittently ponded with repeated applications of 5-cm of water and d. the soil surface continuously sprinkled at a rate less than K_S and equal to 0.3 cm·h^{-1}. The intermittent applications were made weekly when the soil water pressure head at the 30-cm soil depth reached a value of -150 cm. Each method gave rise to different soil water content distributions $\theta(z,t)$ and different water flux density distributions $q(z,t)$. Average values $\overline{\theta(z,t)}$ and $\overline{q(z,t)}$ for each method were a. 0.48 and 0.6, b. 0.44 and 0.09, c. 0.42 and 0.03 and d. 0.42 cm^3·cm^{-3} and 0.3 cm·h^{-1}, respectively. Soil solution samples taken every 30 cm to a depth of 150 cm were frequently extracted and analyzed in the laboratory for their chloride content. From the five plots of each treatment about 3500 samples were analyzed. Additional details regarding the exact nature of the infiltration, redistribution and soil water content profiles are available (Nielsen et al., 1967).

The chloride concentration distributions from the four water application methods manifest significant differences (Fig. 9.28). Although some chloride leached to great depths in the continuously ponded soil after 15 cm water had infiltrated, most all of the solute resided in the top 50 cm of soil. In the other three treatments for the same amount of water infiltrated, large quantities of the solute had been dispersed twice the depth to 100 cm. For 60 cm of infiltration, the center of the solute mass is displaced to greater depths as $\overline{\theta(z,t)}$ decreases.

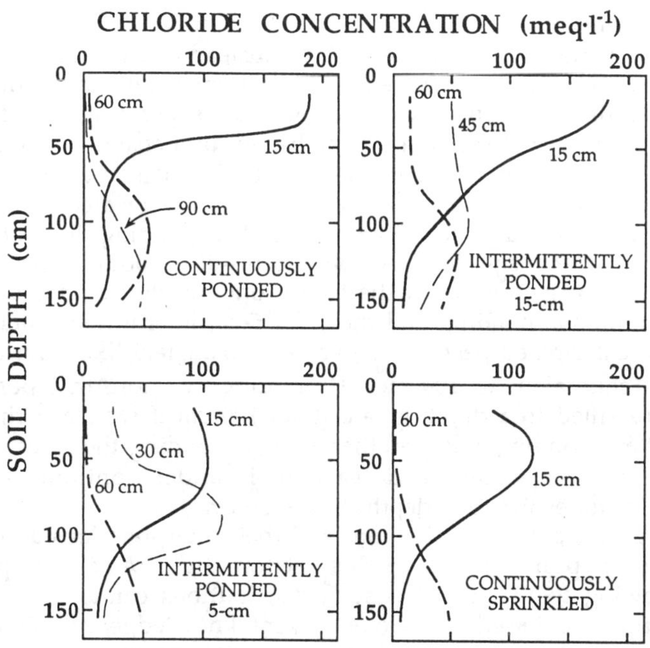

Figure 9.28. Field-measured chloride concentration distributions from a surface application of potassium chloride being leached through the profile as a result of four different water application treatments. Values associated with each curve correspond to cumulative amounts of water infiltrated

That is, for the continuously ponded case, the relative maximum chloride concentration is just below 100 cm while those for the intermittent 5-cm ponding and for continuous sprinkling were both below 150 cm. Indeed, the concentration of solute for the latter two treatments for 60 cm of water infiltrated is everywhere less than that of the continuously ponded case when as much as 90 cm of water had infiltrated. Under the continuously ponded condition, judging the small solute concentration to a depth of 90 cm when a depth of 90 cm of water had been applied, its behavior conformed to the recommendation, "It takes a cm of water to leach a cm of soil". Here we note that for unsaturated, slow leaching conditions, it took only 60 cm of water to leach the same amount of solute as was required for 90 cm of water under near-saturated ponded conditions. The distribution curves for intermittent 5-cm ponding and for continuous sprinkling are remarkably similar. Even though their values of $\overline{q(z,t)}$ differed by one order of magnitude, at such slow flow rates relative differences in mixing by molecular diffusion are apparently insignificant to the spreading of solute at nearly identical $\overline{\theta(z,t)}$.

Two additional comments are appropriate for the data presented in Fig. 9.28. First, consider the 15-cm curve for continuous ponding, the 45-cm curve for 15-cm intermittent ponding and the 30-cm curve for 5-cm intermittent ponding. If we assume that a nitrate fertilizer instead of a chloride salt had been applied to the soil surface, three entirely different solute distributions could have achieved different purposes. For continuous ponding, the fertilizer is leached in a concentrated solution for the benefit of a shallow-rooted crop. For 5-cm intermittent ponding, fertilizer is leached to the roots of a deep-rooted crop (e.g. a fruit tree crop) and not in the vicinity of an associated, shallow-rooted cover crop. The 45-cm curve for 15-cm intermittent ponding illustrates uniform fertilization to a 125-cm soil depth – a desired result for both deeper-rooted and shallow-rooted crops growing in the same community. Second, consider the curves for continuous ponding and those for 5-cm intermittent ponding. If we assume that a chlorinated pesticide moves approximately like chloride and has a lethal dose equivalent to 100 $meq \cdot l^{-1}$ chloride, we note that pests such as nematodes are killed to a depth greater than 100 cm if we leach the pesticide with intermittent ponding. On the other hand, by adding the same quantity of pesticide to the soil surface and leaching it under continuously ponded conditions, nematodes thrive at depths below 60 cm.

Although the solute distributions and their interpretations above support many of the six points made in this section, they remain empirical and speculative owing to the fact that such distributions cannot at this time be accurately predicted based upon our present knowledge of soil hydrology. Different results will no doubt be obtained for different soils, solutes and local conditions. Even for the above experiment, had the intermittent applications of water been made at intervals greater than one week, entirely different results would have occurred. That is, it is known based upon the analysis of the soil water behavior in this study (Nielsen et al., 1967) that at the end of one week, evaporation at the soil surface together with gravitational redistribution of soil water allowed a net vertically upward movement of water from the 90-cm soil depth. Hence, at those smaller soil water contents and water flux densities, the chloride would have started to move more efficiently to the soil surface. In such a case, continually ponding the soil surface would have more efficiently leached the soluble solute per unit water infiltrated from the 150-cm soil profile.

Fig. 9.29 shows an example of the chloride concentrations measured within the soil profiles from which the distributions given in Fig. 9.28 were derived. The figure pertains to the intermittently 15-cm ponded treatment for soil depths of 30, 90 and 150 cm. A distinctive characteristic of the curves for the 30- and 90-cm depths is the relative minimum reached soon after the initiation of each 15-cm water application. Similarly, a relative maximum is reached near 400 h soon after the third 15-cm water application. This behavior is associated with the sampling technique. When soil solution is extracted, the bulk of the flow takes place in the larger water-filled pore sequences. Hence, at the shallower depths, the more solute-free infiltrating water flowing through the larger water-filled pores and extracted by the soil solution samplers during infiltration yields a solution that is, indeed, more dilute than the average solute concentration at that depth. Similarly, the more concentrated solution from the shallower depths temporarily taken into the soil solution samplers at the greater depth is greater than the average solute concentration at that depth.

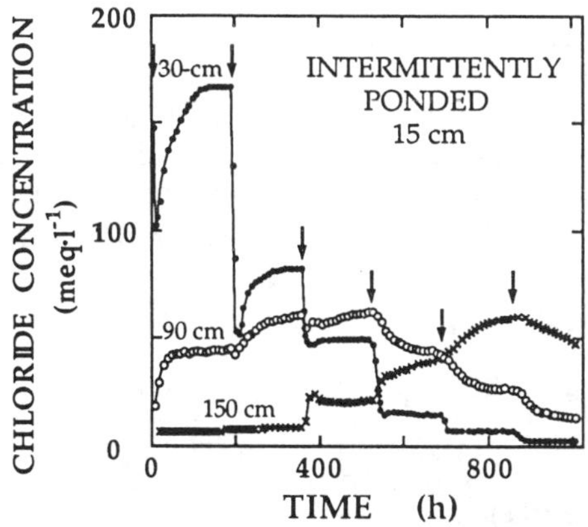

Figure 9.29. Chloride concentration distributions measured at depths of 30, 90 and 150 cm. Zero time represents the time when the first 15-cm ponded water application was initiated. The vertical arrows indicate the other times when additional 15-cm applications were initiated.

Acknowledging that the efficiency of solute leaching obeys the principles discussed here, Dahiya et al. (1985) nevertheless recommend that leaching of saline soils be carried out under continuously ponded conditions based upon their experiments with different plot sizes. They argue that under practical situations in which reclamation is carried out in large field domains, controlled slower, uniform infiltration at smaller soil water contents would require more labor and time.

Except for the post script that follows, we conclude our book with the admonishment that observations and theories of soil hydrology have not yet completely reconciled different scales of space and time in the management of fresh water retained at continental surfaces.

PROBLEMS

1. Considering a soil column in the laboratory to be equivalent to a bundle of independent capillary tubes, calculate breakthrough curves similar to that shown in Fig. 9.5b for a. a column composed of equal-sized capillary tubes and b. a column composed of two tubes of radius 0.1 cm and 4000 tubes of radius 0.001 cm. Discuss your answer in terms of "preferential flow".
2. By changing variables (X, t) of (9.24) to $[(z - vt), t]$ and substituting D_a for D, derive (9.42).
3. Assume that (9.44) describes a pulse of solute (t_1 = 0.4 d) being leached through a homogeneous, water-saturated soil column of length 2 m. For an average pore water velocity $v = 40\ \mathrm{cm \cdot d^{-1}}$, you measured at z = 1 m a

relative maximum concentration $[(C - C_i)(C_o - C_i)^{-1}] = 0.22$. What is the value of D_a? Hint: First obtain $\partial C/\partial z$ and set it equal to zero.

3. a. Experimentally for the above problem, how could you best establish the boundary condition $C = C_o$ $[0, (0 < t \leq t_1)]$? b. Similarly, how could you best experimentally establish the constant solute flux density boundary condition $[vC - D_a\, \partial C/\partial z = \text{constant})$ at $z = 0$? In each case, explain the limitations of alternative techniques.

4. Show that for large values of $v\, D_a^{-1}$, (9.44) reduces to

$$\frac{C-C_i}{C_o-C_i}=\frac{1}{2}\left\{erfc\left[\frac{z-vt}{(4D_at)^{1/2}}\right]-erfc\left[\frac{z-v(t-t_1)}{\left[4D_a(t-t_1)\right]^{1/2}}\right]\right\}.$$

Hint: $\lim_{x\to\infty} \exp(x) erfc(x) = 0$.

5. You are interested in simulating with (9.44) the spreading of a solute pulse (initially of thickness vt_1 = 15 cm in the topsoil) moving below plant roots in the vadose zone of a semi-arid environment. The long-term average net amount of water moving past the zero flux plane in the soil profile is only 0.01 cm·d^{-1}. You first assumed that $D_c \gg D_m$ and approximated D_a in (9.44) by βv using a published field-estimated value of the dispersivity β = 3 cm. Next, you assumed that $D_c = D_m$ = 0.6 cm^2·d^{-1}. a. With your first assumption, calculate the distribution of the solute $\{[C(z) - C_i][C_o - C_i]^{-1}\}$ in the vadose zone for times t = 100 and 1000 y. b. With your second assumption, calculate the distribution of the solute $\{[C(z) - C_i][C_o - C_i]^{-1}\}$ in the vadose zone for the same times. Which assumption do you believe is more realistic, and why?

6. Discuss the implications of using temporal and spatial average values of D_a and v in the above problem.

7. The term ϕ_i in (9.35) sometimes represents the growth of soil microbes with the Michaelis-Menten process having the form

$$\phi_i = \frac{\kappa_m C}{K_m + C}$$

where κ_m is the product of a maximum growth rate constant and the biomass of the microbes and K_m is a "saturation" constant. a. For a concentrated soil solution, show that the above equation reduces to a zero-order equation with a constant value $\phi = k_o$. b. For a dilute soil solution, show that the above equation reduces to a first-order equation with $\phi = kC$ as was assumed in (9.52).

POSTSCRIPT

For our readers who have the intention of getting engaged in modeling hydrological events related to soils, we felt obliged to write this postscript. In the nine chapters of the text, we divided the complex reality of soil hydrology into sections which we considered more lucid for individually dealing with each of the main problems. We simplified reality and distinctly described formulations that can be and are currently used in models. The strict physical interpretation of various phenomena as they were discussed in this text are not only beneficial in modeling hydrological events but without their conceptual understanding the modeling of long term hydrological processes would degenerate into more or less sophisticated computer games.

All of our descriptions of the phenomena in nature are more or less approximate, and if their outputs have a quantitative character, the term models is appropriate. Soil hydrologic models can be defined as systems which describe the reality of hydrologic processes (to a certain degree of approximation) in soils for long periods of time such as those associated with a vegetative season, a hydrologic year or a climatic era. In such cases the spatial scale usually must be greater than that of the pedon and the time scale is extended far beyond the time of one isolated elementary hydrologic process. When compared with the processes and events discussed in the previous chapters, the systems are extended or extrapolated in both space and time. If soil hydrologic models are defined in such a way, our text does not cover all of the essential features of modeling soil hydrology. An instructive, comprehensive text on modeling would inevitably include details on numerical procedures, the theory of systems analysis and formulations applicable to spatial scales greater than that of the pedon. For such content, our book would have increased in volume and themes far beyond our goal to offer basic information on soil hydrology. Hence, this postscript.

We are offering here just some introductory notes for a general orientation in the subject. The notes are accompanied by examples selected from the literature. In the discussion we demonstrate how the complexity of the nature in a watershed modifies the processes described earlier in isolated forms.

MODELS OF SOIL HYDROLOGY

We distinguish between physical, analog and mathematical models. A *physical model* is that of a porous medium, e. g. a Hele-Shaw model or a column of glass beads or soil studied in the laboratory. These models frequently have a smaller scale than reality. *Analog models* are based on the similarity of the flow of water in soil to other transport phenomenon, e. g. the flow of electricity through a conductor. Although somewhat common in groundwater studies, these models presently are rarely applied in soil hydrology. *Mathematical models*, the subject matter of this chapter, commonly require computer methods for investigating a specific problem.

Mathematical models in soil hydrology can be classified either according to the principles of the methods applied or according to the subjects modeled.

Further consideration reveals that the nature of either method tends to be empirical, deterministic or stochastic. Inasmuch as the different kinds of models are frequently merged into larger model structures, sharp boundaries between them do not exist. A broadly used classification of mathematical models in soil hydrology originally proposed for leaching models by Addiscot and Wagenet (1985) is:

- I. Deterministic models
  - A. Mechanistic (usually based on rate parameters)
    - 1. Analytical
    - 2. Numerical
  - B. Functional (usually based on capacity parameters)
    - 1. Partially analytical
    - 2. Soil layers and other simple approaches
- II. Stochastic models
  - A. Mechanistic
  - B. Non-mechanistic

Models of agro-ecosystems have been classified into the following groups (Rohdenburg, 1989):

- I. Empirical/mathematical
- II. Deterministic-analytical

According to Rohdenburg, the stochastic nature of agro-ecosystems is not explicitly included in the formulation of their processes but enters implicitly through the spatial variance structure of the deterministic formulation.

The unique character of each of the two classifications above stems from different conceptualizations. The first starts with a Darcian-scale soil column and subsequent broadening of the model brings new and stochastic features into the model. Accepted at the outset that ecosystem processes are phenomena in spatially heterogeneous media, the stochastic characteristics in the second classification stem just from the initial descriptive part of the system.

If mathematical models are placed into only three categories (empirical, deterministic and stochastic), we must remind ourselves that all three have deterministic origins. Additionally, only the deterministic models adhere strictly to a mathematically rigorous, unique formulation of each of the process rates.

From a pragmatic point of view, we distinguish between research-oriented and practice-oriented models. Here we consider not only a requirement for scientific accuracy, but the entire structure of the model and its aim. Recognizing these requirements, we avoid useless questions (e. g. "How do you imagine that we obtain the required population of soil parameters?" and "Do you expect us to make Swiss cheese of our watershed to obtain enough samples?"). And we also avoid arguments from an opposite group (e. g. "This kind of simplification does not comply with our scientific background"). Recently, Reiniger et al. (1990) found that simpler models (compared with more sophisticated models) were either adequate or even more appropriate for describing the transport of nitrates in agricultural soils. Their results should not suggest that simple models be given higher priority automatically. Our recent uncertainty in the quantification of some detailed processes and the uncertainty in estimating some physical characteristics including their statistical character provide even a greater challenge to develop comprehensive models having either a research-oriented or practice-oriented utility.

Reading the literature on models published during the last decade we are reminded of the "rule of 90 percents". Ninety percent of the models presented in papers are not developed sufficiently for others to use them. Of the remaining 10%, 90% of the models are used by only their creators. And of the remaining 10% which are used by others, 90% of the models are never verified. A similar rule also applies to all scientific publications. Although the "rule of 90 percents" is perhaps sad or cynical, it offends neither innovative modelers nor productive scientists.

Empirical Models

We neglect "completely black box" models which represent the roughest form of empiricism. Inasmuch as they are generally not transferable or applicable on a broader scale either in practice or for research, we do not recommend or discuss them here.

The simplest empirical models are eventually based on the empirical concepts of field capacity and wilting point. The balance of inputs (precipitation or irrigation) and outputs (evapotranspiration) is computed at regular time intervals. Field capacity θ_{FC} is considered the maximum water stored within the soil profile during the time interval. If rainfall or irrigation exceeds the disposable unfilled water storage capacity, the excess water drains from the profile. The disposable water storage DWS is calculated from

$$DWS = \int_0^Z (\theta_{FC} - \theta_i)dz$$

where Z is the soil depth at the lower boundary of the balanced profile. This empirical model is sometimes denoted as the capacitance model (Addiscot and Wagenet, 1985).

Rainfall or irrigation is assumed to either fully or partly infiltrate into the surface. The first assumption is generally fulfilled by small, flat areas. For the second assumption, the ratio of infiltration to runoff is either verified with the calibration of the model or is obtained analytically or semi-analytically (section 6.2.3). With evapotranspiration commonly formulated by Penman's or Penman-Monteith's equations, the soil water regime for a season is obtained. The reduction of potential to actual evapotranspiration is achieved with the principles formulated in (6.153a) through (6.153d). Or, alternatively, the balance of soil water storage is used to determine the actual evapotranspiration. The soil profile is usually subdivided into layers or "compartments". If cumulative infiltration exceeds the DWS of the top layer, the excess is considered as filling the next lower layer. This system of soil water reservoirs occurring in a layered series was first assumed to fill instantaneously without retardation. Later, more realistic rates of soil water movement were formulated using the hydraulic conductivity of each soil layer or "compartment". For example, with each successive compartment first being filled to saturation, the infiltrating water moves through each of water-saturated series of compartments at rate K_S until the cumulative infiltration is accommodated. At that time redistribution of soil water from θ_S to θ_{FC} occurs at a "redistribution rate" $q = K(\theta_{FC})$. Various other more complicated modifications exist. Such refinements have the features of deterministic modeling. Even in its simplest form, this model is being successfully used in irrigated agriculture.

The simplest form without flow rates has an advantage over numerical deterministic models. A numerical solution of partial differential equations with a sink term requires small time steps in order to converge. Unfortunately, routine meteorological data do not provide adequate time-dependent information. Therefore, simple water budget models have real value for practical use and are easily used owing to their simple structure. Because water budget models were in initiated early when computers were not available, they have been comprehensively developed with personal computers during the last two decades (e. g. Baier et al., 1979; Petrovič, 1984; Chopart and Vauclin, 1990).

A different subclass of empirical models are those using correlation analysis between the input and the searched output. For example, adequate soil water storage is merely estimated by correlating meteorological characteristics during periods decisive for optimal crop yields. Such empirical relations are associated with soil mapping units without analyzing the physical processes of water within the soil profile.

Deterministic Models

Deterministic models are based on the assumption that a quantitative physical predictability of processes is attainable when events defined on a boundary or within the profile are considered. The outcome (e. g. the flux to the water table or surface runoff) is thus determined by events occurring earlier than those predicted. For example, if we consider the soil water regime [$\theta(z, t)$ or $\theta(x, y, z, t)$] as an outcome, its prediction is realized through a solution of the Richards' equation with a sink term. Here the upper boundary condition at $z = 0$ is defined from meteorological conditions and the lower boundary at $z = Z$ is defined approximately to match field reality. Three different types of conditions can be found at $z = Z$. (i) The Dirichlet condition with soil water pressure head prescribed at the soil surface and connected to the ground water level. (ii) The Neuman condition as a defined flux at the soil surface usually associated with a unit total potential head gradient when the ground water level is absent or is well below depth Z. (iii) The Cauchy condition as a combination of flux and soil water head relations when the fluxes in the unsaturated soil influence and are influenced by regional groundwater flow.

In a sense, the analytical and approximate solutions of elementary soil hydrological processes (Chapter 6) belong to the simplest class of deterministic models and are classified in the subclass as analytical models. We have already pointed out that analytical models are no doubt instructive for understanding processes but they are restrictive regarding direct application to field situations even for distinct and well-defined 1-dimensional elementary hydrologic processes. Soil profile characteristics as well as θ_i are neither constant with depth nor easily defined analytically.

We describe as Pedon Hydrological Models (PHM) those models which are restricted to the scale of a pedon and related to a time scale larger than just one meteorological event. See publications of field tests for soil water regimes during one or more vegetative seasons (e. g. Renger et al., 1970; Vachaud, 1979). In the vast majority of PHM, the method of finite differences is applied to the solution of Richards' equation with the sink term (Feddes et al., 1978). The model SWATRE (Belmans et al., 1983) is an example of a commonly used PHM.

A numerical PHM applied for large time periods (e. g. a cultivated season or a hydrological year) is troubled with problems related to time discretization. In order to achieve stability and convergence of the numerical scheme, small time steps are required. But this requirement is generally not compatible with routinely available meteorological input data. Feddes et al. (1978) and Belmans et al. (1983) tackled the problem in their early finite difference models. And authors of more recent Galerkin models face the problem by posing different time discretizations in finite difference schemes of time. Some PHMs having no universal or generic character are problem oriented. For example, the Dutch simulation model LAMOS (Bouma et al., 1980) predicts the change of soil water regimes and the occurrence of soil water deficits owing to a lowering of the ground water level.

When the study area is substantially greater than that of a pedon, we must deal with soil heterogeneity. If we assume hypothetically that the heterogeneity is only of a deterministic nature, we could use the finite element method. In such a case, with each node point of the grid theoretically being associated with unique and different values of soil hydraulic parameters, the known heterogeneity is projected into the numerical grid. In practice, this advantage is hardly usable inasmuch as sufficiently detailed data are seldom available to represent the soil spatial heterogeneity. And even when they are available, the scale of the computational grid generally does not correspond with that of the measured soil heterogeneity. Moreover, some soil parameters typically vary by orders of magnitude within distances smaller than the spacing of neighboring grid nodes selected to model a simple hydrologic catchment. This difference in scales lead to the concept of equivalent soils – a concept sometimes not explicitly stated.

Finite element methods are often preferred for the solution of 2- and 3-dimensional soil hydrological problems. In many instances, 2-dimensional models suffice. For example, infiltration coupled with runoff and overland flow on a watershed is typically taken as a 2-dimensional problem over the (x, y)-plane. And the Galerkin finite element method with basic linear functions might be used to obtain the solution of the flow equations subject to imposed initial and boundary conditions. This method, now commonly used, is presented in detail in the literature (e. g. Neuman, 1975; Pinder and Grey, 1977). Because of the heterogeneous nature of field soils, the majority of problems described by a deterministic model requires stochastic entries which are next presented.

Stochastic Models

When the study area is substantially larger than that of a pedon, stochastic properties play an important role inasmuch as the pedons do not appear repetitiously across the landscape even if they belong to the area pedotop at the lowest level of soil taxonomy. Hence, we deal with a polypedon or with the pedotop hydrological model (PPHM).

A watershed or a large region usually consists of more than one pedotop and with lower level taxonomic units grouped into higher level units, we speak of pedochors or pedocomplexes. The gradation of pedotops within the watershed is defined as the land surface catena. And, within a pedocomplex or a catena we find predictable deterministic heterogeneity. When the boundaries of the pedotops are crossed, these deterministic soil hydraulic characteristics manifest different values.

The first approach to model unsaturated flow in spatially variable soils was to discretize the field or watershed into a series of 1-dimensional, non-interacting soil columns. This approach is frequently used to describe water regimes of specific regions (e. g. levels of water deficit within regions mapped by soil surveys). Stein et al. (1991) distinguish two procedures. In the first one, simulations are carried out for all points where soil parameters are available. By means of a statistical interpolation method the model output is predicted for the entire area. They describe the procedure as CI, "Calculate first and interpolate later". The method together with earlier similar procedures is given by Stein et al. (1991).

In the second procedure the variables needed for the model are pedologically and functionally clustered. The methods of clustering range from the traditional representative soil profiles found by detailed soil survey to the application of scaling methods. The simulation model is applied to the reference profile indicative of the pedological or functional cluster. This method is denoted by Stein et al. as IC, "Interpolate first and calculate later". Details can be found in publications of Wösten et al. (1985) and Wösten (1990). If scaling is applied with stochastic evaluation, the "accuracy" of the model output can be assessed.

PPHM consisting of vertical soil columns has been applied in theoretical studies of soil heterogeneity. Because the columns were assumed to be randomly located in the area but uniform in depth, we should speak of areal variation instead of spatial variation. In order to simplify the modeling procedure, it is frequently assumed that K_S is the random variable over the area while parameters of $h(\theta)$ and $K_r(\theta)$ remain constant or are simply shifted with a scale factor (Russo and Bresler, 1981; Binley et al., 1989; Aboujaoude, 1991). This kind of modeling for a simply defined heterogeneous block yields new results. If the concept of an equivalent soil representing the heterogeneous block is accepted, the PPHM of infiltration with Neuman's boundary condition for a flat terrain manifests a concave soil water content profile $\theta_R^*(z)$ rather than the convex deterministic profile $\theta_R(z)$ (Bresler, 1987).

For a sloping terrain, the model of vertical, mutually non-interconnected columns is hardly acceptable. It becomes somewhat more acceptable by considering the assumption of "runon". This term describes the redistribution of excess water on the sloping surface from locations of shorter ponding times (small values of K_S) to those downhill locations having larger ponding times (large values of K_S) (Smith and Hebbert, 1979). In addition to this surface interconnection, the interconnection on the basis of continuous saturated flow is sometimes considered. For example, the interconnection exists in the unsaturated zone component of the physically based distributed catchment model SHE (Système Hydrologique Européen, Abbott et al., 1986).

These types of models used for studying the applicability of an "effective soil" (or "effective soil parameters") in soil hydrology are based on the hypothesis that large scale soil heterogeneity can be successfully lumped into parameters descriptive of an "effective soil". Presently, the heterogeneity of unsaturated soils has only been studied through variations of K_S. Studies have shown that the concept of "effective soil" is restrictive and not generally applicable to all boundary conditions and soils. And, when infiltration and runoff are considered, the concept is applicable only for mildly heterogeneous soils (Aboujaoude, 1991).

Vauclin and Vachaud (1990) found that the "effective soil" concept is acceptable for small rainfall intensities on a simple, uniform slope. On such a soil, increased soil variability is accompanied by increased runoff. However, for large rainfall intensities, the concept completely collapsed. Binley et al. (1989) found similar results studying runoff on rapidly and slowly permeable soils. Moreover, the results from using the "effective soil" concept disagree with those from the PPHM when the distribution of K_S in the watershed follows the rules of geomorphological formation, the existence of soil catenas and the genesis of secondary crust formation (Smith et al., 1990; Aboujaoude, 1991).

In spite of its restrictions, the simplified model of an "effective soil" with its vertical columns accompanied with the runon concept brought new views of elementary runoff hydrology. For example, considering the sloping surface of a heterogeneous soil, the ponding time concept as derived for the pedon-scale elementary hydrological process is no longer a constant characteristic value for a given rainfall intensity. The value of the ponding time can decrease by an order of magnitude when the coefficient of variation of K_S increases from 0 to 0.8 on a short uniform slope (Smith et al., 1990).

Contemporary model studies still deal with uniform slopes even though each concavity or convexity of the surface together with vertical heterogeneity produces a convergence or divergence of flow paths (Zaslavsky, 1970; Zaslavsky and Sinai, 1981). Because the transition to 2- and 3-dimensional modeling is sorely needed, we present a brief review of initial, incomplete results.

The model of vertical columns of random characteristics of soil properties is a special case of Monte Carlo simulation. In order to allow non-vertical flow of water, the mutual "isolation" of the columns must be removed. Unfortunately, smoothing the local mosaic or variation of unsaturated soil properties does not achieve the large scale behavior of a watershed. Nevertheless, Mantoglou and Gelhar (1987) have demonstrated that local variability of soil properties has important large scale effects on hysteresis and anisotropy.

For runoff characteristics, the 3-dimensional Monte Carlo study of Binley et al. (1989) of a heterogeneous inclined soil surface (150 x 150 m) has shown that peak runoff volumes increase with heterogeneity and with the spatial dependence of soil parameters. Sharma et al. (1987) gave a simple explanation of subsurface lateral flow that occurred in their 3-dimensional study. They concluded that hill slopes with a random distribution of hydraulic characteristics provide greater opportunity for soil units with different water capacities to interact than those hill slopes with spatially correlated distributions. Their results improved our understanding of subsurface lateral flow because they ignored models of 1-dimensional vertical columns.

MODELS OF SOLUTE TRANSPORT

The most frequently used models of solute transport in soils are Richards' equation (5.64) for unsaturated water flow and the convection-diffusion equation (9.42) for solute transport. The equations are solved numerically where the Galerkin linear finite element scheme is appropriate. In order to keep the model realistic, a multicomponent system accounts for other flow processes. In addition to water flow, heat transport is often included in the model inasmuch as the temperature influences thermodynamic constants and reaction rates significantly. Moreover, the concentration of CO_2 in soil air fluctuates between 0.035 and 20% in extreme circumstances. The solubility of many mineral salts strongly depends upon CO_2 concentration. Changes in CO_2 concentration are accompanied by changes of O_2 concentration in the soil air. Both concentrations influence pH and there is a direct link between both pH and redox potential with dissolution constants and rates. Hence, multicomponent transport models are preferred nowadays just to get closer to field reality.

The majority of recent procedures has involved two simultaneous equations. One for equilibrium chemical processes such as precipitation, dissolution and cation exchange and a second for transport. At each numerical time step, the transport equation is interfaced with the equilibrium chemical equation. However, iteration between the two equations is neglected. Yeh and Tripathi (1991) have shown that this simplification can cause serious numerical errors. Another limiting factor is the consideration of equilibrium reactions only. Suarez (1985) demonstrated that the chemical system is frequently controlled by kinetic reactions. Hence, both equilibrium and kinetic reactions are nowadays directly incorporated into multicomponent systems of transport (Šimůnek and Suarez, 1993).

POST-POSTSCRIPT

Models of soil hydrology should be related to the reality of a field or watershed by at least two links. The first is a proper characterization of the physical parameters of the domain including appropriate processes and boundary conditions. The second link is related to the scale of the model. One part of this link is statistical evaluation while the other part is the degree of approximation allowed in the model. It follows that a modeler should work in or at least supervise the experimental activity in the field. And, vice versa, effective field experimentation requires the theoretical knowledge of the modeler. Hence, *without properly taken field data all our effort is futile.*

REFERENCES

Abbott, M.B, J.C. Bathurst, J.A. Cunge, P.E. O'Connell and J. Rasmussen. 1986. An introduction to the European hydrological system - Systèrne hydrologique Europeén, SHE.1. History and philosophy of a physically-based distributed modeling system. J. Hydrol. 87:45-59.

Aboujaoude, A. 1991. Modelisation hydrologique de l'infiltration dans les sols encroutés à differents échelles spatialles. Thèse docteur, IMG, Université J. Fourier, Grenoble.

Addiscott, T.M. and R.J. Wagenet. 1985. Concepts of solute leaching in soils: A review of modeling approaches. J. Soil Sci. 36:411-424.

Alemi, M.H., D.R. Nielsen and J.W. Biggar. 1976. Determining the hydraulic conductivity of soil cores by centrifugation. Soil Sci. Soc. Am. J. 40:212-218.

Amanthakrishnan, V., W.N. Gill and A.J. Barduhn. 1965. Laminar dispersion in capillaries: Part I. Mathematical Analysis. Am. Inst. Chem. Eng. J. 11:1063-1072.

Amerman, C.R. 1983. Infiltration measurement. In: Advances of Infiltration, ASAE, St. Joseph, Michigan, p. 201-214.

Amoozegar, A. , D.R. Nielsen and A.W. Warrick. 1982. Soil solute concentration distributions for spatially varying pore water velocities and apparent diffusion coefficients. Soil Sci. Soc. Am. J. 46:3-9.

Amoozegar, A. and A.W. Warrick. 1986. Field measurements of saturated hydraulic conductivity. In: A. Klute (Ed.): Methods of Soil Analysis, Part I. Agronomy Monograph Series No. 9 (2nd edition), ASA and SSSA, p. 735-770, Madison.

Aris, R. 1956. On the dispersion of a solute in a fluid flowing through a tube. Proc. Roy. Soc. London A. 235:67-77.

Arya, L.M., D.A. Farrel and G.R. Blake. 1975. A field study of soil water depletion patterns in presence of growing soybean roots. 1. Determination of hydraulic properties of the soil. Soil Sci. Soc. Am. Proc. 39:424-430.

Arya, L.M. and J.F. Paris. 1981. A physicoempirical model to predict the soil moisture characteristic from particle size distribution and bulk density data. Soil Sci. Soc. Am. J. 45:1023-1030.

Averianov, S.F. 1949. Dependence of water conductivity of soils upon the air content (in Russian) Doklady Akademii Nauk SSSR)Proceedings of Academy of Sciences of USSR) 69, No. 2:141-145.

Baier, W., J.A. Dyer and W.R. Sharp. 1979. The versatile soil moisture budget. Tech. Bull. 87. Land Res. Res. Inst., Ottawa, Canada.

Baver, L.D., W.H. Gardner and W.R. Gardner. 1972. Soil Physics. Wiley, New York.

Bazin, M.J., P.T. Saunders and J.I. Prosser. 1976. Models of microbial interactions in the soil. Chemical Rubber Co. (CRC) Crit. Rev. Microbiol. 5:464-497.

Bear, J. 1969. Hydrodynamic dispersion. In: R.J.M. De Wiest (Ed.): Flow through porous media, Academic Press, New York and London.

Bear, J. 1972. Dynamics of Fluids in Porous Media. Elsevier Science., New York.

Bear, J. and Y. Bachmat. 1967. A generalized theory of hydrodynamic dispersion in porous media. Proc. Intl. Assoc. Sci. Hydrol. Symp. Haifa, Publ. No. 72.

Bear. J., D. Zaslavsky and L. Irmay. 1968. Physical Principles of Water Percolation and Seepage. UNESCO-Paris.

Belmans, C., J.G. Wesseling and R.A. Feddes. 1983. Simulation model of the water balance of a cropped soil: SWATRE. J. Hydrol. 63:271-286.

Bernal, J.D. 1946. Swelling and shrinking. Disc. Faraday Soc., p. 1-17.

Beven, K. and P. Germann. 1982. Macropores and water flow in soils. Water Resour. Res. 18:1311-1325.

Biggar, J.W. and D.R. Nielsen. 1961. Measuring capillary conductivity. Soil Sci. 92:192-193.

Biggar, J.W. and D.R. Nielsen. 1976. Spatial variability of the leaching characteristics of a field soil. Water Resour. Res. 12:78-84.

Binley, A., K. Beven and J. Elgy. 1989. A physically based model of heterogeneous hillslopes. 2. Effective hydraulic conductivities. Water Resour. Res. 25:1227-1233.

Binley, A., J. Elgy and K. Beven. 1989. A physically based model of heterogeneous hillslopes. 1. Runoff production. Water Resour. Res. 25:1219-1226.

Bird, R.B., W.E. Stewart and E.N. Lightfoot. 1960. Transport Phenomenoa. Wiley, New York.

Biswas, T.D., D.R. Nielsen and J.W. Biggar. 1966. Redistribution of soil water after infiltration. Water Resour. Res. 2:513-524.

Boast, C.W. and R.G. Langebartel. 1984. Shape factor for seepage into pits. Soil Sci. Soc. Am. J. 48:10-15.

Bolt, G.H. (Ed.). 1979. Soil Chemistry, Physico-Chemical Models. Elsevier Science, New York.

Bolt, G.H. and P.H. Groenevelt. 1969. Coupling phenomena as a possible cause for non-Darcian behaviour of water in soil. Bull. Intern. Assoc. Sci. Hydrol. 14:17-26.

Bolt, G.H. and P.H. Groenevelt. 1972. Coupling between transport processes in porous media. In: Proc. 2nd Symp. Fundamentals of Transport Phenomena in Porous Media. Univ. of Guelph, p. 630-652.

Boolting, H.W.G., J. Bouma and D. Giménez. 1991. A suction crust infiltrometer for measuring hydraulic conductivity of unsaturated soil near saturation. Soil Sci. Soc. Am. J. 55:566-568.

Bork, H.R. and B. Diekkrüger, 1990. Scale and regionalization problems in soil science. In: Trans. 14th Intl. Congress of Soil Sci., ISSS, Kyoto, I:178-183.

Bosworth, R.C.L. 1948. Distribution of reaction times for laminar flow in cylindrical reactors. Phil. Mag. 39:847-862.

Bouchet, R.J. 1963. Evapotranspiration réele et potentielle, signification climatique. General Assembly, Intl. Assoc. Sci. Hydrol. 62:134-142.

Boulier, J.F., J. Touma and M. Vauclin. 1984. Flux concentration relation-based solution of constant flux infiltration equation: I. Infiltration into nonuniform initial moisture profiles. Soil Sci. Soc. Am. J. 48:245-251.

Bouma, J. 1984. Using soil morphology to develop measurement methods and simulation techniques for water movement in heavy clay soils. In: J. Bouma and P.A.C. Raats (Eds.): Proceedings of the ISSS Symposium on Water and Solute Movement in Heavy Clay Soils, International Institute for Land Reclamation and Improvement (ILRI) Publication 37, Wageningen, p. 298-210.

Bouma, J., P.J.M. de Laat, A.F. van Holst and T.J. von de Nes. 1980. Predicting the effects of changing water table levels and associated soil moisture regimes for soil survey interpretation. Soil Sci. Soc. Am. J. 44:797-802.

Bouwer, H. 1966. Rapid field measurement of air entry value and hydraulic conductivity of soil as significant parameters in flow system analysis. Water Resour. Res. 2:729-738.

Bouwer, H. 1986. Intake rate: Cylinder infiltrometer. In: A. Klute (Ed.): Methods of Soil Analysis, Part I. Agronomy Monograph Series No. 9 (2nd edition), ASA and SSSA, Madison, Wisconsin, p. 825-844.

Bower, C.A., W.R. Gardner and J.O. Goertzen. 1957. Dynamics of cation exchange in soil columns. Soil Sci. Soc. Am. Proc. 21:20-24.

Bradáč, J. and M. Kutílek. 1974. Steady infiltration of water in the homogeneous soil (in Czech). Vodni hosp. 24:74-81.

Bresler, E. and G. Dagan. 1979. Solute dispersion in unsaturated heterogenous soil at field scale: II. Aplications. Soil Sci. Soc. Am. J. 43:467-472.

Bresler, E. and G. Dagan. 1983. Unsaturated flow in spatially variable fields. 2. Application of water flow models to various fields. Water Resour. Res. 19:421-428.

Brewer, R. 1964. Fabric and Mineral Analysis of Soils. John Wiley & Sons, New York.

Brewer, R. 1976. Fabric and Mineral Analysis of Soils. 2nd print. Kreiger Publ. Co. Huntington, New York.

Briggs, L.J. and J.W. McLane. 1907. The moisture equivalents of soils. USDA Bur. of Soils, Bull. 45, pp 23.

Brigham, W.E., P.W. Reed and J.N. Dew. 1961. Experiments on mixing during miscible displacement in porous media. Soc. Petrol. Engrs. J. 1:1-8.

Broadbridge, P. and I. White. 1988. Constant rate rainfall infiltration: A versatile nonlinear model. 1. Analytic solution. Water Resour. Res. 24:145-154.

Bromilow, R.H. and M. Leistra. 1980. Measured and simulated behavior of aldicarb and its oxidation products in fallow soils. Pesticide Sci. 11:389-395.

Brooks, R.H. and A.T. Corey. 1964. Hydraulic properties of porous media. Hydrology paper 3, Colorado St. Univ., Fort Collins.

Brunauer, S., P.H. Emmett and E. Teller. 1938. Adsorption of gases in multimolecular layers. J. Am. Chem. Soc. 60:309-319.

Brutsaert, W. 1966. Probability laws for pore size distributions. Soil. Sci. 101:85-92.

Brutsaert, W. 1977. Vertical infiltration in dry soil. Water Resour. Res. 13:363-368.

Brutsaert, W. and H. Stricker. 1979. An advection-aridity approach to estimate actual regional evapotranspiration. Water Resour. Res. 15:443-450.

Brutsaert, W.H. 1982. Evaporation into the Atmosphere. Reidel Publ. Co. Dordrecht.

Buckingham, E. 1904. Contribution to our knowledge of the aeration of soils. Bull. 25. U.S. Dept. of Agr. Bureau of Soils, Washington, D.C.

Buckingham, E. 1907. Studies on the movement of soil moisture. Bull. 38. U.S. Dept. of Agr. Bureau of Soils, Washington, D.C.

Budagovskij, A.I. 1955. Infiltration of Water in Soil (in Russian). Izdatelstvo Akademii Nauk SSSR (Publishers of Academy of Sciences), Moscow.

Budagovskij, A.I. 1964. Evaporation of Soil Water (in Russian). Izdatelstvo Nauka (Publishers Nauka), Moscow.

Burdine, N.T. 1953. Relative permeability calculations from pore size distribution data. Trans. Am. Inst. Univ. Metall. Pet. Eng. 198:71-87.

Burt, T.P. and P.J. Williams. 1976. Hydraulic conductivity in frozen soils. Earth Surface Processes 11:349-360.

Callebaut, F., D. Gabriels and M. de Boodt (Eds.) 1985. Assessment of Surface Sealing and Crusting. Proc. Symp. in Ghent. Flanders Research Center for Soil Erosion and Soil Conservation, and Department of Soil Physics, University of Ghent.

Cameron, D.A. and A. Klute. 1977. Convective-dispersive solute transport with a combined equilibrium and kinetic adsorption model. Water Resour. Res. 13:183-188.

Carslaw, H.S. and J.C. Jaeger. 1959. Conduction of Heat in Solids. 2nd ed. London, Oxford Univ. Press.

Cassel, D.K. and D.R. Nielsen. 1986. Field capacity and available water capacity. In: A. Klute (Ed.): Methods of Soil Analysis, Part I. Agronomy Monograph Series No. 9 (2nd edition), ASA and SSSA, Madison, Wisconsin, p. 901-926.

Castro, C.L. and D.E. Rolston. 1977. Organic phosphate transport and hydrolysis in soil: Theoretical and experimental evaluation. Soil Sci. Soc. Am. J. 41:1085-1092.

Chatzis, I. and F.A.L. Dullien. 1977. Modelling pore structure by 2-D and 3-D networks of capillary tubes to simulate the drainage capillary pressure of sandstones. J. Can. Pet. Tech. 16:97-104.

Childs, E.C. 1969. The Physical Basis of Soil Water Phenomena. New York, Wiley Interscience.

Childs, E.C. and N. Collis-George. 1950. The permeability of porous materials. Proc. Royal Soc. London A. 210:392-405.

Cho, C.M. 1971. Convective transport of ammonium with nitrification in soil. Can. J. Soil Sci. 51:339-350.

Chong, S.K., R.E. Green and L.R. Ahuja. 1981. Simple *in situ* determination of hydraulic conductivity by power function descriptions of drainage. Water Resour. Res. 17:1109-1114.

Chopart, J. L. and M. Vauclin. 1990. Water balance estimation model: Field test and sensitivity analysis. Soil Sci. Soc. Am. J. 54:1377-1384.

Císler, J. 1969. The solution for maximum velocity of isothermal steady flow of water upward from water table to soil surface. Soil Sci. 108:148.

Císler, J. 1974. Note on the Parlange method for the numerical solution of horizontal infiltration of water into soil. Soil Sci. 117:70-73.

Clausnitzer, V., J.W. Hopmans and D.R. Nielsen, 1992. Simultaneous scaling of soil water retention and hydraulic conductivity curves. Water Resour. Res. 28:19-31.

Clothier, B. and T. Heiler. 1983. Infiltration during sprinkler irrigation: Theory and field results. In: Advances in Infiltration. ASAE Publ. 11-83. St. Joseph, Mich., p. 275-283.

Clothier, B.E., J.H. Knigtht and I. White 1981. Burgers equation: Application to field constant-flux infiltration. Soil Sci. 132:255-261.

Clothier, B.E. and K.R.J. Smettem. 1990. Combining laboratory and field measurements to define the hydraulic properties of soil. Soil Sci. Soc. Am. J. 54:299-304.

Clothier, B.E. and I. White. 1981. Measurement of sorptivity and soil water diffusivity in the field. Soil Sci. Soc. Am. J. 45:241-245.

Coats, K.H. and B.D. Smith. 1964. Dead-end pore volume and dispersion in porous media. Soc. Petrol. Engrs. J. 4:73-84.

Collis-George, N. and D.B. Yates. 1990. The first stage of drainage from ponded soils with encapsulated air. Soil Sci. 149:103-111.

Conca, J.L. and J.V. Wright. 1992. Diffusion and flow in gravel, soil and whole rock. Appl. Hydrogeol. 1:5-24.

Constanz, J. 1982. Temperature dependence of unsaturated hydraulic conductivity of two soils. Soil Sci. Soc. Am. J. 46:466 470.

Corey, A.T. 1977. Mechanics of Heterogenous Fluids in Porous Media. Water Resour. Publ., Fort Collins, Colorado.

Corey, A.T. and A. Klute. 1985. Application of the potential concept to soil water equilibrium and transport. Soil Sci. Soc. Am. J. 49:3-11.

Corey, J.C., D.R. Nielsen and J.W. Biggar. 1963. Miscible displacement in saturated and unsaturated sandstone. Soil Sci. Soc. Am. Proc. 27:258-262.

Crank, J. 1956. The Mathematics of Diffusion. Oxford Univ. Press, London.

Crestana, S., R. Cesareo and S. Mascarenhas. 1986. Using a computed tomography miniscanner in soil science. Soil Sci. 142:56-61.

Currie, J.A. 1960. Gaseous diffusion in porous media. II. Brit. J. Appl. Phys. 11:314-324.

Currie, J.A. and D.A. Rose. 1985. Gas diffusion in structured materials: The effect of trimodal pore size distribution. J. Soil. Sci. 36:487-493.

Cvetkovič, V.D. 1986. A continuum approach to high velocity flow in a porous medium. Transp. Porous Media. 1:63-97.

Cvetkovič, V.D. 1991. mass arrival of reactive solute in single fractures. Water Resour. Res. 27:177-183.

Dagan, G. 1986. Statistical theory of groundwater flow and transport: Pore to laboratory, laboratory to formation and formation to regional scale. Water Resour. Res. 22:102-135.

Dagan, G. and E. Bresler. 1979. Solute dispersion in unsaturated heterogenous soil at field scale: I. Theory. Soil Sci. Soc. Am. J. 43:461-467.

Dagan, G. and E. Bresler. 1983. Unsaturated flow in spatially variable fields. 1. Derivation of models of infiltration and redistribution. Water Resour. Res. 19:413-420.

Dahiya, I.S., K.S. Grewal, R. Anlauf and J. Richter. 1985. Desalinization of a salt-affected soil in plots of various sizes under two modes of water application. J. Agric. Sci., Camb. 104:19-26.

Dalton, F.N. 1992. Development of time-domain reflectometry for measuring soil water content and bulk soil electrical conductivity. In: G.C. Topp, W.D. Reynolds and R.E. Green (Eds.): Advances in Measurement of Soil Physical Properties: Bringing Theory into Practice. SSSA Spec. Publ. No. 30, SSSA, Madison, Wisconsin, p. 143-167.

Danckwerts, P.V. 1953. Continuous flow systems. Chem. Eng. Sci. 2:1-13.

Dane, J.H. and A. Klute. 1977. Salt effects on the hydraulic properties of a swelling soil. Soil Sci. Soc. Am. J. 41:1043-1049.

Darcy, H. 1856. Les fontaines publique de la ville de Dijon. Dalmont, Paris.

Davidson, J.M. and J.R. McDougal. 1973. Experimental and predicted movement of three herbicides in water-saturated soil. J. Environ. Qual. 2:428-433.

Davidson, J.M., D.R. Nielsen and J.W. Biggar. 1963. The measurement and description of water flow through Columbia silt loam and Hesperia sandy loam. Hilgardia 34:601-617.

Davidson, J.M., D.R. Nielsen and J.W. Biggar. 1966. The dependency of soil water uptake and release upon the applied pressure increment. Soil Sci. Soc. Am. Proc. 30:298-304.

Davidson, J.M., L.R. Stone, D.R. Nielsen and M.E. La Rue. 1969. Field measurement and use of soil water properties. Water Resour. Res. 5:1312-1321.

Day, P.R. 1956. Dispersion of a moving salt water boundary advancing through a saturated sand. Trans Am. Geophys. Union 37:595-601.

Dean, T.J., J.P. Bell and A.J.B. Baty. 1987. Soil moisture measurement by an improved capacitance technique. Part I. Sensor design and performance. J. Hydrol. 93:67-78.

de Josselin de Jong, G. 1958. Longitudinal and transverse diffusion in granular deposits. Trans Am. Geophys. Union 39:67-74.

Delesse, A. 1848. Procédé mécanique pour déterminer la composition des roches. Ann. Mines 13:379-388.

Denmead, O.T. and R.H. Shaw. 1962. Availability of soil water to plants as affected by soil moisture content and meteorological conditions. Agron. J. 54:385-390.

Deryaguin, B.V. and N.V. Churaev. 1972. Isotherms of disjoining pressure of water films on the quartz-surface (in Russian). Doklady Akademii Nauk SSSR (DAN - Proceedings of Academy of Sciences of USSR), 207, No. 3:572-575.

Deryaguin, B.V., N.N. Zachovaeva and M.K. Melnikova. 1956. Theory of equilibrium and migration of soil moisture at various moisture contents. Trans. 6th Intl. Congress of Soil Sci., ISSS, Paris, 291-297.

De Vault, D. 1943. The theory of chromatography. J. Am. Chem. Soc. 65:534-540.

Diaz, C.E., I. Chatzis and F.A.L. Dullien. 1987. Simulation of capillary pressure curves using bond correlated site percolation on a simple cubic network. Transp. Porous Media 2:215-240.

Diment, G.A., K.K. Watson and P.J. Blennerhassett. 1982. Stability analysis of water movement in unsaturated porous materials. I. Theoretical consideration. Water Res. Res 18:1248-1254.

Dirksen, C. 1975. Determination of soil water diffusivity by sorptivity measurements. Soil. Sc Soc. Am. Proc. 39:22-27.

Dirksen, C. and S. Matula. 1994. Automatic atomized water spray system for soil hydraulic conductivity measurements. Soil Sci. Soc. Am. J. 58:319-325.

Doering, E.J. 1965. Soil water diffusivity by the one-step method. Soil Sci. 99:322-326.

Doležal, F. and J. Císler. 1969. Discussion of paper by Yu Si Fok. J. Irrig. available [riessen, P.M. and R. Dudal (Eds.) 1989. Lecture Notes on the Geography, Formation, Properties and Use hina (Yang et al., 1992) Agric. University, Wageningen and Katholieke Universiteit, Leuven.

Duffy, C, P.J. Wierenga and R.A. Kselik. 1981. Variations in infiltration rate based on soil survey information and field measurements. New Mexico State Univ., Agr. Experiment Station Bull. 680.

Dullien, F.A.L. 1979. Porous Media: Fluid Transport and Pore Structure. Academic Press, New York.

Dutt, G.R. and P.F. Low. 1962. Diffusion of alkali chlorides in clay-water systems. Soil Sci. 93:233-240.

Dutt, G.R. and K.K. Tanji. 1962. Predicting concentrations of solutes percolated through a column of soil. J.Geophys. Res. 67:3437-3439.

Dyck, S. und G. Peschke. 1983. Grundlagen der Hydrologie. VEB Verlag für Bauwesen (VEB Publisher for Civil Engineering), Berlin.

Dykhuizen, R.C. 1987. Transport of solutes through unsaturated fractured media. Water Research 21:1531-1539.

Eching, S.O., J.W. Hopmans and O. Wendroth. 1994. Unsaturated hydraulic conductivity from transient multistep outflow and soil water pressure data. Soil Sci. Soc. Am. J. 58:687-695.

Edwards, W.M. and W.E. Larson. 1969. Infiltration of water into soils as influenced by surface seal development. Trans. ASAE 12:463-465.

Elrick, D.E. 1961. Transient two-phase capillary flow in porous media. Phys. of Fluids 4:572-575.

Elrick, D.E. and W.D. Reynolds. 1992. Infiltration from constant-head well permeameters and infiltrometers. In: G.C. Topp, W.D. Reynolds and R.E. Green (Eds.): Advances in Measurement of Soil Physical Properties: Bringing Theory into Practice. SSSA Spec. Publ. 30, SSSA, Madison, Wisconsin, p. 1-24.

Everett, D.H. 1954. A general approach to hysteresis, 3. Formal treatment of the independent domain model of hysteresis. Trans. Faraday Soc. 50:1077-1096.

Everett, D.H. 1967. Adsorption hysteresis. In: E.A. Flood (Ed.): Solid gas interface. Vol. 2., Chapt. 36. Marcel Dekker, New York, p. 1055-1113.

Fatt, I. 1956. The network model of porous media I. Capillary pressure characteristics. Pet. Trans. AIME 207:144-164.

Feddes, R.A., P. Kowalik and H. Zaradny. 1978. Simulation of field water use and crop field. John Wiley and Sons, New York.

Federer, C.A. 1982. Transpirational supply and demand: Plant, soil and atmosphere effects evaluated by simulation. Water Resour. Res. 18:355-362.

Fick, A. 1855. Über Diffusion. Ann. der Phys. Leipzig 94:59-86.

Flühler, H., M.S. Ardakani and L.H. Stolzy. 1976. Error propagation in determining hydraulic conductivities from successive water content and pressure head profiles. Soil Sci. Soc. Am. J. 40:830-836.

Freeze, R.A. and J.A. Cherry. 1979. Groundwater. Prentice-Hall, Inc., Englewood, New Jersey.

Freyberg, D.L. 1986 A natural gradient experiment on solute transport in a sand aquifer, 2. Spatial moments and the advection and dispersion of non-reactive solutes. Water Resour. Res.22:3031-2046.

Fried, J.J. and M.A. Combarnous. 1971. Dispersion in porous media. Advances Hydrosci. 7:169-282.

Gardiner, C.W. 1990. Handbook of Stochastic Methods for Physics, Chemistry, and the Natural Science. Springer-Verlag, pp. 442.

Gardner, W.A. 1990. Introduction to Random Process With Applications to Signals and Systems, McGraw-Hill, pp. 546.

Gardner, W.H. 1965. Water content. In: C.A. Black (Ed.): Methods of Soil Analysis, Part 1, Physical and Mineralogical Properties, Agronomy Monograph Series No. 9, ASA, Madison, Wisconsin, p. 82-127.

Gardner, W.R. 1956. Calculation of capillary conducitvity from pressure plate outflow data. Soil. Soc. Am. Proc. 20:317-320.

Gardner, W.R. 1958. Some steady state solutions of the unsaturated moisture flow equation with applications to evaporation from a water table. Soil Sci. 85:228-232.

Gardner, W.R. 1959. Solution of the flow equation for the drying of soils and other porous media. Soil Sci. Soc. Am. Proc. 23:183-187.

Gardner, W.R. 1962. Note on the separation and solution of diffusion type equations. Soil Sci. Soc. Am. Proc. 33:404.

Gardner, W.R. and D.I. Hillel. 1962. The relation of external evaporative conditions to the drying of soils. J. Geophys. Res. 67:4319-4325.

Gardner, W.R., D.I. Hillel and I. Benyamini. 1970. Post irrigation movement of soil water: I. Redistribution. Water Resour. Res. 6:851-861.

Gardner, W.R., D.I. Hillel and I. Benyamini. 1970. Post irrigation movement of soil water: II. Simultaneous redistribution and evaporation. Water Resour. Res. 6:1148-1153.

Gardner, W.R. and M.S. Mayhugh. 1958. Solution and tests of the diffusion equation for the movement of water in soil. Soil Sci. Am. Proc. 22:197-201.

Garrat, J.R. 1978. Flux-profile relationships above tall vegetation. Quart. J. Roy. Meteor. Soc. 105:199-211.

Garrat, J.R. 1992. The Atmospheric Boundary Layer. Cambridge Press, Cambridge.

Gaudet, J.P., H. Jegat, G. Vachaud and P.J. Wierenga. 1977. Solute transfer with diffusion between mobile and stagnant water through unsaturated sand. Soil Sci. Soc. Am. J. 41:665-671.

Gelb, A., J.F. Kasper, R.A. Nash. C.F. Price and A.A. Sutherland. 1974. Applied Optimal Estimation. M.I.T. Press, pp. 374.

Gelhar, L.W., A.L. Gutjahr and R.L. Naff. 1979. Stochastic analysis of macrodispersion in aquifers. Water Resour. Res. 15:1387-1397.

Gerke, H.H. and M.Th. van Genuchten. 1993. A dual-porosity model for simulating the preferential movement of water and solutes in structured porous media. Water Resour. Res. 29:305-319.

Germann, P.F. and K. Beven. 1985. Kinematic wave approximation to infiltration into soils with sorbing macropores. Water Resour. Res. 21:990-996.

Gersevanov, N.M. 1937. Dynamics of Foundations. Strojizdat Publications, Moscow-Leningrad.
Gish, T.J. and J.L. Starr. 1983. Temporal variability of infiltration under field conditions. In: Advances of Infiltration. ASAE Publ. 11-83. St. Joseph, Mich., p. 122-131.
Glueckauf, E. 1949. The theory of chromotography, Part VI. Precision measurements of adsorption and exchange isotherms from column-elution data. J. Chem. Soc. 4:3280-3285.
Gouy, G. 1910. Sur le construction de la charge electrique a la surface a'un electrolyte. J. de Physique 9:657-668.
Gray, D.M. and D.H. Male (Eds.). 1981. Handbook of Snow. Principles, Processes, Management and Use. Pergamon Press, Toronto, Oxford, New York, Sydney.
Greacen, E.L. 1981. Soil Water Assessment by the Neutron Method. CSIRO Australia.
Green, R.E., L.R. Ahuja and S.K. Chong. 1986. Hydraulic conductivity, diffusivity and sorptivity of unsaturated soils. Field methods. In: A. Klute (Ed.): Methods of Soil Analysis, Part I. Agronomy Monograph Series No. 9 (2nd edition), ASA and SSSA, Madison, Wisconsin, p.771-798.
Green, W.H. and G.A. Ampt. 1911. Studies on soil physics: I. Flow of air and water through soils. J. Agric. Sci. 4:1-24.
Greminger, P.J., Y.K. Sud and D.R. Nielsen. 1985. Spatial variability of field measured soil-water characteristics. Soil Sci. Soc. Am. J. 49:1075-1082.
Groenevelt, P.H. and G.H. Bolt. 1972. Water retention in soil. Soil Sci. 113:238-245.
Groenevelt, P.H. and B.D. Kay. 1977. Water and ice potentials in frozen soils. Water Resour. Res. 13:445-449.
Gunzelmann, M., U. Hell and R. Horn. 1987. Die Bestimmung der Wasserspannungs-Wasserleitfähigkeits Beziehung von Bodenaggregaten. Zeitschrift f. Pflanzenernährung und Bodenkunde 150:400-402.
Gutjahr, A. 1985. Spatial variability: Geostatistical methods. In: D.R. Nielsen and J. Bouma (Eds.): Soil Spatial Variability. Pudoc, Wageningen, p. 9-34.
Hadas, A. 1975. Drying of layered columns under non-isothermal conditions. Soil Sci. 119:143-148.
Hadas, A. and D. Hillel. 1972. Steady state evaporation through non-homogeneous soils from a shallow water table. Soil Sci. 113:65-73.
Haines, W.B. 1923. The volume changes associated with variations of water content in soil. J. Agric. Sci. 13:293-310.
Hartge, K.H. 1978. Einführung in die Bodenphysik. Enke, Stuttgart.
Hatano, R. and H.W.G. Boolting. 1992. Using fractal dimensions of stained flow patterns in a clay soil to predict bypass flow. J. Hydrol. 135:121-131.
Hatano, R., H. Nakamoto, T. Sakuma and H. Okajima. 1988. Evapotranspiration in cracked clay field soil. Soil Sci. Plant Nutr. 34:547-555.
Haverkamp, R., M. Kutílek, J.Y. Parlange, L. Rendon and M. Krejča. 1988. Infiltration under ponded conditions: 2. Infiltration equations tested for parameter time-dependence and predictive use. Soil Sci. 145:317-329.

Haverkamp, R., J.Y. Parlange, J.L. Starr, G. Schmitz and C. Fuentes. 1990. Infiltration under ponded conditions. 3. A predictive equation based on physical parameters. Soil Sci. 149:292-300.

Haverkamp, R., M. Vauclin, J. Touma, P.J. Wierenga and G. Vachaud. 1977. A comparison of numerical simulation models for one-dimensional infiltration. Soil Sci. Soc. Am. J. 41:285-294.

Haverkamp, R., M. Vauclin and G. Vachaud. 1984. Error analysis in estimating soil water content from neutron probe measurements: 1. Local standpoint. Soil Sci. 137:78-90.

Heinrich, R. 1886. Über Prüfung der Bodenarten auf Wasserkapacität und Durchlüftbarkeit. In: E. Wollny (Ed.): Forschungen auf Gebiete der Agrikultur-Physik. 9. Band C. Universitätsbuchhandlung C. Winter, Heidelberg, p. 260-274.

Heister, N.K. and T. Vermuelen. 1952. Saturation performance of ion exchange and adsorption columns. Chem. Eng. Prog. 48:505-516.

Helfferich, F. 1962. Ion Exchange. Advances in Chemistry Series. McGraw-Hill, New York.

Hill, D.E. and J.Y. Parlange. 1972. Wetting front instability in layered soils. Soil Sci. Soc. Am. Proc. 36:697-702.

Hillel, D. 1971. Soil and Water. Physical Principles and Processes. Academic Press, New York.

Hillel, D. 1980. Applications of Soil Physics. Academic Press, New York.

Hillel, D. and R.S. Baker. 1988. A descriptive theory of fingering during infiltration into layered soils. Soil Sci. 146:51-56.

Hillel, D. and D.E. Elrick (Eds.). 1990. Scaling in Soil Physics: Principles and Applications. SSSA Spec. Publ. No. 25, SSSA, Madison, Wisconsin.

Hillel, D. and W.R. Gardner. 1970. Transient infiltration into crust-topped profiles. Soil Sci. 109:69-76.

Hillel, D., V. Krentos and Y. Stylianou. 1972. Procedure and test of an internal drainage method for measuring soil hydraulic characteristics *in situ*. Soil Sci. 114:396-400.

Holtan, H.N. 1961. Concept for infiltration estimates in watershed engineering. USDA Res. Ser. Publ.:41-51.

Holtan, H.N. and N.C. Lopez. 1971. USDAHL-70 model of watershed hydrology. USDA Tech. Bull. 1435.

Hopmans, J. W. and J.H. Dane. 1986. Temperature dependence of soil hydraulic properties. Soil Sci. Soc. Am. J. 50:4-19.

Hopmans, J.W. and J.N.M. Stricker. 1989. Stochastic analysis of soil water regime in a watershed. J. Hydrol. 105:57-84.

Hornung, U. 1983. Identification of nonlinear soil physical parameters from an input-output experiment. In: O. Deuflhardt and E. Hairer (Eds.): Workshop on Numerical Treatment of Inverse Problems in Differential Integral Equations. Birkhauser, Boston, p. 227-237.

Horton, R.E. 1940. An approach towards a physical interpretation of infiltration capacity. Soil Sci. Soc. Am. Proc. 5:399-417.

Illangasekare, T.H., R.J. Walter, Jr., M.F. Meier and W.T. Pfeffer. 1990. Modeling of meltwater infiltration in subfreezing snow. Water Resour. Res. 26:1001-1012.

Iwata, S. T. Tabuchi and B.P. Warkentin. 1988. Soil Water Interactions. M. Dekker, New York and Basel.

Jackson, R.D. 1964. Water vapor diffusion in relatively dry soil I. Soil Sci. Soc. Am. Proc. 28:172-176.

Jackson, R.D., S.B. Idso and R.J. Reginato. 1976. Calculation of evaporation rates during the transition from energy-limiting to soil-limiting phases using albedo data. Water Resour. Res. 12:23-26.

Jackson, R.D., B.A. Kimball, R. J. Reginato and F.S. Nakayarna. 1973. Diurnal soil water evaporation: Time-depth-flux patterns. Soil Sci. Soc. Am. Proc. 37:505-509.

Jackson, R.D., D.R. Nielsen and F.S. Nakayama. 1963. On diffusion laws applied to porous materials. U.S. Dept. Agr. ARS 41-86.

Jackson, R.D., R.J. Reginato, B.A. Kimball and F.S. Nakayama. 1974. Diurnal soil water evaporation: Comparison of measured and calculated soil-water fluxes. Soil Sci. Soc. Am. Proc. 38:861-866.

Jackson, T.J. 1993. Measuring surface soil moisture using passive microwave remote sensing. Hydrologic Processes 7:139-152.

Jackson, T.J., D.M. Le Vine, A.J. Griffis, D.C. Goodrich, T.J. Schmugge, C.T. Swift and P.E. O'Neill. 1993. Soil moisture and rainfall estimation over a semiarid environment with the ESTAR microwave radiometer. IEEE Trans. Geosci. Rem. Sens. 31:836-841.

Jalota, S.K. and S.S. Prihar. 1990. Bare-soil evaporation in relation to tillage. In: B.A. Stewart (Ed.): Advances in Soil Science. 12:187-216, Springer Verlag, New York.

Janota, R. 1924. Advances in Soil Reclamation (in Czech). Proceedings, Ministry of Agriculture, No. 34, Prague.

Javandel, J., C. Doughty, Chin-Fu Tsang. 1984. Groundwater transport: Handbook of Mathematical Models. Amer. Geophys. Union, Washingtion, D.C.

Jim, C.Y. 1988. A classification of soil microfabrics. Geoderma 41:315-325.

Journel, A.G. and C.J. Huijbregts. 1978. Mining Geostatistics. Academic Press, New York.

Jury, W.A. 1982. Simulation of solute transport with a transfer function model. Water Resour. Res. 18:363-368.

Jury, W.A. 1983. Chemical transport modeling: Current approaches and unresolved problems. In: SSSA Special Pulication No. 11, SSSA and ASA, Madison, Wisconsin, p. 49-64.

Jury, W.A. 1989. Spatial variability of soil properties. In: S.C. Hern and S.M. Melancon (Eds.): Vadose Zone Modeling of Organic Pollutants. Lewis Publishers, p. 245-269.

Jury, W.A. and G. Sposito. 1985. Field calibration and validation of solute transport models for the unsaturated zone. Soil Sci. Soc. Am. J. 49:1331-1341.

Kalms, J.M., M. Vauclin and G. Vachaud. 1979. Modalites d'alimentation hydrique de deux variétés de riz pluvial en Cote d'Ivoire (Centre). In: Isotopes and Radiation in Research on Soil-Plant Relationships. IAEA-SM-235, Vienna, p. 271-287.

Katchinski, N.A. 1965. Soil physics (in Russian). Vysshaya Shkola (College Publishers), Moscow.
Katul, G.G. and M.B. Parlange. 1992. A Penman-Brutsaert model for wet surface evaporation. Water Resour. Res. 28:121-126.
Katul, G.G., O. Wendroth, M.B. Parlange, C.E. Puente, M.V. Folegatti and D.R. Nielsen. 1993. Estimation of *in situ* hydraulic conductivity function from nonlinear filtering theory. Water Resour. Res. 29:1063-1076.
Kemper, W.D. 1960. Water and ion movement in thin films as influenced by the electrostatic charge and diffuse layer of cations associated with clay mineral surfaces. Soil Sci. Soc. Am. Proc. 24:10-16.
Kinzelbach; W. 1988. The random walk method in pollutant transport simulation. In: E. Custodio, A. Gurgui and J.P.Lobo Ferreira (eds.). Groundwater Flow and Quality Monitoring. D. Reidel, Hingham, MA, p. 227-246.
Kirkham, D. and W. L. Powers. 1972. Advanced Soil Physics. Wiley-Interscience, New York.
Kitagawa, K. 1934. Sur le dispersement et l'écart moyen de l'écoulement des eaux sonterraines. Kyoto Imp. University, Mem. Coll. Sci., Series A, 17; 37-42 and 431-441.
Klute, A. (Ed.). 1986. Methods of Soil Analysis. Part 1. Agronomy Monograph Series No. 9 (2nd edition), ASA and SSSA, Madison, Wisconsin.
Klute, A. and C. Dirksen. 1986. Hydraulic conductivity and diffusivity: Laboratory methods. In: A. Klute (Ed.): Methods of Soil Analysis, Part I. Agronomy Monograph Series No. 9 (2nd edition), ASA and SSSA, Madison, Wisconsin, p. 687-734.
Knighton, R.E. and R.J. Wagenet. 1987. Simulation of solute transport using a continuous time Markov process. 2. Application to transient field conditions. Water Resour. Res. 23:1917-1925.
Kolmogorov, A.N. 1941. Interpolated and extrapolated stationary random sequences (in Russian). Isvestia Akademii Nauk SSSR, seriya matematicheskaya 5 (1).
Kool, J.B. and J.C. Parker. 1988. Analysis of the inverse problem for transient unsaturated flow. Water Resour. Res. 24:817-830.
Kool, J.B.. J.C. Parker and M.T. van Genuchten. 1985. Determining soil hydraulic properties from one-step outflow experiments by parameter estimation. I. Soil Sci. Soc. Am. J. 49:255-293.
Kopecký, J. 1914. Die Physikalischen Eigenschaften des Bodens. Intern. Mittteilungen der Bodenkunde, Berlin 4:138-198.
Kossovitch, P.S. 1904. Soil water properties (in Russian). Zhurnal Opytnoi Agrochimii, vol. 5, No. 2, 3. Quoted according to Rode (1952).
Kovda, V.A. and I. Scabolcs 1979. Modeling of soil salinization and alkalization. Agrokem. Talajt. 28:1-208.
Kostiakov, A.N. 1932. On the dynamics of the coefficient of water-percolation in soils and on the necessity of studying it from a dynamic point of view for purposes of amelioration. Trans. Sixth Comm. Intl. Soc. Soil Sci., Part A:17-21.
Kozeny, I. 1927. Über Kapillarleitfähigkeit des Wassers im Boden. Sitzungsber. Akad. Wiss. Wien 136:303 (quoted according to Zunker, 1930).

Krige, D.G. 1966. Two-dimensional weighted moving average trend surfaces for ore-evaluation. J. South African Inst. Mining Metallurgy 66:13-38.
Kubiena, W.L. 1970. Micromorphological Features of Soil Geography. Rutgers Univ. Press, Brunswick, New Jersey.
Kunze, R.J. and D.R. Nielsen. 1982. Finite-difference solutions of the infiltration equation. Soil Sci. 134:81-88.
Kuráž, V. and J. Matoušek. 1977. A new dielectric soil moisture meter for field measurement of soil moisture. Bull. ICID, 26.1:76-79.
Kuráž, V., M. Kutílek and I. Kašpar. 1970. Resonance-capacitance soil moisture meter. Soil Sci. 110:278-279.
Kutílek, M. 1962a. Hygroscopic soil water (in Czech). Vodohosp. Čas. 10:11-29 and 156-173.
Kutílek, M. 1962b. A new method for determination of the soil specific surface (in Czech). Rostlinná výroba (Plant Production) 33:767-772.
Kutílek, M. 1963. The influence of soil colloids on the value of the contact angle and on the capillary rise of water in soils. Proc. 5 Congr. ICID, Tokyo, p. 1785-1790.
Kutílek, M. 1964. The filtration of water in soils in the region of the laminar flow. Trans. 8th Congress of Soil Sci., ISSS, Bucarest, II:45-51.
Kutílek, M. 1969. Influence of clay minerals upon the statics and dynamics of soil water (in Czech). Research Report, Technical Univ., Prague.
Kutílek, M. 1972. Non-Darcian flow of water in soils-laminar region. In: Fundamentals of Transport Phenomena in Porous Media. Elsevier, Amsterdam, p. 327-340.
Kutílek, M. 1973. The influence of clay minerals and exchangeable cations on soil moisture potential. In: A. Hadas, D. Swartzendruber, P.E. Rijtema, M. Fuchs and B. Yaron (Eds.): Physical Aspects of Soil Water and Salts in Ecosystems. Ecol. Studies 4. Springer, Berlin, p. 153-160.
Kutílek, M. 1978. Soil Science for Water Management (Textbook in Czech). SNTL-ALFA (State Publishers of Technical Literature), Prague.
Kutílek, M. 1980. Constant rainfall infiltration. J. Hydrol. 45:289-303.
Kutílek, M. 1983. Soil physical properties of saline and alkali vertisols. In: Isotope and Radiation Techniques in Soil Physics and Irrigation Studies. IAEA, Vienna, p. 179-190.
Kutílek, M. 1984a. Some theoretical and practical aspects of infiltration in clays with D = const. In: J. Bouma and P.A.C. Raats (Eds.): Water and Solute Movement in Heavy Clay Soils. Proceedings of the ISSS Symposium on Water and Solute Movement in Heavy Clay Soils. Internatioal Institute for Land Reclamation and Improvement (ILRI), Publication 37, Wageningen, p. 114-128.
Kutílek, M. 1984b. Moisture of Porous Materials. SNTL and ČMT (State Publishers of Technical Literature, Prague,.
Kutílek, M. and M. Krejča. 1987. A three-parameters infiltration equation of the Philip's type solution (in Czech). Vodohosp. Čas. 35:52-61.
Kutílek, M., M. Krejča, R. Haverkamp, L.P. Rendon and J.Y. Parlange. 1988. An extrapoladon of algebraic infiltration equations. Soil Tech. 1:47-61.
Kutílek, M. and J. Mls. 1975. Steady evaporation from the layered soil (in Czech). Vodohosp. Čas. 23:163-176.

Kutílek, M. and V. Novák. 1976. The influence of soil cracks upon infiltration and ponding time. In: M. Kutílek and J. Šútor (Eds.): Water in Heavy Soils. Proceedings ICID-ISSS (International Commission on Irrigation and Drainage - International Soil Science Society) Symposium in Bratislava, Czechoslavak Scientific Technical Society Publisher, p. 126-134.
Kutílek, M., R. Rösslerová and H. Othmer. 1992. Models of soil porous systems for unsaturated flow. In: Proc. Colloquium Porous or Fractured Unsaturated Media: Transport and Behavior. Swiss Federal Institute of Technology of Lausanne (EPFL) and University of Neuchatel, p. 71-86.
Kutílek, M. and J. Salingerová. 1966. Flow of water in clay minerals as influenced by adsorbed quinolinium and pyridinium. Soil Sci. 101:385-389.
Kutílek, M. and J. Valentová. 1986. Sorptivity approximations. Transp. Porous Media 1:57-62.
Kutílek, M., T. Vogel and D.R. Nielsen. 1985. Note on the physical characteristics of linear soils. Soil Sci. 139:497-499.
Kutílek, M, K. Zayani, R. Haverkamp, J.Y. Parlange and G. Vachaud. 1991. Scaling of the Richards' equation under invariant flux boundary condittions. Water Resour. Res. 27:2181-2185.
Lapidus, L. and N.R. Amundson. 1952. Mathematics of adsorption in beds. VI. The effect of longitudinal diffusion in ion exchange and chromatographic columns. J. Phys. Chem. 56:984-988.
LaRue. M.E., D.R. Nielsen and R.M. Hagan. 1968. Soil water flux below a ryegrass root zone. Agron. J. 60:625-629.
Lawes, J.B., J.H. Gilbert and R. Warington. 1881. On the amount and composition of the rain and drainage waters collected at Rothamsted. J. R. Agr. Soc. 17 (2nd series):249-279 and 311-350.
Leibenzon, L. S. 1947. Flow of Natural Liquids and Gases in Porous medium (in Russian). Gostekhizdat, Moscow.
Lemon, E.R. 1956. Potentialities for decreasing soil water evaporation loss. Soil Sci. Soc. Am. Proc. 16:160-163.
Letey, J., J. Osborn and R.E. Pelishek. 1962. Measurement of liquid solid contact angles in soil and sand. Soil Sci. 93:149-153.
Libardi, P.L., K. Reichardt, C. Jose, M. Bazza and D.R. Nielsen. 1980. An approximate method of estimating soil water diffusivity for different soil bulk densities. Water Resour. Res. 18:177-181.
Liu, J.-Wen Y., Sam T.S. Wong and L.J. Waldron. 1994. The application of NMR imaging to studying the effect of plant roots on preferential water flow in soil. In: ASA Tomography Workshop 1992, in press.
Low, P.F. 1961. Physical chemistry of clay-water interaction. Advances in Agon. 13:269-327.
Low, P.F. 1962. Effect of quasi-crystalline water on the processes involved in plant nutrition. Soil Sci. 93:6-15.
Lu, T.X., D.R. Nielsen and J.W. Biggar. 1994. Water movement in glass bead porous media III: Theoretical analysis of capillary rise into initially dry media. Water Resour. Res. 29: in press.

Luckner, L., M.Th. van Genuchten and D.R. Nielsen. 1989. A consistent set of parametric models for the two-phase flow of immiscible fluids in the subsurface. Water Resour. Res. 25:2187-2193.

Luckner, L. and W.M. Schestakow. 1991. Migration Processes in the Soil and Groundwater Zone. Lewis Publ. Inc., Chelsea, Michigan.

Lynn, B.H. and T.N. Carlson. 1990. A stomatal resistance model illustrating plant vs. external control of transpiration. Agric. For. Meteor. 52:5-43.

Maasland, M. and H.C. Haskew. 1957. The auger hole method of measuring the hydraulic conductivity of soil and its application to tile drainage problems. Proc. 3rd. Congr., International Commission on Irrigation and Drainage (ICID), p. 8.69-8.114.

Maisel, H. and G. Gnugnoli. 1972, Simulation of Discrete Stochastic Systems. Sci. Res. Assoc., Inc., Chicago.

Malicki, M.A., R. Plagge, M. Renger and R. Walczak. 1992. Application of time domain reflectometry soil moisture microprobe for the determination of unsaturated soil water characteristics from undisturbed soil cores. Irrig. Sci. 13:65-72.

Mansell, R.S., D.R. Nielsen and D. Kirkham. 1968. A method for simultaneous control of aeration and unsaturated water movement in laboratory soil columns. Soil Sci. 106:114-121.

Mantoglou, A. and L.W. Gelhar. 1987. Stochastic modeling of large-scale transient unsaturated flow systems. Water Resour. Res. 23:37-46.

Mantoglou, A. and L.W. Gelhar. 1987. Effective hydraulic conductivities of transient unsaturated flow in stratified soils. Water Resour. Res. 23:57-67.

Marle, C. and P. Defrenne. 1960. La description mathématique du déplacement de fluides miscibles dans un milieu poreux. Rapp. Inst. Fr. Petrole No. 5433.

Marshall, T.J. 1958. A relation between permeability and size distribution of pores. J. Soil Sci. 9:1-8.

Marshall, T.J. and J.W. Holmes. 1979. Soil Physics. Cambridge Univ. Press, Cambridge.

Matheron, G. 1965. Regionalized variables and their estimation (in French). Editions Masson, Paris.

McIntyre, D.S. 1958a. Permeability measurements of soil crusts formed by raindrop impact. Soil Sci. 85:185-189.

McIntyre, D.S. 1958b. Soil splash and the formation of surface crusts by raindrop impact. Soil Sci. 85:261-266.

McLaren, A.D. 1970. Temporal and vectorial reactions of nitrogen in soil: A review. Can. J. Soil Sci. 50:97-109.

McNeal, B.L. 1968. Prediction of the effect of mixed-salt solutions on soil hydraulic conductivity. Soil Sci. Soc. Am. Proc. 32:190-193.

McNeal, B.L., D.A. Layfield, W.A. Norvell and J.D. Rhoades. 1968. Factors influencing hydraulic conductivity of soils in the presence of mixed-salt solutions. Soil Sci. Soc. Am. Proc. 32:187-190.

Means, T.H. and J.G. Holmes. 1901. Soil survey around Fresno, California. Field Operations of the Division of Soils, 1900. U.S. Dept. Agr., House of Representatives, Document No. 526. Washington, D.C., p. 333-384.

Mein, R.G. and C.L. Larson. 1973. Modeling infiltration during a steady state rain. Water Resour. Res. 9:384-394.

M'Ewen M.B. and M.I. Pratt. 1957. The gelation of montmorillonite. Trans. Faraday Soc. 53:535.
Mezencev, V.J. 1948. Theory of formation of the surface runoff (in Russian). Meteorologia i Gidrologia 3:33-40.
Middleboe, V., D.R. Nielsen and D.A. Rennie (Eds.). 1976 Tracer Manual of Crops and Soils. IAEA, Vienna.
Miller, C.W. and L.V. Benson. 1983. Simulation of solute transport in a chemically reactive heterogeneous system: Model development and application. Water Resour. Res. 19:381-391.
Miller, E. E. and R.D. Miller. 1956. Physical theory for capillary flow phenomena. J. Appl. Phys. 27:324-332.
Miller, R.J., J.W. Biggar and D.R. Nielsen. 1965. Chloride displacement in Panoche clay loam in relation to water movement and distribution. Water Resour. Res. 1:63-73.
Millington R.J. and J.P. Quirk. 1959. Permeability of porous media. Nature 183:387-388.
Millington R.J., and J.P. Quirk. 1961. Permeability of porous solids. Trans. Faraday Soc. 57:1200-1207.
Milly, P.C.D. 1984. A linear analysis of thermal effects on evaporation from soil. Water Resour. Res. 20:1075-1086.
Mishra, B.K. and C Misra. 1993. Nitrogen transformation during miscible displacement of ammonium nitrate solution through the root zone of maize. J. Indian Soc. Soil Sci. 41:630-635.
Misra, C., D.R. Nielsen and J.W. Biggar. 1974. Nitrogen transformations in soil during leaching. I. Theoretical considerations. Soil Sci. Soc. Am. Proc. 38:289-293
Mitscherlich, E.A. 1901. Untersuchungen über die physikalischen Bodeneigenschaften. Landw. Jahrb. 30, B:360-445.
Mitchurin, B.N. 1975. Energetics of Soil Water (in Russian). Gidrometeoizdat (Water and Meteorology Publishers), Leningrad.
Miyazaki, T. 1990. Visualization of stream lines in layered soils. In: Trans. 14th Congress of Soil Sci., ISSS, Kyoto, I:272-273.
Mls, J. 1980. Effective rainfall estimation. J. Hydrol. 45:305-311.
Mls, J. 1982. Formulation and solution of fundamental problems of vertical infiltration (in Czech). Vodohosp. Čas. 30:304-313.
Molz, F.J. 1981. Models of water transport in the soil-plant system: A review. Water Resour. Res. 17:1245-1260.
Monin, A.S. and M.A. Obukhov. 1954. Basic laws of turbulent mixing in the ground layer of the atmosphere. Tr. Geofiz. Institut Akad. Nauk SSSR, p. 163-187.
Monteith, J.L. 1981. Evaporation and surface temperature. Quart. J. Roy. Meteor. Soc. 107:1-27.
Morel-Seytoux, H.J. 1976. Derivation of equations for rainfall infiltration. J. Hydrol. 31:203-219.
Morel-Seytoux, H.J. 1982. Analytical results for prediction of variable rainfall infiltration. J. Hydrol. 59:209-230.
Morel-Seytoux, H.J. 1983. Infiltration affected by air, seal, crust, ice and various sources of heterogeneity. In: Advances in Infiltration. ASAE Publ. 11-83, St. Joseph, Michigan, p. 132-146.

Mortland, M.M and W.D. Kemper. 1965. Specific surface. In: C.A. Black (Ed.): Methods of Soil Analysis, Part 1, Physical and Mineralogical Properties, Agronomy Monograph Series No. 9, ASA, Madison, Wisconsin, p. 532-544.

Morton, F.I. 1983. Operational estimates of areal evapotranspiration and their significance to the science and practice of hydrology. J. Hydrol. 66:1-76.

Mualem, Y. 1976. A new model for predicting the hydraulic conductivity of unsaturated porous media. Water Resour. Res. 12:513-522.

Mualem, Y. 1984. A modified dependent domain theory of hysteresis. Soil Sci. 137:283-291.

Mualem, Y., L. Assouline and H. Rohdenburg. 1990. Rainfall induced soil seal. A. A critical review of observations and models. B. Application of a new model to saturated soils. C. A dynamic model with kinetic energy instead of cumulative rainfall as independent variable. Catena 17:185-203, 205-218 and 289-303.

Mualem, Y, and E.E. Miller. 1979. A hysteresis model based on an explicit domain-dependence function. Soil Sci. Soc. Am. J. 43:106-1073.

Muazu, S., J. Skopp and D. Swartzendruber. 1990. Soil water diffusivity determination by a modified one-step outflow method. Soil Sci. Soc. Am. J. 54:1184-1186.

Mushkin, J.G. 1971. Water Subsidized Agricultural Fields (in Russian). Gidrometeoizdat, Leningrad.

Nerpin, S.V. and A.F. Chudnovskij. 1967. Soil Physics (in Russian). Nauka, Moscow.

Neuman, S.P. 1975. Galerkin approach to saturated-unsaturated flow in porous media. In: R.H. Gallagher, J.T. Oden, C. Taylor and O.C. Zienkiewicz (Eds.): Finite Elements in Fluids, Vol. 1: Viscous Flow and Hydrodynamics. John Wiley, London, p. 201-217.

Neuman, S.P. 1976. Wetting front pressure head in the infiltration model of Green and Ampt. Water Resour. Res. 12:564-566.

Nielsen, D.R. and J.W. Biggar. 1961. Miscible displacement in soils. I. Experimantal information. Soil Sci. Soc. Am. Proc. 25:1-5.

Nielsen, D.R. and J.W. Biggar. 1961. Measuring capillary conductivity. Soil Sci. 92:192-193.

Nielsen, D.R. and J.W. Biggar. 1967. The physical characterization of field soils. In: E.F. Bradley and O.T. Denmead (Eds.): The Collection and Processing of Field Data-CSIRO symposium. Wiley Interscience, New York.

Nielsen, D.R., J.W. Biggar and K.T. Erh. 1973. Spatial variability of field measured soil water properties. Hilgardia 42:215-259.

Nielsen, D.R., J.W. Biggar and R.J. Miller. 1967. Field observations of infiltration and soil-water redistribution. Trans. Am. Soc. Agr. Eng. 10:382-387.

Nielsen, D.R., J.M. Davidson, J.W. Biggar and R.J. Miller. 1964. Water movement through Panoche clay loam. Hilgardia 35:491-506.

Nielsen, D.R., D. Kirkham and W.R. van Wijk, 1961. Diffusion equation calculations of field soil water infiltration profiles. Soil Sci. Soc. Am. J. 25:165-168.

Nielsen, D.R., M.T. van Genuchten and J.W. Biggar. 1986. Water flow and solute transport processes in the unsaturated zone. Water Resour. Res. 22:89S-108S.

Nimmo, J.R., J. Rubin and D.P. Hammermeister. 1987. Unsaturated flow in a centrifugal field: measurement of hydraulic conductivity and testing of Darcy's law. Water Resour. Res. 23:124-134.
Njoku, E.G. and J.A. Kong. 1977. Theory for passive microwave remote sensing of near-surface soil moisture. J. Geophys. Res. 82:3108-3118.
Nkedi-Kizza, P. 1979. Ion exchange in aggregated porous media during miscible displacement. Ph.D. dissertation. Univ. of Calif., Davis.
Nkedi-Kizza, P., J.W. Biggar, M.Th. van Genuchten, P.J. Wierenga, H.M. Selim, J.M. Davidson and D.R. Nielsen. 1984. On the equivalence of two conceptual models for describing ion exchange during transport through an aggregated oxisol. Water Resour. Res. 20:1123-1130.
Nofziger, D.L. and D. Swartzendruber. 1974. Material content of binary physical mixtures as measured with a dual energy beam of gamma-rays. J. Appl. Phys. 45:5443-5449.
Norrish, K. 1972. Forces between clay particles. Proc. Intl. Clay Conf., Madrid, p. 375-383.
Novák, V. 1975. Non-isothermal flow of water in unsaturated soils. J. Hydrol. Sci. 2:37-52.
Novák, V. 1981. The structure of evapotranspiration (in Slovak) I. and II. Vodohosp. Čas. 29:476-492 and 581-582.
Novák, V. 1987. Estimation of soil water extraction patterns by roots. Agric. Water Mgt. 12:271-278.
Novák, V. and T. Hurtalová. 1987. Method of estimation of velocity coefficient of turbulent water vapor transport to calculate potential evapotranspiration (in Slovak). Vodohosp. Čas. 35:3-21.
Nye, P.H. and P.B. Tinker. 1977. Solute Movement in the Soil-Root System. Univ. of Calif. Press, Berkeley.
Oatmans, H.D. 1962. Nonlinear theory for frontal stability and viscous fingering in porous media. Soc. Petrol. Engr. J. 2:165-176.
Onsager, L. 1931. Reciprocal relations in irreversible processes, I. and II. Phys. Rev. 37:405-426 and 38:2265-2279.
Oster, J.D. and H. Frenkel. 1980. The chemistry of the reclamation of sodic soils with gypsum and lime. Soil Sci. Soc. Am. J. 44:41-45.
Othmer, H., B. Diekkrüger and M. Kutílek. 1991. Bimodal porosity and unsaturated hydraulic conductivity. Soil Sci. 152:139-150.
Paningbatan, E.P. 1980. Determination of soil moisture characteristic and hydraulic conductivity using a centrifuge. Unpublished Ph.D. dissertation, Univ. Calif., Davis. pp 197.
Parlange, J.Y. 1971. Theory of water movement in soils. 1. One-dimensional absorption. Soil Sci. 111:134-137.
Parlange, J.Y. 1972. Theory of water movement in soils: 8. One-dimensional infiltration with constant flux at the surface. Soil Sci. 114:1-4.
Parlange, J.Y. 1975. On solving the flow equation in unsaturated soils by optimization: Horizontal infiltration. Soil Sci. Soc. Am. Proc. 39:415-418.
Parlange, J.Y.and R.E. Smith. 1976. Ponding time for variable rainfall rates. Can. J. Soil Sci. 56:212-223.

Parlange, J.Y., R. Haverkamp and J. Touma. 1985. Infiltration under ponded conditions: I. Optimal analytical solution and comparison with experimental observations. Soil Sci. 139:305-311.

Parlange, J.Y., I. Lisle,R.D. Braddock and R.E. Smith. 1982. The three parameter infiltration equation. Soil Sci. 133:337-341.

Parlange, M.B. and G.G. Katul. 1992. An advection-aridity evaporation model. Water Resour. Res. 28:127-132.

Parlange, M.B., G.G. Katul and D.R. Nielsen. 1993. Determination of the field scale diffusivity function, soil water storage and evaporation, Water Resour. Res. 29:1279-1286.

Parlange, M.B., G.G. Katul, M.V. Folegatti and D.R. Nielsen. 1993. Evaporation and the field scale soil water diffusivity function. Water Resour. Res. 29:1279-1286.

Parlange, M.B., G.G. Katul, M.L. Kavvas, R.H. Cuenca and D.R. Nielsen. 1992. Physical basis for a time series model of soil water content, Water Resour. Res. 28:2437-2446.

Passerat de Silans, A., L. Bruckler, J.L. Thony and M. Vauclin. 1989. Numerical modeling of coupled heat and water flows during drying in a stratified bare soil – comparison with field obsersvations. J. Hydrol. 105;109-138.

Patterson, D.E. and M.W. Smith. 1981. The measurement of unfrozen water content bt time domain reflectometry: Results from laboratory tests. Can. Geotech. J. 18:131-144.

Peck, A.J. 1984. Field variability of soil physical properties. Adv. in Irrig. 2:189-221.

Pelishek, R.E., J. Osborn and J. Letey. 1962. The effect of wetting agents on infiltration. Soil Sci. Soc. Am. Proc. 26:595-598.

Penka, M. 1985. Transpiration and Consumption of Water by Plants (in Czech). Academia, Prague.

Penman, H.L. 1940. Gas and vapor movements in the soil: I. The diffusion of vapors through porous solids. II. The diffusion of carbon dioxide through porous solids. J. Agr. Sci. 30:436-462 and 570-581.

Penman, H.L. 1948. Natural evaporation from open water, bare soil and grass. Proc. Roy. Soc. London (A) 193;120-145.

Penman, H.L. 1963. Vegetation and Hydrology. Technical Comm. No. 53. Commonwealth Agric. Bur., Farnham Royal, Bucks.

Perroux, K.M. and I. White. 1988. Design for disc permeameter. Soil Sci. Soc. Am. J. 52:1205-1215.

Perroux, K.M., D.E. Smiles and I. White. 1981. Water movement in uniform soils during constant flux infiltration. Soil Sci. Soc. Am. J. 45:237-240.

Peschke, G. and M. Kutílek. 1976. The role of preferential ways in infiltration. In: M. Kutílek and J. Šútor (Eds.): Water in Heavy Soils III. Intl. Symp. Proceedings ICID-ISSS (International Commission on Irrigation and Drainage - International Soil Science Society) Symposium in Brataslava, Czechoslavak Scientific Technical Society Publisher, Prague, p. 70-84.

Peschke, G. and M. Kutílek. 1982. Infiltration model in simulated hydrographs. J. Hydrol. 56:369-379.

Peterson, A.E. and G.D. Bubenzer. 1986. Intake rate: Sprinkler infiltrometer. In: A. Klute (Ed.): Methods of Soil Analysis, Part I. Agronomy Monograph

Series No. 9 (2nd edition), ASA and SSSA, Madison, Wisconsin, p. 845-870.

Petrovič, P. 1984. Vertical hydrological balance and its one-dimensional simulation by the model DAIR. Lab. Hydraulica en Afoerhydr. Delft. Note 65.

Pfannkuch, H.O. 1962. Contribution à l'étude des déplacements de fluides miscibles dans un milieu poreux. Revue de l'Institut Francais du Pétrole 18:215-270.

Phene, C.J., D.A. Clark, G.E. Cardon and R.M. Mead. 1992. Soil matric potential sensor research and applications. In: Advances in Measurement of Soil Physical Properties: Bringing Theory into Practice. SSSA Spec. Publ. No. 30, SSSA, Madison, Wisconsin, p. 263-280.

Philip, J. R. 1955. Numerical solution of equations of the diffusion type with diffusivity concentration-dependent. Trans. Faraday Soc. 51:885-892.

Philip, J.R. 1957a. The theory of infiltration: 1. The infiltration equation and its solution. Soil Sci. 83:345-357.

Philip, J.R. 1957b. The theory of infiltration: 4. Sorptivity and algebraic infiltration equations. Soil Sci. 84:257-267.

Philip, J.R. 1957c. The theory of infiltration: 5. The influence of the initial moisture content. Soil Sci. 84:329-339.

Philip, J.R. 1957d. The physical principles of water movement during the irrigation cycle. Proc. Intl. Congr. ICID 8:124-154.

Philip, J.R. 1958a. The osmotic cell, solute diffusibility and the plant water economy. Plant Physiol. 33:264-271.

Philip, J.R. 1958b. The theory of infiltration: 6. Effect of water depth over soil. Soil Sci. 85:278-286.

Philip, J.R. 1960. General method of exact solution of the concentration-dependent diffusion equation. Aust. J. Phys. 13:1-12.

Philip, J.R. 1969a. Theory of infiltration. Advances Hydrosc. 5:215-305, Academic Press, New York.

Philip, J.R. 1969b. Hydrostatics and hydrodynamics in swelling soils. Water Resour. Res. 5:1070-1077.

Philip, J.R. 1973. On solving the unsaturated flow equation: 1. The flux-concentration relation. Soil Sci. 116:328-335.

Philip, J.R. 1980. Field heterogeneity: Some basic issues. Water Resour. Res. 16:443-448.

Philip, J.R. 1987. The infiltration joining problem. Water Resour. Res. 23:2239-2245.

Philip, J.R. and D.A. de Vries. 1957. Moisture movement in porous materials under temperature gradients. Trans. Am. Geophys. Union 38:222-232.

Philip, J.R. and J.H. Knight. 1974. On solving the unsaturated flow equation 3: New quasi-analytical technique. Soil Sci. 117:1-13.

Pinder, F.G. and W.G. Gray. 1977. Finite Element Simulation in Surface and Subsurface Hydrology. Academic Press, New York.

Poulovassilis, A. 1962. Hysteresis of pore water, an application of the concept of independent domains. Soil Sci. 93:405-412.

Poulovassilis, A., P. Kerkides, S. Elmaloglou and J. Argyrokastritis. 1991. An investigation on the relationship between ponded and constant flux rainfall infiltration. Water Resour. Res. 27:1403-1409.

Prassad, R. 1988. A linear root water uptake model. J. Hydrol. 99:297-306.

Priestly, C.H.B. and R.J. Taylor. 1972. On the assessment of surface heat flux and evaporation using large scale parameters. Monthly Weath. Rev. 100:81-92.

Pullan, A. 1990. The quasilinear approximation for unsaturated porous media flow. Water Resour. Res. 26:1219-1234.

Raats, P.A.C. 1988. Analytical/quasi-analytical approaches to unsaturated flow: Commentary. In: W.L. Steffen and O.T. Denmead (Eds.): Flow and Transport in the Natural Environment: Advances and Applications. Springer Verlag, Berlin, p. 48-58.

Raats, P.A.C. 1990. On the roles of characteristic lengths and times in soil physical processes. In: Trans. 14th Intl. Congress of Soil Sci., ISSS, Kyoto, I:202-207.

Raats, P.A.C. and W.R. Gardner. 1974. Movement of water in the unsaturated zone near a water table. In: J. van Schilfgaarde (Ed.): Drainage for Agriculture. Agronomy Monograph Series No. 17, ASA. Madison, Wisconsin, p. 311-405.

Raschke, R. 1975. Stomatal action. Ann. Rev. Plant Physiol. 26:309-340.

Raupach, M.R. 1988. Canopy transport processes. In: W.L. Steffen and O.T. Denmead (Eds.): Flow and Transport in the Natural Environment: Advances and Applications. Springer Verlag, Berlin, p. 95-127.

Reardon, E.J. 1981. K_d's - Can they be used to describe reversible ion sorption reactions in contaminant migration?. Ground Water 19:279-286.

Reiniger, P., J. Hutson, H. Jansen, J. Kragt, H. Piehler, H. Swarts, H. Vereecken. 1990. Evaluation and testing of models describing nitrogen transport and transformation in soil: A European project. Trans. 14 Intl. Congress of Soil Sci., ISSS, Kyoto, I:156-161.

Remson, J., G.M. Hornberger and F.J. Molz. 1971. Numerical Methods in Subsurface Hydrology with an Introduction to Finite Element Methods. Wiley-Interscience, New York.

Renger, M., W. Giesel, O. Strebel and S. Lorch. 1970. Erste Ergebnisse zur quantitativen Erfassung der Wasserhaushaltskomponenten in der ungesättigten Bodenzone. Z. Pflanzenernährung, Bodenkunde 126:15-35.

Reynolds, W.D. and D.E. Elrick. 1985. *In situ* measurement of field saturated hydraulic conductivity, sorptivity and the α-parameter using the Guelph permeameter. Soil Sci. 140:292-301.

Rible, J.M. and L.E. Davis. 1955. Ion exchange in soil columns. Soil Sci. 79:41-47.

Richards. L.A. 1931. Capillary conduction of liquids through porous media. Physics 1:318-333.

Richards, L.A. 1949. Method of measuring soil moisture tension. Soil Sci. 68:95-112.

Richards, L.A. 1965. Physical condition of water in soil. In: C.A. Black (Ed.): Methods of Soil Analysis, Part 1, Physical and Mineralogical Properties, Agronomy Monograph Series No. 9, ASA, Madison, p. 128-152.

Richards, L.A., W.R. Gardner and G. Ogata. 1956. Physical processes determining water loss from soil. Soil Sci. Soc. Am. Proc. 20:310-314.

Richards, S.J. and L.V. Weeks. 1953. Capillary conductivity values from moisture yield and tension measurements on soil columns. Soil Sci. Soc. Am. Proc. 17:206-209.

Richter, O., P Nörtesheuser and B. Diekkrüger. 1992. Modeling reactions and movement of organic chemicals in soils by coupling of biological and physical processes. Modeling Geo-Biosphere Proc.1:95-114.

Rieu, M. and G. Sposito. 1991. Fractal fragmentation, soil porosity and soil water properties. I. Theory. II. Applications. Soil Sci. Soc. Am. J. 55:1231-1238 and 1239-1244.

Ringrose-Voase, A. and P. Bullock. 1984. The measurement of soil structural parameters by image analysis. In: J. Bouma and P.A.C. Raats (Eds.): Proceedings of the ISSS Symposium on Water and Solute Movement in Heavy Clay Soil. ILPI Publication 37, Wageningen, p. 350-354.

Ritchie, J.T. 1972. Model for predicting evaporation from a row crop with an incomplete cover. Water Resour. Res. 8:1204-1212.

Robert, P.C. , R.H. Rust and W.E. Larson. 1993. Proceedings of Soil Specific Crop Management: A Workshop on Research and Development Issues. Special Pub., Am. Soc. Agron., Crop Sci. Soc. Am. and Soil Sci. Soc. Am., Madison, Wisconsin.

Rode, A.A. 1952. Soil Water (in Russian). Izdatelstvo AN SSSR (Publishers of Academy of Sciences), Moscow.

Rohdenburg, H. 1989. Methods for the analysis of agro-ecosystems in central Europe, with emphasis on geoecological aspects. Catena 16:1-57.

Rose, C.W., W.R. Stern and J.E. Drummond. 1965. Determination of hydraulic conductivity as a function of depth and water content from a soil *in situ*. Aust. J. Soil Res. 3:1-9.

Rosiwal, A. 1898. Über geometrische Gesteinsanalysen. Verhandl. Geol. Reichsamt, Wien, p. 143-175.

Rösslerová, R. 1993. Percolation models: Application for interpretation of soil water retention curves (in Czech). PhD Thesis, Czech Technical University, Prague.

Roth, K. and W.A. Jury. 1993. Modeling the transport of solutes to groundwater using transfer functions. J. Environ. Qual. 22:487-493.

Rubin, J. 1966. Theory of rainfall uptake by soils initially drier than their field capacity and its applications. Water Resour. Res. 2:739-749.

Rubin, J. 1967. Numerical method for analyzing hysteresis-affected, post-infiltration redistribution of soil moisture. Soil Sci. Soc. Am. Proc. 31:13-20.

Rubin, J. and R. Steinhardt, 1964. Soil water relations during rain infiltration: III. Water uptake at incipient ponding. Soil Sci. Soc. Am. Proc. 28:614-619.

Russell, M.B. and L.A. Richards. 1938. The determination of soil moisture energy relations by centrifugation. Soil Sci. Soc. Am. Proc. 3:65-69.

Russo, D. 1993. Stochastic modeling of macrodispersion for solute transport in a heterogeneous unsaturated porous formation. Water Resour. Res. 29:383-397.

Russo, D. and E. Bresler. 1980. Field determinations of soil hydraulic properties for statistical analysis. Soil Sci. Soc. Am. J. 44:682-687.

Russo, D. and E. Bresler, 1981. Effect of field variability in soil hydraulic properties on unsaturated water and salt flows. Soil Sci. Soc. Am. J. 45:675-681.

Saffman, P.G. 1959 A theory of dispersion in a porous medium. J. Fluid Mech. 6:321-349.

Saffman, P.G. 1960. Dispersion due to molecular diffusion and macroscopic mixing in flow through a network of capillaries. J. Fluid Mech.7:194-208.

Sallam, A., W.A. Jury and J. Letey. 1984. Measurement of gas diffusion coefficient under relatively low air-filled porosity. Soil Sci. Soc. Am. J. 48:3-6.

Sardin, m., D. Schweich, F.J. Leij and M.Th. van Genuchten. 1991. Modeling the nonequilibrium transport of linearly interacting solutes in porous media: A review. Water Resour. Res. 27:2287-2307.

Savage, S.M., J. Osborn, J. Letey and C. Heaton. 1972. Substances contributing to fire induced water repellency in soils. Soil Sci. Soc. Am. Proc. 36:674-678.

Schachtschabel, P., H.P. Blume, K.H. Hartge, U. Schwertmann, G. Brümmer and M. Renger. 1984. Lehrbuch der Bodenkunde, 11. Auflage. Enke, Stuttgart.

Scheidegger, A.E. 1954. Statistical hydrodynamics in porous media. J. Appl. Phys. 25:994.

Scheidegger, A.E. 1957. The Physics of Flow through Porous Media. Second edition, 1960. Univ. Toronto Press, Toronto.

Schweikle, V. 1982. Gefügeeigenschaften von Tonböden. Verlag Eugen Ulmer, Stuttgart.

Selim, H.M., J.M. Davidson and R.S. Mansell. 1976. Evaluation of a two-site adsorption-desorption model for describing solute transport in soils. Paper presented at Proceedings, Summer Computer Simulation Conference, Nat. Sci. Found., Washington, D.C., July 12-14.

Shaffer, M.S., R.W. Ribbens, and C.W. Huntley. 1977. Prediction of mineral quality of irrigation return flow, vol. V. Detailed return flow salinity and nutrient simulation model. EPA-600/2-77-179e, U.S. Environ. Prot. Agency, Washington, D.C.

Shainberg, I., J.D. Rhoades and R.J. Prather. 1981. Effect of low electrolyte concentration on clay dispersion and hydraulic conductivity of a sodic soil. Soil Sci. Soc. Am. J. 45:273-277.

Sharma, M.L., R.J. Luxmoore, R. de Angelis, R.C. Ward and G.T. Yeh. 1987. Subsurface water flow simulated for hillslopes with spatially dependent soil hydraulic characteristics. Water Resour. Res. 23:1523-1530.

Simmons. C.S. 1982, A stochastic-convective transport representation of dispersion in one-dimensional porous media systems. Water Resour. Res. 18:1193-1214.

Simmons. C.S., D.R. Nielsen and J.W. Biggar. 1979. Scaling of field-measured soil-water properties. I. Methodology. II. Hydraulic conductivity and flux. Hilgardia 47:74-173.

Simpson, E.S. 1969. Velocity and the longitudinal dispersion coefficient in flow through porous media. In: R.J.M. De Wiest (Ed.): Flow through porous media, Academic Press, New York and London, p. 201-214.

Šimůnek, J. and D.L. Suarez. 1993 UNSATCHEM - 2D Code for Simulating Two-Dimensional Variably Saturated Water Flow, Heat Transport, Carbon

Dioxide Production and Transport and Multicomponent Solute Transport with Major Ion Equilibrium and Kinetic Chemistry. Research Report No. 129, U.S. Salinity Laboratory, USDA, ARS, Riverside, California.

Šimůnek, J. and D.L. Suarez. 1994. Two-dimensional transport model for variably saturated porous media with major ion chemistry. Water Resour. Res. 30:1115-1133.

Šír, M., T. Vogel and M. Cislerová. 1985. The analytical expression of the retention curves and hydraulic conductivities (in Czech). Vodohosp. Čas. 33:74-85.

Šír, M.; M. Kutílek, V. Kuráž, M. Krejča and F. Kubík. 1988. Field estimation of the soil hydraulic characteristics. Soil Tech. 1:63-75.

Skopp, J. 1986. Analysis of time-dependent chemical processes in soils. J. Environ. Qual. 15:205-213.

Slatyer, R.O. 1967. Plant-Water Relationship. Academic Press, London.

Slatyer, R.O. (Ed.). 1973. Plant Response to Climatic Factors. UNESCO, Paris.

Slichter, C.S. 1905. Field measurements of the rate of movement of underground waters. Water Supply and Irrigation, Paper No. 140, Series O, U.S. Dept. Interior, U.S. Geol. Survey, p. 9-122.

Smettem, K.R.J. and B.E. Clothier. 1989. Measuring unsaturated sorptivity and hydraulic conductivity using multiple disc permeameters. J. Soil Sci. 40:563-568.

Smiles, D.E. and M.J. Rosenthal. 1968. The movement of water in swelling materials. Aust. J. Soil Res. 6:237-248.

Smiles, D.E. and E.G. Youngs. 1965. Hydraulic conductivity determinations by several field methods in a sand tank. Soil Sci. 99:83-87.

Smith, R.E. and R.H.B. Hebbert. 1979. A Monte Carlo analysis of the hydrologic effects of spatial variability of infiltration. Water Resour. Res. 15:419-429.

Smith, R.E., D.C. Goodrich and D.A. Woolhiser. 1990. Areal effective infiltration dynamics for runoff of small catchments. Trans. 14 Intl. Congress of Soil Sci. ISSS, Kyoto, I:22-27.

Spanne, P., K.W. Jones, L.D. Prunty and S.H. Anderson. 1994. Potential applications of synchrotron computed microtomography to soil science. In: S.H. Anderson and J.W. Hopmans (eds.): Tomography of Soil-Water-Root Processes. SSSA Special Publ. No. 36, ASA and SSSA, Madison, Wisconsin, p. 43-57.

Sposito, G. 1984. The Surface Chemistry of Soils. New York, Oxford University Press.

Sposito, G. and W.A. Jury. 1990. Miller similitude and generalized scaling analysis. In: D. Hillel and D.E. Elrick (Eds.): Scaling in Soil Physics: Principles and Applications. SSSA Spec. Publ. No. 25, SSSA, Madison, Wisconsin., p. 13-22.

Sposito, G., V.J. Gupta and R.N. Bhattacharya. 1979. Foundation theories of solute transport in porous media: a critical review. Advances in Water Res. 2:59-68.

Srinilta, S.A., D.R. Nielsen and D. Kirkham. 1969. Steady flow of water through a two-layer soil. Water Resour. Res. 5:1053-1063.

Starr, J.L., F.E. Broadbent and D.R. Nielsen. 1974. Nitrogen transformations during continuous leaching. Soil Sci. Soc. Am. Proc. 38:283-289.

Stein, A., I.G. Staritsky, J. Bouma, A.C. van Eijnsbergen and A.K. Bregt. 1991. Simulation of moisture deficits and areal interpolation by universal cokriging. Water Resour. Res. 27:1963-1973.

Steude, J., F. Hopkins and J. Anders. 1994 Industrial X-ray computed tomography applied to soil research. In: S.H. Anderson and J.W. Hopmans (eds.): Tomography of Soil-Water-Root Processes. SSSA Special Publ. No. 36, ASA and SSSA, Madison, Wisconsin, p. 43-57.

Stroosnijder, L. 1976. Cumulative infiltration and infiltration rate in homogeneous soils. Agric. Research Report 847:69-99.

Stull, R.B. 1988. An Introduction to Boundary Layer Meteorology. Kluwer Academic Publishers, Dordrecht.

Sudicky, E.A., R.A. Gillham and E.O. Frind. 1985. Experimental investigation of solute transport in stratified porous media, 1. The non-reactive case. Water Resour. Res. 21:1035-1041.

Swartzendruber, D. 1962. Modification of Darcy's law for the flow of water in soil. Soil Sci. 93:22-29.

Swartzendruber, D. 1963. Non-Darcy behavior and the flow of water in unsaturated soils. Soil Sci. Soc. Am. Proc. 27:491-495.

Swartzendruber, D. 1969. The flow of water in unsaturated soils. In: R.J.M. De Wiest (Ed.): Flow through Porous Media. Academic Press, New York, p. 215-292.

Swartzendruber, D. 1987. A quasi-solution of Richards' equation for the downward infiltration of water into soil. Water Resour. Res. 23:809-817.

Takagi, S. 1960. Analysis of the vertical downward flow of water through a two-layered soil. Soil Sci. 90:98-103.

Talsma, T. 1969. *In situ* measurement of sorptivity. Aust. J. Soil Res. 7:269-276.

Tanji, K.K., L.D. Doneen, G.V. Ferry and R.S. Ayers. 1972. Computer simulation analysis on reclamation of salt-affected soil in San Joaquin Valley, California. Soil Sci. Soc. Am. Proc. 36:127-133.

Taylor, G.I. 1950. The instability of liquid surfaces when accelerated in a direction perpendicular to their planes. Proc. Roy. Soc. London A. 201:192-196.

Taylor, G.I. 1953. Dispersion of soluble matter in solvent flowing slowly through a tube. Proc. Roy. Soc. London A. 219:186-203.

Taylor, S.A. and G.L. Ashcroft. 1972. Physical Edaphology. W. H. Freeman and Co. San Fransisco.

Thom, A.S. 1975. Momentum, mass and heat exchange of plant communities. In: J.L. Monteith (Ed.): Vegetation and Atmosphere. Vol. 1. Academic Press, London, p. 57-109.

Thomas, H.C. 1944. Heterogeneous ino exchange in a flowing system. J. Am. Chem. Soc. 66:1664-1666.

Thony, J.L., J. Hamburger, M. Vauclin and G. Vachaud. 1979. Etude détaillée de l'évaporation sur sol nu. In: Isotopes and Radiation in Research on Soil-Plant Relationships. IAEA-SM-235, Vienna, p. 247-258.

Thony, J.L., G. Vachaud, B.E. Clothier and R. Angulo-Jaramillo. 1991. Field measurement of the hydraulic properties of soil. Soil Tech. 4:111-123.

Thornthwaite, C.W. 1948. An approach toward a rational classification of climate. Geograph. Rev. 38:55-94.

Tillotson, P.M. and D.R. Nielsen. 1984. Scale factors in soil science. Soil Sci. Soc. Am. J. 48:953-959.

Tokunaga, K.I. 1988. X-ray stereoradiographs using new contrast media on soil macropores. Soil Sci. 146:199-206.

Toledo, P.G., R.A. Novy, H.T. Davis and L.E. Scriver. 1990. Hydraulic conductivity of porous media at low water content. Soil Sci. Soc. Am. J. 54:673-679.

Topp. G.C., J.L. Davis and A.P. Annan. 1980 Electromagnetic determination of soil water content: Measurements in coaxial transmission lines. Water Resour. Res. 16:574-582.

Topp, G.C., W.D. Reynolds and R.E. Green (Eds.). 1992. Advances in Measurement of Soil Physical Properties: Bringing Theory into Practice. SSSA Special Publ. No. 30. SSSA, Madison, Wisconsin.

Touma, J. and M. Vauclin. 1986. Experimental and numerical analysis of two-phase infiltration in a partially saturated soil. Transp. Porous Media 1:27-55.

Travis, C.C. and E.C. Etnier. 1981. A survey of sorption relationships for reactive solutes in soil J. Environ. Qual. 10:219-225.

Tyler, S.W. and S.W. Wheatcraft. 1989. Application of fractal mathematics to soil water retention estimation. Soil Sci. Soc. Am. J. 53:987-996.

Tyler, S.W., J.B. Chapman, S.H. Conrad and D. Hammermeister. 1994. Paleoclimatic resonse of a deep vadose zone in southern Nevada, U.S.A. as inferred from soil water tracers. Proc. 1st Intl. Symp., Tracers in Arid Zones. Intl. Atomic Energy Agency, Vienna. (In press).

Uehara, G., B.B. Trangmar and R.S. Yost. 1985. Spatial variability of soil parameters. In D.R. Nielsen and J. Bouma (eds.) Soil Spatial Variability. Pudoc, Wageningen, The Netherlands, p. 61-95.

Ulaby, F.T. and C. Elachi. 1990. Radar Polarimetry for Geoscience Applications. Artech House.

Vachaud, G. 1979. Research related to agricultural development in the semiarid tropics. Proc. 14th Collog. Intl. Potash Inst.: Soils in Mediterranean Type Climates and their Field Potential. Bern. 59-78.

Vachaud, G., M. Vauclin and R. Laty. 1990b. CARHYD: A computer aided system for characterization of the hydraulic conductivity of field soils. Soil Tech. 3:131-144.

Vachaud, G., Z. Chaaboumi, L. El Amani and M. Vauclin. 1983. Methodologie d'etude du bilan hydrique d'une culture a l'echelle de la parcelle. In: Isotope and Radiation Techniques in Soil Physics and Irrigation Studies, IAEA, Vienna, p. 79-101.

Valocchi, A.J. 1984. Describing the transport of ion-exchanging contaminants using an effective K_d approach. Water Resour. Res. 20:499-503.

Van der Molen, W.H. 1956. Desalinization of soils as a column process. Soil Sci. 81:19-27.

van Genuchten, M.Th. 1980. A closed form equation for predicting the hydraulic conductivity of unsaturated soils. Soil Sci. Soc. Am. J. 44:892-898.

van Genuchten, M.Th. 1981. A comparison of numerical solutions of the one-dimensional unsaturated-saturated flow and mass transport equations. Advances Water Res. 5:47-55.

van Genuchten, M.Th. 1985. A general approach for modeling solute transport in structured soils. Memoires Intl. Assoc. Hydrogeol. 17:513-526.

van Genuchten, M.Th. and W.J. Alves. 1982. Analytic solutions of the one-dimensional convective-dispersive solute transport equation. Tech. Bull. 1661, U.S. Dept. of Agr., Washigton, D.C.

van Genuchten, M.Th. and D.R. Nielsen. 1985. On describing and predicting the hydraulic properties of unsaturated soils. Annales Geophysicae 3:615-628.

van Genuchten, M.Th. and P.J. Wierenga. 1976. Mass transfer studies in sorbing porous media. I. Analytical solutions. Soil Sci. Soc. Am. J. 40:473-480.

van Genuchten, M.Th., J.M. Davidson and P.J. Wierenga. 1974. An evaluation of kinetic and equiliibrium equations for the prediction of pesticide movement in porous media. Soil Sci. Soc. Am. Proc. 38:29-35.

van Genuchten, M.Th., F. Kavel, W.B. Russell and S.R. Yates. 1989. Direct and indirect methods for estimating the hydraulic properties of unsaturated soils. In: J. Bouma and A.K. Bregt (Eds.): Land Qualities in Space and Time. Pudoc, Wageningen, p. 61-72.

Vauclin, M. and G. Vachaud. 1990. Deterministic versus stochastic modelings of soil water balance components. Trans. 14th Intl. Congress of Soil Sci., ISSS, Kyoto, I:16-19.

Vauclin, M., R. Haverkamp, G. Vachaud. 1979. Résolution numerique d'une équation dė diffusion non-linèaire. Presse Universitaires, Grenoble.

Vauclin, M., R. Haverkamp, and G. Vachaud. 1984. Error analysis in estimating soil water content from neutron probe measurements: 2. Spatial standpoint. Soil Sci. 137:141-148.

Vauclin, M., S.R. Vieira, G. Vachaud and D.R. Nielsen. 1983. The use of cokriging with limited field soil observations. Soil Sci. Soc. Am. J. 47:175-184.

Vauclin, M., J.L. Thony, G. Vachaud, P. Revol, P. Ruelle, D.E. Elrick and B.E. Clothier. 1991. Field determination of soil hydraulic properties by suction and pressure infiltrometers Paper presented at EGS XVI Meeting. April, Wiesbaden.

Veihmeyer, F.J., O.W. Israelsen and J.P. Conrad. 1924. The moisture equivalent as influenced by the amount of soil used in its determination. Tech. Paper 16, Univ. of Calif., Coll. of Agric., Berkeley.

Vogel, T. and M. Císlerová. 1988. On the reliability of unsaturated hydraulic conductivity calculated from the moisture retention curve. Transp. Porous Media 3:1-16.

Vogel, T., M. Císlerová and J.W. Hopmans. 1991. Porous media with linearly variable hydraulic properties. Water Resour. Res. 27:2735-2741.

Wagenet, R.J. and J.L. Starr. 1977. A method for simultaneous control of the water regime and gaseous atmosphere in soil columns. Soil Sci. Soc. Am. J. 41:658-659.

Wagenet, R.J., J.W. Biggar and D.R. Nielsen. 1977. Tracing the transformations of urea fertilizer during leaching. Soil Sci. Soc. Am. J. 41:896-902.

Wang, J.H. 1952. Tracer-diffusion in liquids. III. The self-diffusion of chloride

ion in aqueous sodium chloride solutions. J. Am. Chem. Soc. 74:1612-1615.
Wang, J.R., P.E. O'Neill, T.J. Jackson and E.T. Engman. 1983. Multifrequency measurements of the effects of soil moisture, soil texture and surface roughness. IEEE Geosci. Rem. Sens. 21:44-51.
Wardlow, N.C, Y. Li and D. Forbes. 1987. Pore-throat size correlation from capillary pressure curves. Transp. Porous Media 2:597-614.
Warrick, A.W. 1988. Additional solutions for steady-state evaporation from a shallow water table. Soil Sci. 146:63-66.
Warrick, A.W. 1990. Application of scaling to the characterization of spatial variability in soils. In: D. Hillel and D.E. Elrick (Eds.): Scaling in Soil Physics: Principles and Applications. SSSA Spec. Publ. No. 25, SSSA, Madison, Wisconsin, p. 39-57.
Warrick, A.W. 1992. Models for disc infiltrometers. Water Resour. Res. 28:1319-1327.
Warrick, A.W. 1993. Inverse estimation of soil hydraulic properties with scaling: one dimensional infiltration. Soil Sci. Soc. Am. J. 578:631-636.
Warrick, A.W. and A.A. Hussen. 1993. Scaling of Richards' equation for infiltration and drainage. Soil Sci. Soc. Am. J. 57:15-18.
Warrick, A.W., D.O. Lomen and A. Islas. 1990. An analytical solution to Richards' equation for a draining soil profile. Water Resour. Res. 26:253-258.
Warrick, A.W., G.J. Mullen and D.R. Nielsen. 1977. Scaling field properties using a similar media concept. Water Resour. Res. 13:355-362.
Watson, K.K. 1966. An instantaneous profile method for determining the hydraulic conductivity of unsaturated porous materials. Water Resour. Res. 2:709-715.
Webster, R. 1985. Quantitative spatial analysis of soil in the field. Advances in Soil Sci. Vol. 3:1-70. Springer Verlag, New York.
Webster, R. and M.A. Oliver. 1990. Statistical Methods in Soil and Land Resource Survey. Oxford Univ. Press.
Wendroth, O., G.G. Katul, M.B. Parlange, C.E. Puente and D.R. Nielsen. 1993. A nonlinear filtering approach for determining hydraulic conductivity functions. Soil Sci. 156:293-301.
Wendroth, O., W. Ehlers, J.W. Hopmans, H. Kage, J. Halbertsma and J.H.M. Wösten. 1993. Reevaluation of the evaporation method for determining hydraulic functions in unsaturated soils. Soil Sci. Soc. Am. J. 57:1436-1443.
Wheatcraft, S.W. and J.H. Cushman. 1991 Hierarchical approaches to transport in heterogeneous porous media. Rev. Geophys. 29 Supplement:261-267.
Whisler, F.D., A. Klute and R.J. Millington. 1968. Analysis of steady state evapotranspiration from soil column. Soil Sci. Soc. Am. Proc. 32:167-174.
Whitaker, S. 1986. Flow in porous media. I: A theoretical derivation of Darcy's law. Transp. Porous Media 1:3-25.
White, I. 1988. Measurement of soil physical properties in the field. In: W.L. Steffen and O.T. Denmead (Eds.): Flow and Transport in the Natural Environment: Advances and Applications. Springer, Berlin.
White, I. 1989. Comments on sorptivity approximations by Kutílek and Valentová. Transp. Porous Media 2:317-321.

White, I. and P. Broadbridge. 1988. Constant rate rainfall infiltration: A versatile nonlinear model. 2. Application of solutions. Water Resour. Res. 24:155-162.

White, I. and K.M. Perroux. 1987. Use of sorptivity to determine field soil hydraulic properties. Soil Sci. Soc. Am. J. 51:1093-1101.

White, I. and K.M. Perroux. 1989. Estimation of unsaturated hydraulic conductivity from field sorptivity measurements. Soil Sci. Soc. Am. J. 53:324-329.

White, I., M.J. Sully and M.D. Melville. 1989. Use and hydrological robustness of time-to-incipient ponding. Soil Sci. Soc. Am. J. 53:1343-1346.

Wilding, L.P. 1985. Spatial variability: its documentation, accomodation and implication to soil surveys. In: D.R. Nielsen and J. Bouma (Eds.): Soil Spatial Variability. Pudoc, Wageningen.

Wilding, L.P. and C.T. Hallmark. 1984. Development of structural and microfabric properties in shrinking and swelling clays. In: J. Bouma and P.A.C. Raats: Water and Solute Movement in Heavy Clay Soils. ISSS Symp. Proceedings of the ISSS Symposium on Water and Solute Movement in Heavy Clay Soils. International Institute for Land Reclamation and Improvement (ILRI) Publication 37, Wageningen, p. 1-18.

Willis, W.O. 1960. Evaporation from layered soils in the presence of a water table. Soil Sci. Soc. Am. Proc. 24:239-242.

Wilson, J.N. 1940. A theory of chromatography. J. Am. Chem. Soc. 63:1583-1591.

Wind, G.P. 1955. A field experiment concerning capillary rise of moisture in a heavy clay soil. Neth. J. Agr. Sci. 3:60-69.

Wooding, R.A. 1959. The stability of a viscous liquid in a vertical tube containing porous material. Proc. Roy. Soc. London A. 252:120-134.

Wooding, R.A. 1968. Steady infiltration from a shallow circular pond. Water Resour. Res. 4:1259-1273.

Wösten, J.H.M. 1990. Quantification of soil hydraulic variability to allow accuracy estimation of model output. Trans. 14th Intl. Congress of Soil Sci., ISSS, Kyoto, I:134-139.

Wösten, J.H.M. and M.Th. van Genuchten. 1988. Using texture and other soil properties to predict the unsaturated soil hydraulic functions. Soil Sci. Soc. Am. J. 52:1762-1770.

Wösten, J.H.M., J. Bouma and G.H. Stoffelsen. 1985. Use of soil survey data for regional soil water simulation models. Soil Sci. Soc. Am. J. 49:175-184.

Yang, S., Z. Lei and Y Xue. 1992. An experimental study of local recharge in the thick unsaturated zone. J. Hydraulic Engineering 1:495-502.

Yeh, G.T. and S.V. Tripathi. 1991. A model for simulating transport of reactive multispecies components: Model development and demonstration. Water Resour. Res. 27:3075-3091.

Yeh, T.-C., L.W. Gelhar and A.L. Gutjahr. 1985a. Stochastic analysis of unsaturated flow in heterogeneous soils, 1. Statistically anisotropic media with variable α. Water Resour. Res. 21:447-456.

Yeh, T.-C., L.W. Gelhar and A.L. Gutjahr. 1985b. Stochastic analysis of unsaturated flow in heterogeneous soils, 2. Statistically isotropic media. Water Resour. Res. 21:457-464.

Yeh, T.-C., L.W. Gelhar and A.L. Gutjahr. 1985c. Stochastic analysis of unsaturated flow in heterogeneous soils, 3. Observations and applications. Water Resour. Res. 21:465-472.

Youngs, E.G. 1964. An infiltration method of measuring the hydraulic conductivity of unsaturated porous materials. Soil Sci. 97:307-311.

Youngs, E.G. 1983. The use of similar media theory in the consideration of soil water redistribution in infiltrated soils. In: Advances in Infiltration. ASAE Publication 1183, St. Joseph, Mich., p. 48-54.

Yule, D.F. and J.T. Ritchie. 1980. Soil shrinkage relationships of Texas vertisols: Small cores. Soil Sci. Soc. Am. J. 44:1285-1291.

Zaslavsky, D. 1970. Some Aspects of Watershed Hydrology. USDA-ARS No. 41-157.

Zaslavsky, D. and G. Sinai. 1981. Surface hydrology: IV. Flow in sloping layered soil. ASCE 107, No. HY 1:1-93.

Zegelin, S.J., I. White and G.F. Russell. 1992. A critique of the time domain reflectometry technique for determining field soil-water content. In: Advances in Measurement of Soil Physical Properties: Bringing Theory into Practice. SSSA Spec. Publ. No. 30, SSSA, Madison, Wisconsin, p. 187-208.

Zeiger, H. 1983. The biology of stomatal quard cells. Ann. Rev. Plant Physiol. 34:441-475.

Zunker, F. 1930. Das Verhalten des Bodens zum Wasser. In: E. Blanck: Handbuch der Bodenlehre. Bd.VI Berlin, p. 66-220.

FAO. 1977. Crop Water Requirement. FAO Irrigation and Drainage Paper 24, Rome.

FAO/UNESCO. 1985. Soil Map of the World. Third Draft. FAO-UN.

Keys to Soil Taxonomy. Fourth Ed. 1990. SMSS Technical Monograph No. 19. Virginia Polytechnic Institute and State University.

Soil Taxonomy. 1975. Agriculture Handbook No. 436, SCS, USDA.

Terminology in Soil Physics. 1976. ISSS Bull. 49:26-36.

U.S. Department of Agriculture. 1951. Handbook No. 18. Soil Survey Manual, Washigton, D. C.

INDEX